Up Close and Personal with the Urantia Book

EXPANDED EDITION

JJ Johnson

Library of Congress Control Number: 2008934637
ISBN 978-0-9795926-2-1

Dedication

Part Two of this book is respectfully dedicated to those cosmic conscious citizens who recognize the time is at hand to "Remove the Apron" and be about our Father's Business.

Part Three goes out to those forward-looking men and women of spiritual insight who will dare to construct a religious philosophy of living based upon the highest concepts of cosmic truth, universe beauty and divine goodness as revealed in The Urantia Book.

Personally, this book is dedicated to my two sons **Michael Andrew** and **David Matthew** who have blessed their mom and dad with such Love and Joy that it continually overflows from our hearts. Also, our beloved grandchildren who are a source of love and inspiration in recent years.

Note on Page and Paper References from *The Urantia Book*

Passages from *The Urantia Book* are referenced by Page numbers and Paragraph number of the Urantia Foundation edition on the left margin.

Indicated on the right margin is the Paper, Section, and Paragraph for use with the Uversa Press edition of *The Urantia Book*.

See example below.

Page 1091:8   *Paper 99:5.9*

    ...It is high time that man had a religious experience so personal and so sublime that it could be realized and expressed only by "feelings that lie too deep for words."

The quote above is the 8th full paragraph on page 1091 of the Foundation edition. It is in Paper 99, Section 5, Paragraph 9 in the Uversa Press edition.

Note on Use of **Bold**

Throughout this book, certain words, phrases, or sentences are in **bold** in order to assist the reader in locating portions of the text that are relevant to the theme. This use of bold is entirely my own, and not in the original text of *The Urantia Book*.

Contents

Foreword

Philosopher Mortimer Adler framed a significant academic and cultural problem when he declared in 1940:

"One of the troubles [with universities] is that scientists, philosophers, and theologians, or teachers of religion, have long failed to communicate with one another. The structure of a modern university, with its departmental separations, and its total lack of order among specialized disciplines, represent perfectly the disunity and chaos of modern culture ... one might even hope for communication to lead to mutual understanding, and thence to agreement about the truths that could unify our culture. "

Personally, I have always considered the unified relations of science, philosophy, and religion in *The Urantia Book* as the most important aspect of the work. JJ Johnson has completed his philosophical description of *The Urantia Book*'s holistic character by adding the new chapter in Up Close and Personal with *The Urantia Book*, relating science with religion (COSAR). It is an impressive achievement.

I congratulate JJ on his extensive research in producing this new chapter. He is one of the most significant students of *The Urantia Book* who is defining its revelatory insights and balanced approach toward breaking down the barriers between the disciplines of religion and science. Perhaps such tireless efforts will one day lead to the mutual understanding that Adler believed would unify our culture.

—*Dr. Meredith Justin Sprunger*

Dr. Meredith Justin Sprunger has had a remarkable career as a college president, professor, and an ordained minister. He has served various United Church of Christ congregations in the Midwest. He taught psychology at Elmherst College and then at the Indiana Institute of Technology, where he was head of the Department of Psychology, Chairman of the Division of Liberal Arts, and as President. A student of *The Urantia Book* for over fifty years, Dr. Sprunger is the author and contributor of several *Urantia Book*-related works.

Acknowledgements

The author gratefully acknowledges the assistance of **Beth Wenger**. She has been of immeasurable help in bringing the draft manuscript to fruition. As associate editor, Beth has given lavishly of her time and helped make this labor of love much more than it would have been without her. Beth and her husband Bruce are long-time students of *The Urantia Book*. They reside in Phoenix, Arizona, and are founding members of the Grand Canyon Society for Readers of *The Urantia Book*.

To **Larry Mullins**, whose door and heart have always been open to me and other Urantians who do not hesitate to call upon his untiring willingness to help see a worthy Urantia project through. Larry is perhaps the most prolific writer of secondary works of *The Urantia Book*. One only has to look at the list of fourteen other books authored by Larry noted in his book *A History of the Urantia Papers*, published in 2000, to get a feel of his dedication to the Urantia community for over three decades. Larry and his lovely wife Joan live in Florida.

To **Bob Debold**, who has been a source of inspiration during my assignment in the Washington, D.C. area. As associate editor, Bob also left his thumbprint throughout these pages. Bob's recent paper "Crouching Deity—Hidden Supreme" exemplifies his deep thinking approach to such topics and his willingness to share with all of us.

PART I

SCIENCE, RELIGION, AND THE TRUTH OF REALITY

The Urantia Book disclosed Plate Tectonics and a dozen other facts decades before scientists confirmed it. This new chapter will cover them all.

While on assignment in the Sinai in late 2007, I bolted up in bed one day with an impress on my mind: "The COSAR Principle—A Revelatory Proposition." The following chapter is the fruition of this impress.

Impress as used above: To urge, as something to be remembered or done

—JJ

Chapter One

The COSAR Principle

—A Revelatory Proposition—

The Coordination of Science and Religion (COSAR) with the truth of reality is best achieved by revelation. *The Urantia Book*, a revelation of epochal significance to our strife torn planet, serves this *function* with a high degree of satisfaction for the reader. The manner in which the coordination of both science and religion is so exquisitely embedded throughout *The Urantia Book*, I call the COSAR Principle. The role that the COSAR Principle performs can be valued by approaching *The Urantia Book*'s development of the truth of reality from either the scientific content or the religious truths that are revealed within its pages. Yet to have a deeper appreciation and fuller awareness of the truth of reality that is revealed in *The Urantia Book* one must be able to integrate in a meaningful way both the "S" (Science) and "R" factors in the COSAR Principle into his or her personal picture of reality. I address the "R" (Religious/Spiritual) approach in Chapter Two. In fact, my personal motivation for *Up Close* is expressed in Chapter Two in general and culminates in the "Wake Up Call".

The "S" approach, in COSAR, is offered as a rebuttal to Scientism—a hard "S" position. Scientism is defined as the denial of any reality that science cannot detect and analyze. It perpetuates an offensive against Western religious tradition by marginalizing religious claims that have little chance of empirical support, such as the existence of the soul. To be fair, at the polar opposite, theologies make preposterous claims (like the belief that the earth is 6000 years old) of which science has solid empirical data proving otherwise. This hard "R" position is not acceptable to the deep thinkers and those who are thirsty for truth. Both extremes exhibit a viewpoint that creates an unhealthy confrontation. These S and R extremes encourage a volatile imbalance that produces truth blinders and is mutually destructive.

Students of *The Urantia Book* have been successful in bridging the gap of these two apparently divergent aspects of their intellectual and spiritual nature. Sincere students recognize that this can best be resolved with the COSAR Principle that produces a self-evident elegance throughout *The Urantia Book*. This revelatory proposition—the Truth of Reality—offered in *The Urantia Book* creates a harmony of mind and satisfaction of spirit which answers in human experience those cravings on how the Infinite functions in both the "S" and "R" arenas. *The Urantia Book* synthesizes the sciences of nature and the theology of religion into a consistent and logical universe philosophy that metaphysics has failed to achieve.

The Coordination of S and R provides the fundamentalist "S-minded" types insight into *The Urantia Book*'s intriguing teachings about nature—science—that help one to discard the assertions of Scientism that there is nothing in our world except matter and energy, that is, only those things that can be quantified and measured. The truth of reality presented in *The Urantia Book* opens a window of opportunity to further expand the COSAR Principle. The first three parts of *The Urantia Book* were completed in 1934-35. Since that time the *Urantia Papers* have been found to contain an extraordinarily high percentage of correct assertions about the cosmos as well as scientific facts that were either not known or accepted by the scientific mainstream until decades later. This alone should leave one either quite amazed or thoughtfully curious. In either case, if we suspend judgment and let the truths come to light by scientifically observing and analyzing the hard core facts and concrete data contained in the *Urantia Papers* themselves, a different scientific perspective would emerge. This promising intellectual unity can then be strengthened by revelation, moving the hard "S" over to shake hands with the "R".

I feel it is only fair that in this expanded edition of *Up Close* I provide a few examples of the scientific content revealed in *The Urantia Book*. They show both the accuracy and precision of their assertions well before the facts supporting such concepts were acknowledged as correct by the majority of scientists and received into the scientific mainstream. This is for the scientific community, Doubting Thomas's, and those that are thirsty for truth but who, as of yet, have not personally experienced or discerned the truths covered in Chapter Two. Even the orthodox scientists can be won over to the COSAR Principle, provided they recognize and agree with the facts of the science content in *The Urantia Book* that is supported by reputable independent sources.

Ultimately this chapter is an attempt to pique the curiosity of the scientifically minded with bold examples that will validate the *Urantia Papers'* historically unbelievable odds of "getting it right". To have known beyond any doubt these claims were right in 1934-35 and teach them in an emphatic way is statistically impossible through guesswork. Again, the *Urantia Papers*, an epochal revelation, accomplish COSAR by coordinating both S and R with the truth of reality. It provides this coordinating *function* by bridging the gap between S and R with the truth of reality. The authors of *The Urantia Book* had to carefully measure their approach to achieve this goal. They were well aware that while "the historic facts and religious truths will stand on the records of the ages to come, within a few short years many of our statements regarding the physical sciences will stand in need of revision in consequence of additional scientific developments and new discoveries." (Paper 101).

The Urantia Book provides cosmic revelation of a scientific nature by employing two techniques: 1) recalling long-lost truths and facts, and 2) advancing new truths and facts, some which are given well in advance of our discovering and accepting them into the scientific mainstream. The papers tell us in quite ardent terms that they reveal the reality of nature (science) in ways that illuminate the relationship to the spiritual truths contained in *The Urantia Book*. Their descriptions provide insights that coordinate the nature of Deity with the reality of the cosmos under the COSAR Principle. *The Urantia Book* provides us with specific, precise and noteworthy corrections which, in turn, are confirmed by the established scientific record. We then become obliged to trust in the coordinated scientific assertions and religious truths (COSAR) that humans have a high destiny in the universe. Our personality survival and ascension career is eloquently portrayed throughout the *Urantia Papers*.

Before I support my contention about the scientific veracity of the revelators' astounding claims, allow me to provide a few thoughts about just what this extraordinary accuracy and precision may or may not mean. It has been said that if I can make my pen fly in the air and land directly back in my hand, this does not prove the truth of my *next* statement. I don't want to give the impression that if one looks at the scientific correctness found throughout the papers that they are in some way prophetic, mystical, or inspired from the hand of God. However, I do claim that the science content in the papers are *instructive* and will lead one—anyone—to further integrate in a meaningful way both the "S" and "R" factors into a more truthful picture of reality.

It requires an intellectual discipline and thirst for truth to acknowledge and accept the accuracy of the scientific facts and cosmology revealed in the *Urantia Papers*. For example, geologists have revised and pushed back to 600-800 million years the *scientific fact* of plate tectonics and the date of commencement of breakup. This was long after *The Urantia Book* was published in 1955. These numbers now agree with the dates in *The Urantia Book*. This should bring about a pregnant pause and leave any reasonable scientist with the desire to explore further. It will require *scientists*, not religionists, to plumb the cosmos and creatively energize their minds for appropriate answers.

Consider the following personal comments from the *Garden of Ediacara* by Mark A.S. McMenamin (McMenamin 1998) "They [Urantia Papers] embraced continental drift at a time when it was decidedly out of vogue in the scientific community...However, the concept of a billion-year-old supercontinent split apart, forming gradually widening ocean basins in which early marine life flourished, is unquestionably in this book [*The Urantia Book*]...Furthermore, they even got the timing of *that* approximately correct at 650 to 600 million years ago ('These inland seas of olden times were truly the cradle of evolution')...Cases such as this one (which is by no means unique) are an exercise in humility for me as a scientist."

Students of *The Urantia Book* have the utmost admiration and respect for scientists like Professor McMenamin who boldly accept scientific facts and truths from whatever source provided it will stand up to a disciplined scientific peer process. The examples of science content in *The Urantia Book* provided in this chapter coincide with and predate the validation of our science. These examples will stand up to scientific scrutiny; such scrutiny can only lead to the acceptance that these concepts presented in *The Urantia Book* predate their discovery and/or verification by the scientific community. If you are a scientist or scientifically inclined and do not care to seriously experience and confirm for yourself the accuracy of the examples below I respect that and wish you well on your continued quest of new discoveries. However, for those so-called scientists who lash out at the McMenamin's of the world and deny or ignore the facts that can be independently confirmed by reliable sources just because the truths were revealed "in a work of religious revelation decades before scientists find out *anything* about the subject"— shame on you! To careless scientists who would be so presumptuous: I am reminded of a quote that was passed onto me years ago..."I would take great pride in my intelligence were it not for the fact that that attribute is shared by some of the stupidest men on earth"—Albert Camus in *The Fall*.

The fact that the revelators almost nonchalantly enlighten the student of *The Urantia Book* of the conjunction of Jupiter and Saturn in the constellation of Pisces as having occurred on May 29, September 29, and December 5 of the year 7 B.C. is now irrefutable in its precision. This is related to the legend of the star of Bethlehem and the idea that the Magi were directed to find the baby Jesus via an ancient astrological GPS mechanism. In 1976, at California's Jet Propulsion Laboratory, in close association with the U.S. Naval Observatory, a team of JPL scientists calculated the positions of all major bodies in the solar system throughout a span of forty-four centuries, from 1411 B.C. to A.D. 3002. To meet rigorous scientific methodology, they omitted all previous analytical theories of motion for individual objects. This new method embraced a technique of simultaneous numerical integration on a Univac 1100/81 (a calculation method), inconceivable just a few decades before the 1976 calculations. The task required nine days of computer time resulting in a magnetic tape output officially dubbed DE 102 and now known as the Long Ephemeris Tape. The point being made here is one of *accuracy*. Roger Sinnott admits in his Sky and Telescope article that the differences deduced from Tuckerman's tables are hardly dramatic—but "The differences are real enough." The dates taken from the JPL output in 1976 and published in 1986 by Bretagnon and Simon in Paris after carefully comparing their work with DE 102, match with uncanny accuracy *The Urantia Book* dates published in 1955. What was refutable up until 1976 and not brought to light until 1986 was the *precision* of the three dates in the same year (7 B.C.) almost two thousand years ago (see Example 4 below). As of 1976 the precision of the triple, three-times-in-one-year conjunction no longer can be discredited. These conjunctions never could lead the Magi to the manger. The star of Bethlehem was not a "star" visible on the night of Jesus' birth, but rather an extrapolation of three astrological events into a reification of a religious myth. There are a myriad of scenarios on "The Star" dates. Just Google 'The Star of Bethlehem', Wikipedia, and other sites to see the various projected dates and how each arrives at their own conclusion. What is most noteworthy is that the three precise dates mentioned above in *The Urantia Book* that was published in 1955 and substantiated in 1976 by the Jet Propulsion Laboratory in California are now completely in agreement with but a single day (see reference links for explanation of this one day difference).

Mercier in 1871 is the only other known "Star of Bethlehem" account that can make this claim. His conjunctions were identical with *The Urantia Book* dates. It is interesting to report that C. Pritchard in 1856 arrived at three dates in 7 B.C. that were closer to JPLs dates than the Tuckerman tables which were discovered in 1981 to have certain systematic flaws that resulted in the tables being several days off. Pritchard's 04 December date does not agree with *The Urantia Book*, Mercier's and JPLs date of 05 December. However, *The Urantia Book*, Mercier and C. Pritchard agree with the 29 September 7 B.C. date compared to the one day difference of the

JPL date of 30 September 7 B.C. In summary, Pritchard in 1856 is off 2 days from JPL. *The Urantia Book* and the Mercier dates in 1871 are one day off of the JPL dates. *The Urantia Book*, Mercier, and Pritchard happen to agree with the date of 29 September 7 B.C. that accounts for the one day difference with the JPL date.

What is equally if not most important about this scientific revelation (now proven both precise and accurate), is the religious significance and how our science supports the dates given in *The Urantia Book*. Here is a perfect example of science aiding religion—COSAR in Action. Each individual pondering this fact and reading further in the paper about the birth of Jesus will notice one is reminded that myths become traditions and traditions become accepted as fact. This is especially true during a time when information travels by word of mouth.

My point to the hard "S" types is the credibility of the science content in these examples from the *Urantia Papers* can now be firmly established in that they predate the proof of science. But this trustworthiness is paradoxically attenuated by the revelators themselves. They appear not to be really concerned with our revering their accuracy. All scientists know measurement is always an approximation. So with regard to knowledge, as opposed to truth, there can never be absolute certainty, only increasing probability of approximation. This proven reliability of the authors of the *Urantia Papers* coupled with the unassuming manner in which the science content is convincingly presented is a combination that is most desirable and enduring.

What we have in the following examples is the depiction of irrefutable evidence that the authors are in an authoritative position to provide cosmic revelation by the two techniques described earlier in this chapter. They tell us it is a requirement of revelation to not provide "unearned or premature knowledge"—and they don't. This Revelatory Proposition that provides these corrections and statements of cosmology are *not inspired* but are of immense value as they reduce confusion and transiently clarify scientific knowledge that affords the COSAR Principle an opportunity to *function* as a Coordinating conduit for the "S" and "R" factors in COSAR.

In the following pages I provide four detailed examples and nine additional scientific concepts in *The Urantia Book* that predate the validation of our science. I have provided reputable independent references for the interested scientist and those that are thirsty for truth to begin further scrutiny. It is worth noting that the science content in *The Urantia Book* has taken aback scientifically minded readers as well as scientists who are not affiliated with readers of the *Urantia Papers*. There is a growing widespread acknowledgement among the scientific community that certain parts of the scientific data and cosmology that are revealed could not have been known prior to 1955 when *The Urantia Book* was published. Kary Mullis is another example of a scientist boldly reporting the science content in *The Urantia Book* that he observed predates our science. Mullis was awarded the Nobel Prize in Chemistry in 1993. On his website (http://www.karymullis.com/) under his subtopic "Books" he states:

The Urantia Book was purportedly written by extraterrestrials and published in 1955. It has been freely available on the internet since 2001. Several scientific developments, unexpected in 1955, reported in 2005 in Science and Nature, and referenced below, were somehow, described rather precisely already in The Urantia Book. I have documented three cases here, but there are many contemporary scientific discoveries which were first posited as far as I can tell, in this rather large tome. There is much in here, the truth of which cannot be judged from the apparent truth of these several instances. The book claims a large number of authors. Much of it would be considered "politically incorrect" and might infuriate some people. I suggest that you not be shooting at the messenger; I am just reporting what I have observed. Striking Coincidences Between The Urantia Book (Copyright 1955) three articles in Science: 309 (2005), and one in Nature, (2005).

In the examples that follow I will use The Urantia Book publication date of 1955 but would like to reiterate that Parts I and II were completed in 1934 and at the end of Part III the date of 1935 is given. Even in Part III the date of 1934 is noted on four different occasions (pp 707:7, 710:6, 716:7, 828:1 Papers 62:5.1, 62:7.7, 63:6.8, 74:0.1—see page iv regarding the use of page references). Whichever date you choose to use as your starting point, all of the eventual scientific verification and acceptance for each example wasn't available until well after the Urantia Papers revealed these scientific facts and discoveries.

These examples merely sample the large litany of science The Urantia Book provides. It is my hope the reader will explore further the tremendous wealth of coordination the Urantia Papers deliver related to the COSAR Principle. It will prove to be worth your while and a great investment of your time.

Scientific concepts in *The Urantia Book* published in 1955 that predate the validation of our science and before accepted into the scientific mainstream:

NOTE: Discussion of the following examples can be found at: http://www.truthbook.org/index.cfm?linkID=101.

Example 1:

Collapse of a Star Due to Massive Neutrino Out-flux During a Supernova

The Urantia Book unequivocally states the following:

Page 464:5   *Paper 41:8.3*

In large suns—small circular nebulae—when hydrogen is exhausted and gravity contraction ensues, if such a body is not sufficiently opaque to retain the internal pressure of support for the outer gas regions, then a sudden collapse occurs. The gravity-electric changes give origin to vast quantities of tiny particles devoid of electric potential, and such particles readily escape from the solar interior, thus bringing about the collapse of a gigantic sun within a few days. It was such an emigration of these "runaway particles" that occasioned the collapse of the giant nova of the Andromeda nebula about fifty years ago. This vast stellar body collapsed in forty minutes of Urantia time.

The supernova collapse of a star due to neutrino out-flux was not proven until the SN1987A supernova was monitored by instruments that could detect the neutrinos reaching the earth from the supernova. The "tiny particles devoid of electric potential" can only be "neutrinos" as it was later named.

References:

Glasziou, K. 1996. Neutrinos, Neutrons and Neutron Stars. *Science, Anthropology and Archeology in The Urantia Book.* http://www.urantiabook.org/archive/readers/doc182.htm#neutrinos.

Hoyle and Narlikar. 1980. *The Physics Astronomy Frontier.* San Francisco: W.H. Freeman & Co.

Novikov, I. 1990. *Black Holes and the Universe.* Cambridge, UK: Cambridge University Press.

Sutton, C. 1992. *Spaceship Neutrino.* Cambridge, UK: Cambridge University Press.

EXAMPLE 2:

CONTINENTAL DRIFT AND PLATE TECTONICS

The Urantia Book indisputably reveals the following details:

Page 663:1   *Paper 57:8:23*

750,000,000 years ago the first breaks in the continental land mass began as the great north-and-south cracking, which later admitted the ocean waters and prepared the way for the westward drift of the continents of North and South America, including Greenland. The long east-and-west cleavage separated Africa from Europe and severed the land masses of Australia, the Pacific Islands, and Antarctica from the Asiatic continent.

While continental drift had been proposed by Wegner in 1912, the concept was almost universally rejected by geologists up until the 1960's. In the 1960's, measurements of the ocean floor showed that sea floor spreading was taking place. This was recognized by geologists as the mechanism that was pushing continents apart and thus causing continental drift. It was recognized that the continents are giant floating plates drifting about on the earth's surface. This phenomenon was labeled "plate tectonics."

References:

Glasziou, K. Continental Drift. *Science, Anthropology and Archeology in The Urantia Book.* http://www.urantiabook.org/archive/readers/doc182.htm #drift.

McMenamin, M. A. S. 1998. *The Garden of Ediacara: Discovering the First Complex Life,* 173-176, 267. New York: Columbia University Press.

Scientific American 250 (2), 41, 1984.

Ibid 256 (4), 84, 1987.

Topics and Projects. http://www.ubthenews.com.

H.E. Le Grand. 1988. *Drifting Continents and Shifting Theories.* Cambridge, UK: Cambridge University Press.

EXAMPLE 3:

BLACK HOLES

The Urantia Book explicitly presents the following:

Page 173:1   *Paper 15:6.6*

The Dark Islands of Space. These are the dead suns and other large aggregations of matter devoid of light and heat. The dark islands are sometimes enormous in mass and exert a powerful influence in universe equilibrium and energy manipulation. The density of some of these large masses is well-nigh unbelievable.

The existence of black holes was theoretically proven in 1939: "But in 1939, Robert Oppenheimer published papers (with various co-authors) which predicted that stars above about three solar masses would collapse into black holes for the reasons presented by Chandrasekhar." Reference: Wikipedia article on black holes. But the existence of black holes was in doubt until astronomers determined in 1971 that the star in Cygnus X1 has a companion smaller than the earth but with more mass than a neutron star; this companion would have to be a black hole. The "dark islands of space" fit the concept of black holes. The name "Black Holes" was coined in 1968.

References:

Pathway to Discovery: Is a Black Hole Really a Hole? http://amazing -space.stsci.edu/resources/explorations/blackholes/lesson/whatisit/ history.html.

Glasziou, K. Black Holes. *Science, Anthropology and Archeology in The Urantia Book.* http://www.urantiabook.org/archive/readers/doc184.htm#black %20holes.

EXAMPLE 4:

DATES FOR CONJUNCTIONS—THE STAR OF BETHLEHEM ACCOUNT

The Urantia Book reveals the following facts:

Page 1352:3   *Paper 122:8.7*

On May 29, 7 B.C., there occurred an extraordinary conjunction of Jupiter and Saturn in the constellation of Pisces. And it is a remarkable astronomic fact that similar conjunctions occurred on September 29 and December 5 of the same year.

These three conjunctions listed in *The Urantia Book* at last provides a reasonable basis for the legend of the Star of Bethlehem which astronomers and Biblical scholars alike have been searching for and both can wholeheartedly accept.

Until the advent of digital computers and sophisticated computer programs, the Star dates being put forth did not have the validity that is now supported by the JPL dates. In 1976, at the Jet Propulsion Laboratory in California, new planetary ephemerides were computed. Using data from a table of these ephemerides, Roger W. Sinnott found that the dates for the three conjunctions occur on May 29, September 30, and December 5 in 7 B.C. Two of these dates are *exactly* the same as revealed in *The Urantia Book*; one is off by only one day. (Note: a one day difference may be as little as a fraction of a second or as much as 24 hours, depending upon conventions used to define the date to which a particular night belongs). The probability for attaining this result in 1955 from random guesswork is astronomical. The Mercier dates of 1871 demonstrate that this knowledge was earned and authoritatively brought to light in exquisite detail in *The Urantia Book*. *The Urantia Book* ignored all the other Star accounts that are out there and unequivocally supported the Mercier dates that in turn were confirmed in the 1986 report from the results of the JPL dates.

References:

Sinnott, R. December 1986. Star of Bethlehem. *Sky and Telescope*.

Bain, R. and K. Glasziou, M. Neibaur, and F. Wright, eds. The Star of Bethlehem. *Science Content of The Urantia Book*. http://www.truthbook.org/index.cfm?linkID=101.

Mercier, L.P. 1871. *Outlines of the life of the Lord Jesus Christ*. Oxford University.

Additional Scientific Concepts in *The Urantia Book* that Coincide with and Predate the Validation of Science or Before Being Accepted into the Scientific Mainstream

The following list contains scientific concepts revealed in *The Urantia Book*. Some were not verified by our science until after the first three parts of the *Urantia Papers* were received in the mid 1930's; most were not verified by our science until after the book was published in 1955.

Planetary Atmospheres: The density of the atmospheres of Venus and Mars were not confirmed until after the advent of spacecraft that could visit these planets and measure the density. *The Urantia Book* states that Venus has an atmosphere considerably denser than that of our planet, while Mars has an atmosphere considerably less dense than ours.

Urantia Book Reference: Page 561:5; Paper 49:2.6

Motions of the Moon: *The Urantia Book* revealed that the moon is moving away from the earth. This was not confirmed until the 1970's when lasers were bounced off reflectors left on the moon by US astronauts to measure the change in distance over time between the moon and the earth.

Urantia Book Reference: Page 657:6; Paper 57:6.3

Scientific American 249 (6), 71.

Scharringhausen, B. 2002. Is the Moon Moving Away from the Earth? *Curious about astronomy?* http://curious.astro.cornell.edu/question.php?number =124.

Tycho Brahe's Nova of 1572: the *Urantia Papers* state that this nova was due to the explosion of a double star. This was not confirmed until 1967 by data from the Einstein X-ray observatory.

Urantia Book Reference: Page 458:5; Paper 41:3:5

Urantia Brotherhood Bulletin. Nova of 1572 Explained.

SN 1572 or Tycho's Nova. Wikipedia. http://en.wikipedia.org/wiki/SN_1572.

Crab Nebula: *The Urantia Book* speaks of a small remnant star at the center of the nebula. This star, which is a pulsar, was first detected in 1967.

Urantia Book Reference: Page 464:6; Paper 41:8.4

Kaufmann, W. and R. Freeman. *The Universe*. W.H. Freeman and Company.

SN 1054 (Crab Supernova). Wikipedia. http://en.wikipedia.org/wiki/SN_1054.

AGE OF THE SOLAR SYSTEM: the *Urantia Papers* give the age of the earth as 4.5 billion years. As late as 1950, the age of the earth was accepted by scientists as about 2 billion years. More recent radioactive element dating has established the age of the earth as 4.55 billion years.

Urantia Book References: Pages 655:9, 656:1-7, 657:1-3; Paper 57:5.4-14

Kaufmann, W. and R. Freeman. *The Universe*. W.H. Freeman and Company.

THE GREAT KENTUCKY VOLCANIC ERUPTION: the *Urantia Papers* inform us of a great volcanic eruption about 325 million years ago in the area of Kentucky. In the 1980's, volcanic ash in the area was dated to about 310 million years in age.

Urantia Book Reference: Page 675:4; Paper 59:2.5

Keiser, A.F., Blake, B.M. Jr., and Grady, W.C. Spring 1999. Volcanic Ash in West Virginia. *Geoscience Education in the Mountain State: CATS Geology Telecourse*. http://www.wvgs.wvnet.edu/www/geoeduc/edcs99a2.htm.

X-RAYS FROM THE SUN: *The Urantia Book* let us know that our sun and other stars generate and radiate x-rays. Scientists did not detect these x-rays until 1948.

Urantia Book References: Pages 460:6-461:1, 465:1.4; Paper 41:5.1.3-5 41:9.1.4

Hoyle, F. and Narlikar, J. 1980. *The Physics-Astronomy Frontier*. p. 173. San Francisco: W.H. Freeman and Co.

Clark, D.H. 1987. *The Cosmos from Space*. New York: Crown.

Keller, C.U. July 1995. X-rays from the Sun. *Cellular and Molecular Life Sciences*. Volume 51, Number 7, Pages710-720.

THE MYSTERY OF THE MEDITERRANEAN BASIN: The Papers state that the Mediterranean Sea was isolated from the Atlantic ocean for millions of years. This fact was discovered by geologists in 1970.

Urantia Book References: Pages 697:3, 698:4, 721:8, 728:2, 826:6-827:0, 889:3-4, 890:6-8, 891:0-1; Papers 61:3.8, 61:4.2, 64:4.10, 64:7.13, 73:7.1, 80:1-2, 80:2.2-5

Morrison, P. 1987. *The Ring of Truth: An Inquiry into How We Know What We Know*. Random House.

TEMPERATURE OF DEEP SPACE: *The Urantia Book* tells us that space has a temperature; it is not devoid of heat. This was not confirmed until 1965 when noise due to the temperature of space was first detected by Arno Penzias and Robert Wilson of Bell Telephone.

Urantia Book Reference: Page 473:4; Paper 42:4.6.

NASA. Discovery of the Cosmic Microwave Background. *Universe 101: Our Universe.* http://map.gsfc.nasa.gov/m_uni/uni_101bbtest3.html.

Harwit, M. 1981. *Cosmic Discovery: The Search, Scope, and Heritage of Astronomy.* New York: Basic Books.

Merken, M. 1985. *Physical Science with Modem Applications.* Philadelphia: Saunders Pub.

RESEARCH PROJECTS THAT DIRECTLY OR INDIRECTLY RELATE TO *THE URANTIA BOOK* WHICH WARRANTS KEEPING ON YOUR RADAR SCREEN

Allow me to now draw attention to three interesting projects that stand to have further significant bearing on the validity and value of the *Urantia Papers* should these projects come to fruition.

ATLANTIS AS IT RELATES TO THE GARDEN OF EDEN ACCOUNT IN PAPER 73:

Sarmast, R. Cyprus-Atlantis Project. http://www.discoveryofatlantis.com/.

DNA SEQUENCE STRUCTURE OF THE GENE MICROCEPHALIN.

Modern copies of D Alleles arose from a single progenitor copy about 37,000 years ago. Non-D Alleles likely evolved in a separate lineage about 1.1 million years ago. This should grab your attention!

Howard Hughes Medical Institute. 2006. Could Interbreeding Between Humans and Neanderthalls Have Led to an Enhanced Human Brain? http://www.hhmi.org/news/lahn20061006.html.

The dates of 37,000 and 1.1 million years in Lahn's DNA research as it relates to Papers 62, 74, and 78 in *The Urantia Book* are remarkable. On 17 November 2006 I initiated my first of many emails to Professor Lahn regarding his article. Professor Lahn is on the right track with the DNA analysis but the wrong train of thought with his speculation that the Neanderthals are the source of the original copy of the D Allele.

Adam and Eve arrived on our planet 37,848 years ago from the year A.D. 1934 (Paper 74). The purpose of their arrival was to biologically uplift the evolutionary races. The genetic exchange between the biologic uplifters and the mortal races 37,000 years ago as revealed in *The Urantia Book* satisfies the two prominent competing theories regarding interbreeding. Also, religious scholars now have a consistent history of Adam and Eve that science has confirmed through DNA research 60 years after it was revealed in *The Urantia Book*. Once again COSAR in Action!

The Urantia Book accurately supports the non-D alleles of 1.1 million years with the same detailing consistency as it did with the D allele of 37,000 years ago.

This research was brought to my attention by Lee Rector, a long time student of *The Urantia Book*, on 09 November 2006 and merits ongoing investigation.

APPARATUS FOR GYROSCOPIC PROPULSION EXPLAINED:

Ganid Productions LLC. 2007. www.ganid.com. http://www.youtube.com/watch?v =7Lka6d6DDBs &feature=email.

Antigravity is mentioned twenty-one times in *The Urantia Book*. The following passage from Paper 9 in *The Urantia Book* is insightful to the understanding of this project.

Page 101.3   *Paper 9:3.3*

Antigravity can annul gravity within a local frame; it does so by the exercise of equal force presence. It operates only with reference to material gravity, and it is not the action of mind. The gravity-resistant phenomenon of a gyroscope is a fair illustration of the *effect* of antigravity but of no value to illustrate the cause of antigravity.

EVOLUTION OF SCIENCE AND THE SCIENTIST

In the following two paragraphs Dick Bain offers a historical perspective on the evolution of science and the scientist:

"Before the rise of the scientific method and the advent of the so-called Age of Reason in the 17th century, there was a great deal of pseudoscience that was widely accepted. When genuine scientists began to apply reason and the scientific method to their investigations, it was necessary to distance themselves from all the unverifiable mysticism and hokum that passed as science. Unfortunately, the church was as much an impediment to the advancement of science as was all the mystical garbage that passed for truth in those days. Sad to relate, the rift that developed between science and religion has not been fully healed to this day.

"It was necessary for scientists to distance themselves from all the pseudoscience, but did they throw the baby out with the bath water? The methods of science are not the methods of religion; because science can only measure material realities does not mean that no other realities exist. A real scientist should admit that there are realities even in his or her own life that cannot be measured. Can anyone measure love rather than just the physiological effects of it? Much of the content of *The Urantia Book* can only be validated in our souls, but some of the science can be shown to predate our science and this provides assurance that a large amount of science content in *The Urantia Book* is what it says it is. This assurance can encourage us to seriously consider the spiritual truths in the book as well. Each of us must study *The Urantia Book* with an open heart and critical mind and let the Spirit of Truth confirm its validity supported by the facts of the cosmology that is part of this revelation."

Let us now view the COSAR Principle as it is revealed in *The Urantia Book* in light of Dick's insight into the above history of the reflective scientist and the evolution of science. "A logical and consistent philosophic concept of the universe cannot be built up on the postulations of either materialism or spiritism, for both of these systems of thinking, when universally applied, are compelled to view the cosmos in distortion, the former contacting with a universe turned inside out, the latter realizing the nature of a universe turned outside in. Never, then, can either science or religion, in and of themselves, standing alone, hope to gain an adequate understanding of universal truths and relationships without the guidance of human philosophy and the illumination of divine revelation." (Paper 103)

Philosophy is a great tool for us as we examine and study science and religion; it helps harmonize these two diametrically opposite avenues of approaching the impersonal and personal natures of the cosmos and Deity. The art of philosophy associated with an epochal revelation of truth is a mighty leveler when it comes to coordinating the scientific and religious aspects of the COSAR Principle. So, as the so-called "Age of Reason" continues to evolve more and more scientists will discover for themselves the cosmic truth, universe beauty and divine goodness revealed in *The Urantia Book.*

Progress has been made since 1940 when philosopher Mortimer Adler (See Foreword) drew attention to the dilemma of lack of communication between scientists, philosophers, and theologians. We are by no means out of the woods yet. However, with the likes of the Templeton Foundation's "Science and Religion Dialogue" programs (www.Templeton.org) and the work of Paul Davies and his group (http://cosmos.asu.edu/) we are advancing toward the COSAR Principle as outlined in this chapter.

CONCLUSIONS

It is impossible to believe that all the correct scientific concepts in *The Urantia Book* that predate the discoveries of our science are simply lucky guesses. If the truths revealed in this book are not from the authors that take credit for these papers, then how else can we account for this amazing fact? The astronomical odds of "getting it right" by sheer guesswork should make the hair stand up on the back of your neck. If not, frankly I'm puzzled.

Though *The Urantia Book* addresses primarily our spiritual natures, it also illuminates the relationship between science and religion. The *Urantia Papers* emphasize that religion and science are not two antagonistic views of reality; they are views of two aspects of one reality, those aspects being the personal and impersonal natures of God. The bumper sticker you may have seen lately "In the Beginning, God Created Evolution" demonstrates the increasing consciousness of our Cosmic Citizenship and a healthy respect for science (COSAR expressed on a bumper sticker).

When all is said and done, I have come up with the COSAR Principle—A Revelatory Proposition to demonstrate and restate the manner by which *The Urantia Book* Coordinates Science and Religion with the truth of reality. New and long time readers alike of *The Urantia Book* should be mindful that this is the *function* of an epochal revelation (Pause to reflect on the front cover picture with the quote below it). For those that fail to accept or grasp this, I will simply say have a nice day. Your approach to the truth of reality will continue to be an ineffective method at metaphysics. Also, you are not being fair to yourself if you embrace only one side of COSAR. The deep, independent and free thinkers now have available to them within the pages of *The Urantia Book*, an "accurate geology, profound scientific truths" and a cosmology worthy of the 21st century, something that a growing number of the spiritual minded and the "skeptics alike find lacking" in much of the so-called sacred scriptures (McMenamin pp. 173-174).

These are indeed stirring times! Not in 2000 years has our world been gifted with a Revelation that ascends to a level of cosmic dignity as does *The Urantia Book*. *The Urantia Book* is not a mere revelation of facts. *The Urantia Book* is an Epochal Revelation of Truth. This Revelatory Proposition (The Truth of Reality) in the *Urantia Papers* coordinate science and religion (COSAR) in such a way that is vastly superior to any other written source available on our planet.

I will leave you with the following quote from the Foreword of *The Urantia Book*:

Page 17   *Foreword: Acknowledgement.3*

We are fully cognizant of the difficulties of our assignment; we recognize the impossibility of fully translating the language of the concepts of divinity and eternity into the symbols of the language of the finite concepts of the mortal mind. But we know that there dwells within the human mind a fragment of God, and that there sojourns with the human soul the Spirit of Truth; and we further know that these spirit forces conspire to enable material man to grasp the reality of spiritual values and to comprehend the philosophy of universe meanings. But even more certainly we know that these spirits of the Divine Presence are able to assist man in the spiritual appropriation of all truth contributory to the enhancement of the ever-progressing reality of personal religious experience—God-consciousness.

Acknowledgment

I would not attempt to provide the examples of the science content of *The Urantia Book* in this chapter to support the COSAR Principle without the decades of research in this area by Dr. Matt Neibaur, Ken Glasziou, Richard Bain, & Frank Wright. These dedicated Urantians exemplify the passage in Paper 25:4.5 – "In the universal regime you are not reckoned as having possessed yourself of knowledge and truth until you have demonstrated your ability and willingness to impart this knowledge and truth to others." I most gratefully acknowledge my indebtedness to Matt Neibaur and the team of scientists who carry on in keeping all of us abreast of the science content in the *Urantia Papers* as it continues to validate our natural science. I express my deep gratitude to Dick Johnson for the personal touch that he contributed to this chapter as only Dick can do.

The spiritual forward urge is the most powerful driving force present in this world; the truth-learning believer is the one progressive and aggressive soul on earth.
—Urantia Book, Paper 194:3.4

PART II

EVANGELISM

There is a solemn, heartfelt, message in Part II
for those who are ready to
"Take off the apron"
and be about our Father's business.

The Urantia Book—You've Got To Read It

How does one go about summarizing a two thousand ninety seven-page letter from God? Sometime during my first year of reading *The Urantia Book* in 1975, I decided to put to paper my personal insights gleamed from our fifth epochal revelation based on my religious experiences up to that time. Being thirty years younger, considerable more brain cells, and having read *The Urantia Book* from cover to cover three times the first year the following newspaper summary is basically what unfolded. Some paragraphs have been rearranged and a couple added about six years ago. The original summary appeared in *The Honolulu Advertiser* on Tuesday, February 14, 1978. (*The Captain Cook* edition even made Mainland distribution).

I had not planned on publishing it, but when I wrote the Urantia Foundation for approval to give a talk along these lines, they responded favorably but said a disclaimer would have to be included at the bottom if I placed it in writing. That prompted me to publish it in the *Honolulu Advertiser*. The original version did not have the word Urantia or any other term that would point to *The Urantia Book*. I did not leave my name or contact address either, just basically what follows with the exceptions noted above. The original title was "Man's Gift to God (Whole Hearted Desire To Do His Will)." Because of the disclaimer the Foundation insisted I insert, individuals were able to find the source of my inspiration because the Foundation's disclaimer noted *The Urantia Book*. I personally know of a long-time reader who was on vacation in Hawaii who found the book from that article. I did not hear about that until a year or so later. I had already departed Hawaii a month before the article came out as I accepted a position with an Aerospace related company in Saudi Arabia.

"Individuals of all faiths should become familiar with the truths revealed in The Urantia Book to discern if these are of any value. Will it bring you closer to God and God closer to you? Once you discern that these truths are of value to you, allow the will of our Heavenly Father to become dominant and transcendent in your heart. And then, as transformed spirit-led faith sons and daughters of God, let us join hands in spiritual unity with all other Urantia Book believers, and as universe cosmic-conscious citizens let us go forth to transform the world...."

This paragraph above was one of the paragraphs I included in the Arizona Republic, when I published the current summary below in the later part of the 90s. I did not feel that the 70s was the appropriate time to include this paragraph. I point this out, as I will dwell into this deeper in the Epilogue of Part II of this book.

The Urantia Book - You've Got To Read It!

**It is high time that man had a religious experience
so personal and so sublime that it could be realized and expressed
only by "feelings that lie too deep for words."**

Did that move you? It did me. I found it in *The Urantia Book*. The truth should so illuminate your heart that you would readily accept it from whatever source. *The Urantia Book* is a religious revelation of epochal significance to our planet. This collection of papers coordinates both science and religion with the truth of reality. It provides a coordinated and unbroken explanation of both science and religion.

Based upon my personal religious experiences and guided by the truths revealed in *The Urantia Book* I offer the following insights:

Those who know God have experienced the fact of his presence.

Such God-knowing mortals hold in their personal experience the only positive proof of the existence of the living God which one human being can offer to another.

The experience of faith is the assurance of God- consciousness.

The Urantia Book is the desire of truth seekers of all ages. This epochal revelation of Truth derives its authority from the fruits of its acceptance in the hearts of those individuals who choose to accept these truths based upon the leading of the divine fragment that indwells each of us and guided by the spirit of truth. It reveals the origin and destiny of man and everything in-between, our relationship to the cosmos and our Universal Father, a detailed description of the origin and history of this planet, and our ascension career. The papers culminate in a restatement of the life and teachings of Jesus the likes of which you have never seen before—every single year of his life, before he was incarnated to after he was crucified.

Pause to consider the following truths revealed in *The Urantia Book*: Being finite evolutionary free will creatures makes it impossible for us to fully and completely understand God, the infinite being. Although God has superhuman qualities such as omnipotence, omnipresence, omniscience and other divine attributes, it does not lessen the fact that he is a personal loving father to each and every one of us. The Father idea is the highest human concept of God we have.

Among the many attributes of God love is second to none. Love is the desire to do good to others. *If we desire to do good to others then we desire to do the Father's will.*

God, our Universal Father bestowed upon each and every one of us our own unique and individual personality along with giving us free will within our own mind domain. God, the first source and center of all things, has also bestowed within us a fragment of himself. This divine fragment that indwells our mind is not our mind; it is not our personality; and it is not our conscious. It is pre-mind, pre-personal and super-conscious. We must choose with our own free will to become God-like, to become perfect as a creature as God is perfect as a creator. To the extent that we desire to become God-like and desire to do the Father's will, it is to that extent that this divine fragment can give us insight of a spiritual nature and cosmic meaning.

God, our Heavenly Father, has made available to us a divine plan. We have the opportunity, potential and capacity to ascend all the way to where God personally resides. If we accept this ascension career, this divine plan, and desire to do the Father's will, upon mortal death we will receive another form on another sphere. It will be more than the material flesh and blood that we have now but it will be less than a spiritual form. Our personality will survive and indwell this new form along with the divine fragment that is in us now, which is what makes it possible for us to survive mortal death and is what sets us apart from the rest of the animal kingdom. Everything that we have experienced that is of survival value, which is the soul, will also indwell this new form along with our memory.

Nothing can take the place of experience. The only thing that we gain from death itself is the fact that we experience survival. There are no short cuts. What we do not learn here we must learn sooner or later.

If we reject this ascension career, this divine plan, and make a knowing and conscious decision not to do the Father's will, upon mortal death the personality will not survive. We will cease to exist and be as if we never were. I am not referring to those who are ignorant or those who have sincere and honest doubts. Since anything of survival value is never lost, whatever that personality experienced that was worthwhile (which consists of truth, beauty and goodness) will be absorbed into the oversoul of creation, the Supreme Being.

Once we recognize our personal status as a truly loved child of God, our Spirit Father, we will become increasingly aware that we are all spiritual brothers and sisters. Having found God within ourselves, we cannot help but go forth sharing the Father's love with our spiritual family. The personal teachings and life example of Jesus of Nazareth as restated in *The Urantia Book* inspires us to share his religious faith and enter into the spirit of the Master's life of unselfish service for humankind.

There indwells within our heart the Spirit of Truth that Jesus of Nazareth, the Son of Man and Son of God, has bestowed upon us. This Spirit of Truth, also known as the comforter, will validate so that we can recognize in our mind that these truths are of a spiritual origin. Once we discern

that these truths are of a spiritual origin we still must choose with our own free will either to accept or reject them. To the extent that we accept these truths and desire to do the Father's will, to that extent we are in harmony with the universe and our Universal Father.

The religious challenge of this day and age is to those far seeing and forward-looking men and women of spiritual insight who would dare to construct a new and appealing religious philosophy of living out of the enlarged and exquisitely integrated modern concepts of cosmic truth, universal beauty and divine goodness revealed in *The Urantia Book*.

Individuals of all faiths should become familiar with the truths revealed in *The Urantia Book* to discern if these are of any value. Will it bring you closer to God and God closer to you? Once you discern that these truths are of value to you, allow the will of our Heavenly Father to become dominant and transcendent in your heart. And then, as transformed spirit-led faith sons and daughters of God, let us join hands in spiritual unity with all other Urantia Book believers, and as universe cosmic-conscious citizens let us go forth to transform the world.

Those who confess their faith in the Fatherhood of God certainly recognize the spiritual brotherhood of all humankind. With *The Urantia Book* as our guide along with our personal religious experiences we have the opportunity and potential to spiritually uplift our planet and establish worldwide peace under law.

We are all sons and daughters of God and through faith we can actually realize and daily experience this ennobling truth.

May the peace of our Universal Father be with you.

(In Memory of John Haines)

—JJ Johnson
www.azurantia.org
www.urantiabook.org

Chapter Two

Evangelism and The Urantia Book

Page 1780:5   *Paper 160:5.3*

If something has become a religion in your experience, it is self-evident that you already have become an active evangel of that religion since you deem the supreme concept of your religion as being worthy of the worship of all mankind, all universe intelligences. If you are not a **positive and missionary evangel of your religion,** you are self-deceived in that what you call a religion is only a traditional belief or a mere system of intellectual philosophy. If your religion is a spiritual experience, your object of worship must be the universal spirit reality and ideal of all your spiritualized concepts. All religions based on fear, emotion, tradition, and philosophy I term the intellectual religions, while those based on true spirit experience I would term the true religions. The object of religious devotion may be material or spiritual, true or false, real or unreal, human or divine. Religions can therefore be either good or evil.

Over three decades ago, when I first encountered the above passage, I naturally embraced and immediately identified with the evangelic message. As sincere students of *the Urantia Papers*, let's not lose sight of our loyalty, obligation, privilege and joy of being a "positive and missionary evangel" of our religion.

The following passages are provided for additional guidance and insight that relate directly or indirectly to the evangelical call within all of us who desire to serve in this manner. The quotes below cover all four parts of *The Urantia Book* and encompass the expressions of numerous authors.

Page 193:8   *Paper 16:7.9*

Morality can never be advanced by law or by force. It is a personal and freewill matter and must be **disseminated by the contagion of the contact of morally fragrant persons** with those who are less morally responsive, but who are also in some measure desirous of doing the Father's will.

Page 279:13   *Paper 25:4.5*

Those mortals and midwayers who serve transiently with the advisers are chosen for such work because of their expertness in the concept of universal law and supreme justice. As you journey toward your Paradise goal, constantly acquiring added knowledge and enhanced skill, you are continuously afforded the opportunity to give out to others the wisdom and experience you have already accumulated; all the way in to Havona you enact the role of a pupil-teacher. You will work your way through the ascending levels of this vast experiential university by imparting to those just below you the new-found knowledge of your advancing career. **In the universal regime you are not reckoned as having possessed yourself of knowledge and truth until you have demonstrated your ability and your willingness to impart this knowledge and truth to others.**

Page 381:1   *Paper 34:6.7*

Those who have received and recognized the indwelling of God have been born of the Spirit. "You are the temple of God, and the spirit of God dwells in you." It is not enough that this spirit be poured out upon you; **the divine Spirit must dominate and control every phase of human experience.**

Page 1017:3–6   *Paper 93:4.1–4*

The ceremonies of the Salem worship were very simple. Every person who signed or marked the clay-tablet rolls of the Melchizedek church committed to memory, and subscribed to, the following belief:
1. I believe in El Elyon, the Most High God, the only Universal Father and Creator of all things.
2. I accept the Melchizedek covenant with the Most High, which bestows the favor of God on my faith, not on sacrifices and burnt offerings.
3. I promise to obey the seven commandments of Melchizedek and to **tell the good news of this covenant with the Most High to all men.**

Page 1041:5   *Paper 94:12.7*

All Urantia is waiting for the proclamation of the ennobling message of Michael, unencumbered by the accumulated doctrines and dogmas of nineteen centuries of contact with the religions of evolutionary origin. The hour is striking for presenting to Buddhism, to Christianity, to Hinduism, even to the peoples of all faiths, not the gospel about Jesus, but the living, spiritual reality of the gospel of Jesus.

Page 1051:1   *Paper 95:7.3*

Here and there throughout Arabia were families and clans that held on to the hazy idea of the one God. Such groups treasured the traditions of Melchizedek, Abraham, Moses, and Zoroaster. There were numerous centers that might have responded to the Jesusonian gospel, but the Christian missionaries of the desert lands were an austere and unyielding group in contrast with the compromisers and innovators who functioned as missionaries in the Mediterranean countries. Had the followers of Jesus taken more seriously his injunction to **"go into all the world and preach the gospel,"** and had they been more gracious in that preaching, less stringent in collateral social requirements of their own devising, then many lands would gladly have received the simple gospel of the carpenter's son, Arabia among them.

Page 1673:4   *Paper 149:4.4*

And then Jesus discoursed on the dangers of courage and faith, how they sometimes lead unthinking souls on to recklessness and presumption. He also showed how **prudence and discretion, when carried too far, lead to cowardice and failure.** He exhorted his hearers to strive for originality while they shunned all tendency toward eccentricity. He pleaded for sympathy without sentimentality, piety without sanctimoniousness. He taught reverence free from fear and superstition.

Page 1725:4   *Paper 155:1.3*

Jesus continued to teach the twenty-four, saying: "The heathen are not without excuse when they rage at us. Because their outlook is small and narrow, they are able to concentrate their energies enthusiastically. Their goal is near and more or less visible; wherefore do they strive with valiant and effective execution. You who have professed entrance into the kingdom of heaven are **altogether too vacillating and indefinite in your teaching conduct.** The heathen strike directly for their objectives; you are guilty of too much chronic yearning. If you desire to enter the kingdom, why do you not take it by spiritual assault even as the heathen take a city they lay siege to? You are hardly worthy of the kingdom when your service consists so largely in an attitude of regretting the past, whining over the present, and vainly hoping for the future. Why do the heathen rage? Because they know not the truth. Why do you languish in futile yearning? Because you obey not the truth. Cease your useless yearning and **go forth bravely doing that which concerns the establishment of the kingdom.**

Page 1733:1   *Paper 155:6.13*

When you once begin to find God in your soul, presently you will begin to discover him in other men's souls and eventually in all the creatures and creations of a mighty universe. **But what chance does the Father have to appear as a God of supreme loyalties and divine ideals in the souls of men who give little or no time to the thoughtful contemplation of such eternal realities?** While the mind is not the seat of the spiritual nature, it is indeed the gateway thereto.

Page 1745:1   *Paper 157:2.2*

When Jesus had thus spoken, he withdrew and prepared for the evening conference with his followers. At this conference it was decided to undertake a united mission throughout all the cities and villages of the Decapolis as soon as Jesus and the twelve should return from their proposed visit to Caesarea Philippi. The Master participated in planning for the Decapolis mission and, in dismissing the company, said: "I say to you, beware of the leaven of the Pharisees and the Sadducees. Be not deceived by their show of much learning and by their profound loyalty to the forms of religion. Be only concerned with the spirit of living truth and the power of true religion. It is not the fear of a dead religion that will save you but rather your faith in a living experience in the spiritual realities of the kingdom. Do not allow yourselves to become blinded by prejudice and paralyzed by fear. Neither permit reverence for the traditions so to pervert your understanding that your eyes see not and your ears hear not. **It is not the purpose of true religion merely to bring peace but rather to insure progress.** And there can be no peace in the heart or progress in the mind unless you fall wholeheartedly in love with truth, the ideals of eternal realities. The issues of life and death are being set before you—the sinful pleasures of time against the righteous realities of eternity. Even now you should begin to find deliverance from the bondage of fear and doubt as you enter upon the living of the new life of faith and hope. And when the feelings of service for your fellow men arise within your soul, do not stifle them; when the emotions of love for your neighbor well up within your heart, give expression to such urges of affection in intelligent ministry to the real needs of your fellows."

Page 1930:6   *Paper 178:1.9*

So long as the rulers of earthly governments seek to exercise the authority of religious dictators, you who believe this gospel can expect only trouble, persecution, and even death. But the very light which you bear to the world, and even the very manner in which you will suffer and die for this gospel of the kingdom, will, in themselves, eventually enlighten the whole world and result in the gradual divorcement of politics and religion. **The persistent preaching of this gospel of the kingdom will some day bring to all nations a new and unbelievable liberation, intellectual freedom, and religious liberty.**

Page 1931:2   *Paper 178:1.11*

Remember that you are commissioned to preach this gospel of the kingdom— the supreme desire to do the Father's will coupled with the supreme joy of the faith realization of sonship with God—**and you must not allow anything to divert your devotion to this one duty.** Let all mankind benefit from the overflow of your loving spiritual ministry, enlightening intellectual communion, and uplifting social service; but none of these humanitarian labors, nor all of them, should be permitted to take the place of proclaiming the gospel. These mighty ministrations are the social by-products of the still more mighty and sublime ministrations and transformations wrought in the heart of the kingdom believer by the living Spirit of Truth and by the personal realization that the faith of a spirit-born man confers the assurance of living fellowship with the eternal God.

Page 1931:5   *Paper 178:1.14*

You are not to be passive mystics or colorless ascetics; you should not become dreamers and drifters, supinely trusting in a fictitious Providence to provide even the necessities of life. You are indeed to be gentle in your dealings with erring mortals, patient in your intercourse with ignorant men, and forbearing under provocation; **but you are also to be valiant in defense of righteousness, mighty in the promulgation of truth, and aggressive in the preaching of this gospel of the kingdom, even to the ends of the earth.**

Page 2082:9   *Paper 195:9.4*

Religion does need new leaders, spiritual men and women who will dare to depend solely on Jesus and his incomparable teachings. If Christianity persists in neglecting its spiritual mission while it continues to busy itself with social and material problems, **the spiritual renaissance must await the coming of these new teachers of Jesus' religion who will be exclusively devoted to the spiritual regeneration of men.** And then will these spirit-born souls quickly supply the leadership and inspiration requisite for the social, moral, economic, and political reorganization of the world.

31

Epilogue

Individuals of all faiths should become familiar with the truths revealed in The Urantia Book to discern if these are of any value. Will it bring you closer to God and God closer to you? Once you discern that these truths are of value to you, allow the will of our Heavenly Father to become dominant and transcendent in your heart. And then, as transformed spirit-led faith sons and daughters of God, let us join hands in spiritual unity with all other Urantia Book believers, and as universe cosmic-conscious citizens let us go forth to transform the world.

The above paragraph, included in the Prologue, was drawn from the many reinforcements below. These passages are insightful and uplifting. Nothing but good can come from a sincere study of these references covering Parts II, III, and IV, including 14 Papers with over 1700 pages between the first and last reference.

Page 381:1   *Paper 34:6.7*

Those who have received and recognized the indwelling of God have been born of the Spirit. "You are the temple of God, and the spirit of God dwells in you." It is not enough that this spirit be poured out upon you; **the divine Spirit must dominate and control every phase of human experience.**

Page 804:15   *Paper 71:4.3*

And this progress in the arts of civilization leads directly to the realization of the highest human and divine goals of mortal endeavor—the social achievement of the brotherhood of man and the personal status of God-consciousness, which becomes revealed in **the supreme desire of every individual to do the will of the Father in heaven.**

Page 1124:3   *Paper 102:6.1*

The philosophic elimination of religious fear and the steady progress of science add greatly to the mortality of false gods; and even though these casualties of man-made deities may momentarily befog the spiritual vision, they eventually destroy that ignorance and superstition which so long obscured the living God of eternal love. The relation between the creature and the Creator is a living experience, a dynamic religious faith, which is not subject to precise definition. To isolate part of life and call it religion is to disintegrate life and to distort religion. **And this is just why the God of worship claims all allegiance or none.**

Page 1175:1   *Paper 106:9.12*

To material, evolutionary, finite creatures, **a life predicated on the living of the Father's will leads directly to the attainment of spirit supremacy in the personality arena and brings such creatures one step nearer the comprehension of the Father-Infinite.** Such a Father life is one predicated on truth, sensitive to beauty, and dominated by goodness. Such a God-knowing person is inwardly illuminated by worship and outwardly devoted to the wholehearted service of the universal brotherhood of all personalities, a service ministry which is filled with mercy and motivated by love, while all these life qualities are unified in the evolving personality on ever-ascending levels of cosmic wisdom, self-realization, God-finding, and Father worship.

Page 1434:2   *Paper 130:4.3*

The highest level to which a finite creature can progress is the recognition of the Universal Father and the knowing of the Supreme. And even then such beings of finality destiny go on experiencing change in the motions of the physical world and in its material phenomena. Likewise do they remain aware of selfhood progression in their continuing ascension of the spiritual universe and of growing consciousness in their deepening appreciation of, and response to, the intellectual cosmos. Only in the perfection, harmony, and unanimity of will can the creature become as one with the Creator; and such a state of divinity is attained and maintained only by the creature's continuing to live in time and eternity by consistently conforming his finite personal will to the divine will of the Creator. **Always must the desire to do the Father's will be supreme in the soul and dominant over the mind of an ascending son of God.**

Page 1521:1   *Paper 136:8.6*

Jesus chose to establish the kingdom of heaven in the hearts of mankind by natural, ordinary, difficult, and trying methods, just such procedures as his earth children must subsequently follow in their work of enlarging and extending that heavenly kingdom. For well did the Son of Man know that it would be "through much tribulation that many of the children of all ages would enter into the kingdom." Jesus was now passing through the great test of civilized man, to have power and steadfastly refuse to use it for purely selfish or personal purposes.

Page 1522:1   *Paper 136:9.2*

The Jews envisaged a deliverer who would come in miraculous power to cast down Israel's enemies and establish the Jews as world rulers, free from want and oppression. Jesus knew that this hope would never be realized. He knew that the kingdom of heaven had to do with the overthrow of evil in the hearts of men, and that it was **purely a matter of spiritual concern.** He thought out the advisability of inaugurating the spiritual kingdom with a brilliant and dazzling display of power—and such a course would have been permissible and wholly within the jurisdiction of Michael—but he fully decided against such a plan. He would not compromise with the revolutionary techniques of Caligastia. He had won the world in potential by submission to the Father's will, and he proposed to finish his work as he had begun it, and as the Son of Man.

Page 1522:5   *Paper 136:9.6*

Rome was mistress of the Western world. The Son of Man, now in isolation and achieving these momentous decisions, with the hosts of heaven at his command, represented the last chance of the Jews to attain world dominion; but this earthborn Jew, who possessed such tremendous wisdom and power, declined to use his universe endowments either for the aggrandizement of himself or for the enthronement of his people. He saw, as it were, "the kingdoms of this world," and he possessed the power to take them. The Most Highs of Edentia had resigned all these powers into his hands, but he did not want them. The kingdoms of earth were paltry things to interest the Creator and Ruler of a universe. **He had only one objective, the further revelation of God to man, the establishment of the kingdom, the rule of the heavenly Father in the hearts of mankind.**

Page 1535:5   *Paper 137:7.13*

While Jesus later directed that the apostles should go forth, as John had, preaching the gospel and instructing believers, he laid emphasis on the proclamation of the "good tidings of the kingdom of heaven." He unfailingly impressed upon his associates that they must "show forth love, compassion, and sympathy." He early taught his followers that **the kingdom of heaven was a spiritual experience having to do with the enthronement of God in the hearts of men.**

Page 1537:2   *Paper 137:8.15*

"Entrance into the Father's kingdom waits not upon marching armies, upon overturned kingdoms of this world, nor upon the breaking of captive yokes. The kingdom of heaven is at hand, and **all who enter therein shall find abundant liberty and joyous salvation.**"

Page 1568:5   *Paper 140:1.2*

"**The new kingdom which my Father is about to set up in the hearts of his earth children is to be an everlasting dominion.** There shall be no end of this rule of my Father in the hearts of those who desire to do his divine will. I declare to you that my Father is not the God of Jew or gentile. Many shall come from the east and from the west to sit down with us in the Father's kingdom, while many of the children of Abraham will refuse to enter this new brotherhood of the rule of the Father's spirit in the hearts of the children of men."

Page 1588:4   *Paper 141:2.1*

The night before they left Pella, Jesus gave the apostles some further instruction with regard to the new kingdom. Said the Master: "You have been taught to look for the coming of the kingdom of God, and now I come announcing that this long-looked-for kingdom is near at hand, even that it is already here and in our midst. In every kingdom there must be a king seated upon his throne and decreeing the laws of the realm. And so have you developed a concept of the kingdom of heaven as a glorified rule of the Jewish people over all the peoples of the earth with Messiah sitting on David's throne and from this place of miraculous power promulgating the laws of all the world. But, my children, you see not with the eye of faith, and you hear not with the understanding of the spirit. I declare that **the kingdom of heaven is the realization and acknowledgment of God's rule within the hearts of men.** True, there is a King in this kingdom, and that King is my Father and your Father. We are indeed his loyal subjects, but far transcending that fact is the transforming truth that we are his sons. In my life this truth is to become manifest to all. Our Father also sits upon a throne, but not one made with hands. The throne of the Infinite is the eternal dwelling place of the Father in the heaven of heavens; he fills all things and proclaims his laws to universes upon universes. And the Father also rules within the hearts of his children on earth by the spirit which he has sent to live within the souls of mortal men.

Page 1609:5   *Paper 143:2.4*

"By the old way you seek to suppress, obey, and conform to the rules of living; by the new way **you are first transformed by the Spirit of Truth** and thereby strengthened in your inner soul by the constant spiritual renewing of your mind, and so are you endowed with the power of the certain and joyous performance of the gracious, acceptable, and perfect will of God. Forget not—it is your personal faith in the exceedingly great and precious promises of God that ensures your becoming partakers of the divine nature. Thus by your faith and the spirit's transformation, you become in reality the temples of God, and his spirit actually dwells within you. If, then, the spirit dwells within you, you are no longer bondslaves of the flesh but free and liberated sons of the spirit. The new law of the spirit endows you with the liberty of self-mastery in place of the old law of the fear of self-bondage and the slavery of self-denial."

Page 1865:1   *Paper 170:5.11*

The kingdom, to the Jews, was the Israelite *community*; to the gentiles it became the Christian *church*. **To Jesus the kingdom was the sum of those** *individuals* **who had confessed their faith in the fatherhood of God,** thereby declaring their wholehearted dedication to the doing of the will of God, thus becoming members of the spiritual brotherhood of man.

Page 1866:2   *Paper 170:5.19*

Sooner or later another and greater John the Baptist is due to arise proclaiming "the kingdom of God is at hand"—meaning a return to the high spiritual concept of Jesus, who proclaimed that **the kingdom is the will of his heavenly Father dominant and transcendent in the heart of the believer**—and doing all this without in any way referring either to the visible church on earth or to the anticipated second coming of Christ. There must come a revival of the actual teachings of Jesus, such a restatement as will undo the work of his early followers who went about to create a sociophilosophical system of belief regarding the fact of Michael's sojourn on earth. In a short time the teaching of this story about Jesus nearly supplanted the preaching of Jesus' gospel of the kingdom. In this way a historical religion displaced that teaching in which Jesus had blended man's highest moral ideas and spiritual ideals with man's most sublime hope for the future—eternal life. And that was the gospel of the kingdom.

Page 1951:2   *Paper 180:6.1*

After Peter, James, John, and Matthew had asked the Master numerous questions, he continued his farewell discourse by saying: "And I am telling you about all this before I leave you in order that you may be so prepared for what is coming upon you that you will not stumble into serious error. The authorities will not be content with merely putting you out of the synagogues; I warn you the hour draws near when they who kill you will think they are doing a service to God. And all of these things they will do to you and to those whom you lead into the kingdom of heaven because they do not know the Father. They have refused to know the Father by refusing to receive me; and they refuse to receive me when they reject you, provided you have kept my new commandment that you love one another even as I have loved you. I am telling you in advance about these things so that, when your hour comes, as mine now has, you may be strengthened in the knowledge that all was known to me, and that my spirit shall be with you in all your sufferings for my sake and the gospel's. It was for this purpose that I have been talking so plainly to you from the very beginning. I have even warned you that a man's foes may be those of his own household. Although this gospel of the kingdom never fails to bring great peace to the soul of the individual believer, it will not bring peace on earth until man is **willing to believe my teaching wholeheartedly and to establish the practice of doing the Father's will as the chief purpose in living the mortal life.**

Page 2061:6   *Paper 194:2.8*

Jesus lived a life which is a revelation of man submitted to the Father's will, not an example for any man literally to attempt to follow. This life in the flesh, together with his death on the cross and subsequent resurrection, presently became a new gospel of the ransom which had thus been paid in order to purchase man back from the clutch of the evil one—from the condemnation of an offended God. Nevertheless, even though the gospel did become greatly distorted, it remains a fact that this new message about Jesus carried along with it many of the fundamental truths and teachings of his earlier gospel of the kingdom. And, sooner or later, **these concealed truths of the fatherhood of God and the brotherhood of men will emerge to effectually transform the civilization of all mankind.**

Page 2088:4   *Paper 196:0.9*

The Master's entire life was consistently conditioned by this living faith, this sublime religious experience. **This spiritual attitude wholly dominated his thinking and feeling, his believing and praying, his teaching and preaching.** This personal faith of a son in the certainty and security of the guidance and protection of the heavenly Father imparted to his unique life a profound endowment of spiritual reality. And yet, despite this very deep consciousness of close relationship with divinity, this Galilean, God's Galilean, when addressed as Good Teacher, instantly replied, "Why do you call me good?" When we stand confronted by such splendid self-forgetfulness, we begin to understand how the Universal Father found it possible so fully to manifest himself to him and reveal himself through him to the mortals of the realms.

Wake Up Call

As religious teachers of a new order, as positive and missionary evangels of our religion, as transformed spirit-led faith sons and daughters of God, let us join hands in spiritual unity with all other Urantia Book believers, and as universe cosmic-conscious citizens let us go forth to transform the world. With the wholehearted desire to allow the will of our Heavenly Father to become dominant and transcendent in our hearts these spiritual fruits shall become manifest. Being on the ground floor of an epochal revelation, we have a unique opportunity to make our contribution to the Supreme. To the extent that we elect not to, to that extent will the Supreme have to wait for the new order of teachers and evangels to experience the joy and satisfaction of sharing the Father's love with their spiritual brothers and sisters—but that will be their contribution, not ours.

Are we to be recognized as a fellowship of believers in *The Urantia Book*, or readily identified as a fellowship of spiritual sisters and brothers wholeheartedly dedicated to doing the will of our Heavenly Father based upon our personal spiritual experiences and guided by the truths revealed in *The Urantia Book*?

...As transformed spirit-led faith sons and daughters of God, let us join hands in spiritual unity with all other Urantia Book believers, and as universe cosmic-conscious citizens let us go forth to transform the world.

The Spiritual Fellowship, with the principles outlined in their website at www.TheSpiritualFellowship.org, is one resource for linking up with other believers to fulfill this call to service. *The Urantia Book* Fellowship website at www.urantiabook.org is another one I go to from time to time. There are a number of other websites that are available; decide for yourself which ones match your interests and your approach to outreach service and learning.

—JJ Johnson

Potpourri

Many of the seven prayers in Paper 144:5 that Jesus presented to the twelve Apostles for illustration of other matters, came from other inhabited planets. Though the apostles were not at liberty to present these "Parable Prayers" in their public teachings, they profited much from all these revelations in their personal religious experiences. Jesus incorporated these prayer models as illustrations in connection with the intimate instructions of the twelve. He did not inform them that many of the seven prayers were from other inhabited plants. With the passing of two millenniums and another epochal revelation in our midst, the following "Form of Prayer" is submitted from a student of *The Urantia Book.*

Heavenly Father we desire to be dominated and led by our Divine
　　Fragment that you have so lovingly bestowed within us.
We have been blessed with the Spirit of Truth that Master Michael of Nebadon
　　Has left us and pray that our fruits will be worthy of this insight.
We feel at home in the Local Universe of Nebadon and especially in the System of Satania
　　With all the wonderful personalities created by our Local universe Mother Spirit
Who are dedicated to helping us ascend to Paradise.
　　We desire to work hand in hand with the spiritual personalities in accordance with
Your will to further reveal the Fatherhood of God and the Spiritual Brotherhood of humankind
　　To our spiritual brothers and sisters who have yet to discover your Love.
We pray that we will continue to be receptive to cosmic strength, spiritual insight and wisdom
　　So that we may become perfect as a creature as you are perfect as a creator.
We also pray that we will continue to become more spiritually mature so that we can place
　　Ourselves in line for increase service to our brothers and sisters on Urantia.
As we become one with our Divine Fragment we look forward to continue sharing
　　Your love with our spiritual brothers and sisters as we ascend to Paradise.
Heavenly Father based on our personal spiritual experiences and
　　Guided by the truths revealed in The Urantia Book,
We have come to know you better as the days and years go by.
　　We recognize that you are the first source and center of all things and
Though you are infinitely beyond our finite understanding,
　　You are a personal loving Father to each and every one of us.
Because of this we are thankful and grateful for the opportunity and
　　Privilege as a mortal of the realm to be able to wholeheartedly worship you.

It is the desire of the author of the above prayer that we abide by the same instructions that the apostles were enjoined to follow regarding these "Forms of Prayers." I request that we honor this wish.

The spiritual status of any religion may be determined
by the nature of its prayers.

Divine Counselor, Page 67, Paper 5:4.8

Before you embark on Part III with the Study Aids, I would like to turn your attention to Paper 139 and Paper 192:2. When Jesus made his thirteenth appearance, as reported in Paper 192, it was his third manifestation to the apostles as a group. I invite you to study the admonition Jesus passed on to the apostles during this visit with the ten, two by two. It is a wonderful testament to see what the apostles did after the Prince of Peace completed his mission on Urantia based on what Jesus personally told them during their strolls. A few examples to look for in Paper 139 are: "Andrew's great service to the kingdom was in advising Peter, James, and John concerning the choice of the first missionaries who were sent out to proclaim the gospel, and also counseling these early leaders about the organization of the administrative affairs of the kingdom." Peter—"When he was fully assured that Jesus had forgiven him...the fires of the kingdom burned so brightly within his soul that he became a great and saving light to thousands who sat in darkness." Alpheus Twins—Paper 192:2—"Go on believing and remembering your association with me, when I am gone, and after you have, perchance, returned to the work you used to do before you came to live with me."

It is insightful to take a look at these two papers and see what Jesus said to the ten apostles during their two by two walks, and then observe in Paper 139 how each one took His personal talk to heart and acted accordingly. Each apostle received their final instructions from the Son of Man as a serious injunction to carry them out in accordance with their own spiritual growth and insight. Each of us can take the Fifth Epochal Revelation in the same vein. *The Urantia Book* reveals cosmic truth, universe beauty and divine goodness in such a way that allows us to act with loving service in accordance with our personal spiritual growth and insight. As cosmic conscious citizens and students of *The Urantia Book* no less should be expected of us than was expected of the apostles with the fourth epochal revelation.

PART III

STUDY AIDS

*Let us **always remember** that an Epochal Revelation
confirms, validates and is consistent
with our personal revelations.*

—JJ

Chapter Three

Always Remember And Never Forget

Page 2001:2   *Paper 186:3.4*

This peculiar-minded David Zebedee was the only one of the leading disciples of Jesus who was inclined to take a literal and plain matter-of-fact view of the Master's assertion that he would die and "rise again on the third day." David had once heard him make this prediction and, being of a literal turn of mind, now proposed to assemble his messengers early Sunday morning at the home of Nicodemus so that they would be on hand to spread the news in case Jesus rose from the dead. David soon discovered that none of Jesus' followers were looking for him to return so soon from the grave; therefore did he say little about his belief and nothing about the mobilization of all his messenger force on early Sunday morning except to the runners who had been dispatched on Friday forenoon to distant cities and believer centers.

The *Urantia Book* reveals truth in a somewhat authoritative way, and rightfully so—but always with respect for our free will. Each one of us can grow at our own pace consistent with the cosmic truth, universe beauty, and divine goodness revealed in *the Urantia Papers*. In one of my readings I decided to apply David Zebedee's "peculiar-minded" trait, as portrayed in the above paragraph, to the extent of taking the passages that stated "always remember" and "never forget" quite seriously with a "plain matter-of-fact view" of the authors that chose to use these phrases in expressing their views.

The following are samples in the Foreword and over twenty papers in Parts I, II, III, and IV of authors who elected to express themselves with such strong conviction. If one is of the opinion the authors chose their words carefully, wouldn't it seem prudent to treat their comments with considerable reflection? Although David Zebedee came to mind when reading the following passages I have not "literally" committed them to memory. Again, as in other parts of this book, this is by no means an exhaustive compilation of such phrases. For example, "keep in mind" is another selective phrase used in Paper 180 in addition to an "always remember." Some of these teachings were directed towards a particular apostle, or others, but that shouldn't prevent us from gleaning as much insight as possible as it relates to our own personal life and experiences.

Page 15:2   *Foreword:11.12*

Always remember: Potential infinity is absolute and inseparable from eternity. Actual infinity in time can never be anything but partial and must therefore be nonabsolute; neither can infinity of actual personality be absolute except in unqualified Deity. And it is the differential of infinity potential in the Unqualified Absolute and the Deity Absolute that eternalizes the Universal Absolute, thereby making it cosmically possible to have material universes in space and spiritually possible to have finite personalities in time.

Page 31:8   *Paper 1:7.8*

The fact of the Paradise Trinity in no manner violates the truth of the divine unity. The three personalities of Paradise Deity are, in all universe reality reactions and in all creature relations, as one. Neither does the existence of these three eternal persons violate the truth of the indivisibility of Deity. I am fully aware that I have at my command no language adequate to make clear to the mortal mind how these universe problems appear to us. But you should not become discouraged; not all of these things are wholly clear to even the high personalities belonging to my group of Paradise beings. **Ever bear in mind** that these profound truths pertaining to Deity will increasingly clarify as your minds become progressively spiritualized during the successive epochs of the long mortal ascent to Paradise.

Page 63:3   *Paper 5:1.5*

However Urantia mortals may differ in their intellectual, social, economic, and even moral opportunities and endowments, **forget not** that their spiritual endowment is uniform and unique. They all enjoy the same divine presence of the gift from the Father, and they are all equally privileged to seek intimate personal communion with this indwelling spirit of divine origin, while they may all equally choose to accept the uniform spiritual leading of these Mystery Monitors.

Page 78:2   *Paper 6:5.6*

Ever remember, the Eternal Son is the personal portrayal of the spirit Father to all creation. The Son is personal and nothing but personal in the Deity sense; such a divine and absolute personality cannot be disintegrated or fragmentized. God the Father and God the Spirit are truly personal, but they are also everything else in addition to being such Deity personalities.

Page 89:4   *Paper 7:7.5*

In all these widespread activities of the far-flung spiritual administration of the Eternal Son, **do not forget** that the Son is a person just as truly and actually as the Father is a person. Indeed, to beings of the onetime human order the Eternal Son will be more easy to approach than the Universal Father. In the progress of the pilgrims of time through the circuits of Havona, you will be competent to attain the Son long before you are prepared to discern the Father.

Page 96:1 *Paper 8:5.5*

Ever remember that the Infinite Spirit is the *Conjoint* Actor; both the Father and the Son are functioning in and through him; he is present not only as himself but also as the Father and as the Son and as the Father-Son. In recognition of this and for many additional reasons the spirit presence of the Infinite Spirit is often referred to as "the spirit of God."

Page 112:8 *Paper 10:4.5*

Ever remember that what the Infinite Spirit does is the function of the Conjoint Actor. Both the Father and the Son are functioning in and through and as him. But it would be futile to attempt to elucidate the Trinity mystery: three as one and in one, and one as two and acting for two.

Page 189:2 *Paper 16:4.1*

The Seven Master Spirits are the full representation of the Infinite Spirit to the evolutionary universes. They represent the Third Source and Center in the relationships of energy, mind, and spirit. While they function as the co-ordinating heads of the universal administrative control of the Conjoint Actor, **do not forget** that they have their origin in the creative acts of the Paradise Deities. It is literally true that these Seven Spirits are the personalized physical power, cosmic mind, and spiritual presence of the triune Deity, "the Seven Spirits of God sent forth to all the universe."

Page 508:2 *Paper 44:8.4*

But **every human being should remember:** Many ambitions to excel which tantalize mortals in the flesh will not persist with these same mortals in the morontia and spirit careers. The ascending morontians learn to socialize their former purely selfish longings and egoistic ambitions. Nevertheless, those things which you so earnestly longed to do on earth and which circumstances so persistently denied you, if, after acquiring true mota insight in the morontia career, you still desire to do, then will you most certainly be granted every opportunity fully to satisfy your long-cherished desires.

Page 578:5 *Paper 50:6.5*

You should not forget that for two hundred thousand years all the worlds of Satania have rested under the spiritual ban of Norlatiadek in consequence of the Lucifer rebellion. And it will require age upon age to retrieve the resultant handicaps of sin and secession. Your world still continues to pursue an irregular and checkered career as a result of the double tragedy of a rebellious Planetary Prince and a defaulting Material Son. Even the bestowal of Christ Michael on Urantia did not immediately set aside the temporal consequences of these serious blunders in the earlier administration of the world.

Page 855:4   *Paper 77:1.2*

It is well always to bear in mind that the successive bestowals of the Sons of God on an evolving planet produce marked changes in the spiritual economy of the realm and sometimes so modify the workings of the interassociation of spiritual and material agencies on a planet as to create situations indeed difficult of understanding. The status of the one hundred corporeal members of Prince Caligastia's staff illustrates just such a unique interassociation: As ascendant morontia citizens of Jerusem they were supermaterial creatures without reproductive prerogatives. As descendant planetary ministers on Urantia they were material sex creatures capable of procreating material offspring (as some of them later did). What we cannot satisfactorily explain is how these one hundred could function in the parental role on a supermaterial level, but that is exactly what happened. A supermaterial (nonsexual) liaison of a male and a female member of the corporeal staff resulted in the appearance of the first-born of the primary midwayers.

Page 948:7   *Paper 85:7.2*

You must remember that feeling, not thinking, was the guiding and controlling influence in all evolutionary development. To the primitive mind there is little difference between fearing, shunning, honoring, and worshiping.

Page 999:6   *Paper 91:6.3*

No matter how difficult it may be to reconcile the scientific doubtings regarding the efficacy of prayer with the ever-present urge to seek help and guidance from divine sources, **never forget** that the sincere prayer of faith is a mighty force for the promotion of personal happiness, individual self-control, social harmony, moral progress, and spiritual attainment.

Page 1005:3   *Paper 92:3.1*

The study of human religion is the examination of the fossil-bearing social strata of past ages. The mores of the anthropomorphic gods are a truthful reflection of the morals of the men who first conceived such deities. Ancient religions and mythology faithfully portray the beliefs and traditions of peoples long since lost in obscurity. These olden cult practices persist alongside newer economic customs and social evolutions and, of course, appear grossly inconsistent. The remnants of the cult present a true picture of the racial religions of the past. **Always remember,** the cults are formed, not to discover truth, but rather to promulgate their creeds.

Page 1091:1   *Paper 99:5.2*

Always keep in mind: True religion is to know God as your Father and man as your brother. Religion is not a slavish belief in threats of punishment or magical promises of future mystical rewards.

Page 1115:3   *Paper 101:9.2*

When you presume to sit in critical judgment on the primitive religion of man (or on the religion of primitive man), **you should remember** to judge such savages and to evaluate their religious experience in accordance with their enlightenment and status of conscience. Do not make the mistake of judging another's religion by your own standards of knowledge and truth.

Page 1153:3   *Paper 105:1.6*

Ever remember that man's comprehension of the Universal Father is a personal experience. God, as your spiritual Father, is comprehensible to you and to all other mortals; but *your experiential worshipful concept of the Universal Father must always be less than your philosophic postulate of the infinity of the First Source and Center, the I AM.* When we speak of the Father, we mean God as he is understandable by his creatures both high and low, but there is much more of Deity which is not comprehensible to universe creatures. God, your Father and my Father, is that phase of the Infinite which we perceive in our personalities as an actual experiential reality, but the I AM ever remains as our hypothesis of all that we feel is unknowable of the First Source and Center. And even that hypothesis probably falls far short of the unfathomed infinity of original reality.

Page 1153:6   *Paper 105:2.1*

In considering the genesis of reality, **ever bear in mind** that all absolute reality is from eternity and is without beginning of existence. By absolute reality we refer to the three existential persons of Deity, the Isle of Paradise, and the three Absolutes. These seven realities are co-ordinately eternal, notwithstanding that we resort to time-space language in presenting their sequential origins to human beings.

Page 1296:3   *Paper 118:2.1*

The ubiquity of Deity must not be confused with the ultimacy of the divine omnipresence. It is volitional with the Universal Father that the Supreme, the Ultimate, and the Absolute should compensate, co-ordinate, and unify his time-space ubiquity and his time-space-transcended omnipresence with his timeless and spaceless universal and absolute presence. **And you should remember** that, while Deity ubiquity may be so often space associated, it is not necessarily time conditioned.

Page 1298:1   *Paper 118:4.1*

Many of the theologic difficulties and the metaphysical dilemmas of mortal man are due to man's mislocation of Deity personality and consequent assignment of infinite and absolute attributes to subordinate Divinity and to evolutionary Deity. **You must not forget** that, while there is indeed a true First Cause, there are also a host of co-ordinate and subordinate causes, both associate and secondary causes.

Page 1444:6   *Paper 131:2.6*

"God has made man a little less than divine and has crowned him with love and mercy. The Lord knows the way of the righteous, but the way of the ungodly shall perish. The fear of the Lord is the beginning of wisdom; the knowledge of the Supreme is understanding. Says the Almighty God: `Walk before me and be perfect.' **Forget not** that pride goes before destruction and a haughty spirit before a fall. He who rules his own spirit is mightier than he who takes a city. Says the Lord God, the Holy One: `In returning to your spiritual rest shall you be saved; in quietness and confidence shall be your strength.' They who wait upon the Lord shall renew their strength; they shall mount up with wings like eagles. They shall run and not be weary; they shall walk and not be faint. The Lord shall give you rest from your fear. Says the Lord: `Fear not, for I am with you. Be not dismayed, for I am your God. I will strengthen you; I will help you; yes, I will uphold you with the right hand of my righteousness.'"

Page 1452:3   *Paper 131:8.5*

"The Great Supreme is all-pervading; he is on the left hand and on the right; he supports all creation and indwells all true beings. You cannot find the Supreme, neither can you go to a place where he is not. If a man recognizes the evil of his ways and repents of sin from the heart, then may he seek forgiveness; he may escape the penalty; he may change calamity into blessing. The Supreme is the secure refuge for all creation; he is the guardian and savior of mankind. If you seek for him daily, you shall find him. Since he can forgive sins, he is indeed most precious to all men. **Always remember** that God does not reward man for what he does but for what he is; therefore should you extend help to your fellows without the thought of rewards. Do good without thought of benefit to the self."

Page 1464:4   *Paper 132:5.10*

"6. If you chance to secure wealth by flights of genius, if your riches are derived from the rewards of inventive endowment, do not lay claim to an unfair portion of such rewards. The genius owes something to both his ancestors and his progeny; likewise is he under obligation to the race, nation, and circumstances of his inventive discoveries; **he should also remember** that it was as man among men that he labored and wrought out his inventions. It would be equally unjust to deprive the genius of all his increment of wealth. And it will ever be impossible for men to establish rules and regulations applicable equally to all these problems of the equitable distribution of wealth. You must first recognize man as your brother, and if you honestly desire to do by him as you would have him do by you, the commonplace dictates of justice, honesty, and fairness will guide you in the just and impartial settlement of every recurring problem of economic rewards and social justice."

And then, in bidding him farewell, Jesus said: "My brother, **always remember** that man has no rightful authority over woman unless the woman has willingly and voluntarily given him such authority. Your wife has engaged to go through life with you, to help you fight its battles, and to assume the far greater share of the burden of bearing and rearing your children; and in return for this special service it is only fair that she receive from you that special protection which man can give to woman as the partner who must carry, bear, and nurture the children. The loving care and consideration which a man is willing to bestow upon his wife and their children are the measure of that man's attainment of the higher levels of creative and spiritual self-consciousness. Do you not know that men and women are partners with God in that they co-operate to create beings who grow up to possess themselves of the potential of immortal souls? The Father in heaven treats the Spirit Mother of the children of the universe as one equal to himself. It is Godlike to share your life and all that relates thereto on equal terms with the mother partner who so fully shares with you that divine experience of reproducing yourselves in the lives of your children. If you can only love your children as God loves you, you will love and cherish your wife as the Father in heaven honors and exalts the Infinite Spirit, the mother of all the spirit children of a vast universe."

To the earnest leader of the Mithraic cult he said: "You do well to seek for a religion of eternal salvation, but you err to go in quest of such a glorious truth among man-made mysteries and human philosophies. Know you not that the mystery of eternal salvation dwells within your own soul? Do you not know that the God of heaven has sent his spirit to live within you, and that this spirit will lead all truth-loving and God-serving mortals out of this life and through the portals of death up to the eternal heights of light where God waits to receive his children? **And never forget:** You who know God are the sons of God if you truly yearn to be like him."

To the Greek contractor and builder he said: "My friend, as you build the material structures of men, grow a spiritual character in the similitude of the divine spirit within your soul. Do not let your achievement as a temporal builder outrun your attainment as a spiritual son of the kingdom of heaven. While you build the mansions of time for another, neglect not to secure your title to the mansions of eternity for yourself. **Ever remember,** there is a city whose foundations are righteousness and truth, and whose builder and maker is God."

Reality of material existence attaches to unrecognized energy as well as to visible matter. When the energies of the universe are so slowed down that they acquire the requisite degree of motion, then, under favorable conditions, these same energies become mass. And **forget not,** the mind which can alone perceive the presence of apparent realities is itself also real. And the fundamental cause of this universe of energy-mass, mind, and spirit, is eternal—it exists and consists in the nature and reactions of the Universal Father and his absolute co-ordinates.

Page 1488:1   *Paper 134:5.3*

Religious teachers must always remember that the spiritual sovereignty of God overrides all intervening and intermediate spiritual loyalties. Someday civil rulers will learn that the Most Highs rule in the kingdoms of men.

Page 1490:4   *Paper 134:6.1*

If one man craves freedom—liberty—**he must remember** that all other men long for the same freedom. Groups of such liberty-loving mortals cannot live together in peace without becoming subservient to such laws, rules, and regulations as will grant each person the same degree of freedom while at the same time safeguarding an equal degree of freedom for all of his fellow mortals. If one man is to be absolutely free, then another must become an absolute slave. And the relative nature of freedom is true socially, economically, and politically. Freedom is the gift of civilization made possible by the enforcement of LAW.

Page 1577:2   *Paper 140:6.8*

And then said Jesus: "But you will stumble over my teaching because you are wont to interpret my message literally; you are slow to discern the spirit of my teaching. **Again must you remember** that you are my messengers; you are beholden to live your lives as I have in spirit lived mine. You are my personal representatives; but do not err in expecting all men to live as you do in every particular. **Also must you remember** that I have sheep not of this flock, and that I am beholden to them also, to the end that I must provide for them the pattern of doing the will of God while living the life of the mortal nature."

Page 1577:5   *Paper 140:6.11*

When Jesus heard this, he said: "Be willing, then, to take up your responsibilities and follow me. Do your good deeds in secret; when you give alms, let not the left hand know what the right hand does. And when you pray, go apart by yourselves and use not vain repetitions and meaningless phrases. **Always remember** that the Father knows what you need even before you ask him. And be not given to fasting with a sad countenance to be seen by men. As my chosen apostles, now set apart for the service of the kingdom, lay not up for yourselves treasures on earth, but by your unselfish service lay up for yourselves treasures in heaven, for where your treasures are, there will your hearts be also.

Page 1584:5   *Paper 140:10.2*

Another great handicap in this work of teaching the twelve was their tendency to take highly idealistic and spiritual principles of religious truth and remake them into concrete rules of personal conduct. Jesus would present to them the beautiful spirit of the soul's attitude, but they insisted on translating such teachings into rules of personal behavior. **Many times, when they did make sure to remember what the Master said, they were almost certain to forget what he did not say.** But they slowly assimilated his teaching because Jesus was all that he taught. What they could not gain from his verbal instruction, they gradually acquired by living with him.

Page 1609:5   *Paper 143:2.4*

"By the old way you seek to suppress, obey, and conform to the rules of living; by the new way you are first *transformed* by the Spirit of Truth and thereby strengthened in your inner soul by the constant spiritual renewing of your mind, and so are you endowed with the power of the certain and joyous performance of the gracious, acceptable, and perfect will of God. **Forget not**—it is your personal faith in the exceedingly great and precious promises of God that ensures your becoming partakers of the divine nature. Thus by your faith and the spirit's transformation, you become in reality the temples of God, and his spirit actually dwells within you. If, then, the spirit dwells within you, you are no longer bondslaves of the flesh but free and liberated sons of the spirit. The new law of the spirit endows you with the liberty of self-mastery in place of the old law of the fear of self-bondage and the slavery of self-denial."

Page 1641:4   *Paper 146:3.2*

The apostles were a bit disconcerted by the open manner of Jesus' assent to many of the Greek's propositions, but Jesus afterward privately said to them: "My children, marvel not that I was tolerant of the Greek's philosophy. True and genuine inward certainty does not in the least fear outward analysis, nor does truth resent honest criticism. **You should never forget** that intolerance is the mask covering up the entertainment of secret doubts as to the trueness of one's belief. No man is at any time disturbed by his neighbor's attitude when he has perfect confidence in the truth of that which he wholeheartedly believes. Courage is the confidence of thoroughgoing honesty about those things which one professes to believe. Sincere men are unafraid of the critical examination of their true convictions and noble ideals."

Page 1660:1–8, 1661:1   *Paper 148:4.2–10*

"Do not make the mistake of confusing *evil* with the *evil one*, more correctly the *iniquitous one*. He whom you call the evil one is the son of self-love, the high administrator who knowingly went into deliberate rebellion against the rule of my

Father and his loyal Sons. But I have already vanquished these sinful rebels. Make clear in your mind these different attitudes toward the Father and his universe. **Never forget** these laws of relation to the Father's will:

"Evil is the unconscious or unintended transgression of the divine law, the Father's will. Evil is likewise the measure of the imperfectness of obedience to the Father's will.

"Sin is the conscious, knowing, and deliberate transgression of the divine law, the Father's will. Sin is the measure of unwillingness to be divinely led and spiritually directed.

"Iniquity is the willful, determined, and persistent transgression of the divine law, the Father's will. Iniquity is the measure of the continued rejection of the Father's loving plan of personality survival and the Sons' merciful ministry of salvation.

"By nature, before the rebirth of the spirit, mortal man is subject to inherent evil tendencies, but such natural imperfections of behavior are neither sin nor iniquity. Mortal man is just beginning his long ascent to the perfection of the Father in Paradise. To be imperfect or partial in natural endowment is not sinful. Man is indeed subject to evil, but he is in no sense the child of the evil one unless he has knowingly and deliberately chosen the paths of sin and the life of iniquity. Evil is inherent in the natural order of this world, but sin is an attitude of conscious rebellion which was brought to this world by those who fell from spiritual light into gross darkness.

"You are confused, Thomas, by the doctrines of the Greeks and the errors of the Persians. You do not understand the relationships of evil and sin because you view mankind as beginning on earth with a perfect Adam and rapidly degenerating, through sin, to man's present deplorable estate. But why do you refuse to comprehend the meaning of the record which discloses how Cain, the son of Adam, went over into the land of Nod and there got himself a wife? And why do you refuse to interpret the meaning of the record which portrays the sons of God finding wives for themselves among the daughters of men?

"Men are, indeed, by nature evil, but not necessarily sinful. The new birth—the baptism of the spirit—is essential to deliverance from evil and necessary for entrance into the kingdom of heaven, but none of this detracts from the fact that man is the son of God. Neither does this inherent presence of potential evil mean that man is in some mysterious way estranged from the Father in heaven so that, as an alien, foreigner, or stepchild, he must in some manner seek for legal adoption by the Father. All such notions are born, first, of your misunderstanding of the Father and, second, of your ignorance of the origin, nature, and destiny of man.

"The Greeks and others have taught you that man is descending from godly perfection steadily down toward oblivion or destruction; I have come to show that man, by entrance into the kingdom, is ascending certainly and surely up to God and divine perfection. Any being who in any manner falls short of the divine and spiritual ideals of the eternal Father's will is potentially evil, but such beings are in no sense sinful, much less iniquitous.

"Thomas, have you not read about this in the Scriptures, where it is written: `You are the children of the Lord your God.' `I will be his Father and he shall be my son.' `I have chosen him to be my son—I will be his Father.' `Bring my sons from far and my daughters from the ends of the earth; even every one who is called by my name, for I have created them for my glory.' `You are the sons of the living God.' `They who have the spirit of God are indeed the sons of God.' While there is a material part of the human father in the natural child, there is a spiritual part of the heavenly Father in every faith son of the kingdom."

Page 1672:6   *Paper 149:3.3*

When Jesus first met with the evangelists at the Bethsaida camp, in concluding his address, he said: "**You should remember** that in body and mind—emotionally—men react individually. The only *uniform* thing about men is the indwelling spirit. Though divine spirits may vary somewhat in the nature and extent of their experience, they react uniformly to all spiritual appeals. Only through, and by appeal to, this spirit can mankind ever attain unity and brotherhood." But many of the leaders of the Jews had closed the doors of their hearts to the spiritual appeal of the gospel. From this day on they ceased not to plan and plot for the Master's destruction. They were convinced that Jesus must be apprehended, convicted, and executed as a religious offender, a violator of the cardinal teachings of the Jewish sacred law.

Page 1697:1   *Paper 151:6.8*

As they were about to depart, Amos besought Jesus to permit him to go back with them, but the Master would not consent. Said Jesus to Amos: "**Forget not** that you are a son of God. Return to your own people and show them what great things God has done for you." And Amos went about publishing that Jesus had cast a legion of devils out of his troubled soul, and that these evil spirits had entered into a herd of swine, driving them to quick destruction. And he did not stop until he had gone into all the cities of the Decapolis, declaring what great things Jesus had done for him.

Page 1715:4   *Paper 153:5.4*

"My beloved, **you must remember** that it is the spirit that quickens; the flesh and all that pertains thereto is of little profit. The words which I have spoken to you are spirit and life. Be of good cheer! I have not deserted you. Many shall be offended by the plain speaking of these days. Already you have heard that many of my disciples have turned back; they walk no more with me. From the beginning I knew that these halfhearted believers would fall out by the way. Did I not choose you twelve men and set you apart as ambassadors of the kingdom? And now at such a time as this would you also desert? Let each of you look to his own faith, for one of you stands in grave danger." And when Jesus had finished speaking, Simon Peter said: "Yes, Lord, we are sad and perplexed, but we will never forsake you. You have taught us the words of eternal life. We have believed in you and followed with you all this time. We will not turn back, for we know that you are sent by God." And as Peter ceased speaking, they all with one accord nodded their approval of his pledge of loyalty."

Page 1732:4   *Paper 155:6.11*

Never forget there is only one adventure which is more satisfying and thrilling than the attempt to discover the will of the living God, and that is the supreme experience of honestly trying to do that divine will. And fail not to remember that the will of God can be done in any earthly occupation. Some callings are not holy and others secular. All things are sacred in the lives of those who are spirit led; that is, subordinated to truth, ennobled by love, dominated by mercy, and restrained by fairness—justice. The spirit which my Father and I shall send into the world is not only the Spirit of Truth but also the spirit of idealistic beauty.

Page 1765:5   *Paper 159:3.3*

In bringing men into the kingdom, do not lessen or destroy their self-respect. While overmuch self-respect may destroy proper humility and end in pride, conceit, and arrogance, the loss of self-respect often ends in paralysis of the will. It is the purpose of this gospel to restore self-respect to those who have lost it and to restrain it in those who have it. Make not the mistake of only condemning the wrongs in the lives of your pupils; remember also to accord generous recognition for the most praiseworthy things in their lives. **Forget not** that I will stop at nothing to restore self-respect to those who have lost it, and who really desire to regain it.

Page 1768:1–2   *Paper 159:4.5–6*

"Nathaniel, never permit yourself for one moment to believe the Scripture records which tell you that the God of love directed your forefathers to go forth in battle to slay all their enemies—men, women, and children. Such records are the words of men, not very holy men, and they are not the word of God. The Scriptures always have, and always will, reflect the intellectual, moral, and spiritual status of those who create them. Have you not noted that the concepts of Yahweh grow in beauty and glory as the prophets make their records from Samuel to Isaiah? And **you should remember** that the Scriptures are intended for religious instruction and spiritual guidance. They are not the works of either historians or philosophers.

"The thing most deplorable is not merely this erroneous idea of the absolute perfection of the Scripture record and the infallibility of its teachings, but rather the confusing misinterpretation of these sacred writings by the tradition-enslaved scribes and Pharisees at Jerusalem. And now will they employ both the doctrine of the inspiration of the Scriptures and their misinterpretations thereof in their determined effort to withstand these newer teachings of the gospel of the kingdom. Nathaniel, **never forget**, the Father does not limit the revelation of truth to any one generation or to any one people. Many earnest seekers after the truth have been, and will continue to be, confused and disheartened by these doctrines of the perfection of the Scriptures."

Page 1770:2 *Paper 159:5.10*

Jesus did not advocate the practice of negative submission to the indignities of those who might purposely seek to impose upon the practitioners of nonresistance to evil, but rather that his followers should be wise and alert in the quick and positive reaction of good to evil to the end that they might effectively overcome evil with good. **Forget not**, the truly good is invariably more powerful than the most malignant evil. The Master taught a positive standard of righteousness: "Whosoever wishes to be my disciple, let him disregard himself and take up the full measure of his responsibilities daily to follow me." And he so lived himself in that "he went about doing good." And this aspect of the gospel was well illustrated by many parables which he later spoke to his followers. He never exhorted his followers patiently to bear their obligations but rather with energy and enthusiasm to live up to the full measure of their human responsibilities and divine privileges in the kingdom of God.

Page 1822:1 *Paper 165:4.6*

"**But never forget** that, after all, wealth is unenduring. The love of riches all too often obscures and even destroys the spiritual vision. Fail not to recognize the danger of wealth's becoming, not your servant, but your master."

Page 1834:3 *Paper 167:1.5*

Then went Jesus over to where the sick man sat and, taking him by the hand, said: "Arise and go your way. You have not asked to be healed, but I know the desire of your heart and the faith of your soul." Before the man left the room, Jesus returned to his seat and, addressing those at the table, said: "Such works my Father does, not to tempt you into the kingdom, but to reveal himself to those who are already in the kingdom. You can perceive that it would be like the Father to do just such things because which one of you, having a favorite animal that fell in the well on the Sabbath day, would not go right out and draw him up?" And since no one would answer him, and inasmuch as his host evidently approved of what was going on, Jesus stood up and spoke to all present: "My brethren, when you are bidden to a marriage feast, sit not down in the chief seat, lest, perchance, a more honored man than you has been invited, and the host will have to come to you and request that you give your place to this other and honored guest. In this event, with shame you will be required to take a lower place at the table. When you are bidden to a feast, it would be the part of wisdom, on arriving at the festive table, to seek for the lowest place and take your seat therein, so that, when the host looks over the guests, he may say to you: `My friend, why sit in the seat of the least? come up higher'; and thus will such a one have glory in the presence of his fellow guests. **Forget not**, every one who exalts himself shall be humbled, while he

57

who truly humbles himself shall be exalted. Therefore, when you entertain at dinner or give a supper, invite not always your friends, your brethren, your kinsmen, or your rich neighbors that they in return may bid you to their feasts, and thus will you be recompensed. When you give a banquet, sometimes bid the poor, the maimed, and the blind. In this way you shall be blessed in your heart, for you well know that the lame and the halt cannot repay you for your loving ministry."

Page 1849:4   *Paper 168:4.13*

10. All genuine spirit-born petitions are certain of an answer. Ask and you shall receive. **But you should remember** that you are progressive creatures of time and space; therefore must you constantly reckon with the time-space factor in the experience of your personal reception of the full answers to your manifold prayers and petitions.

Page 1891:4   *Paper 173:2.3*

It was altogether proper that the temple rulers and the officers of the Jewish Sanhedrin should ask this question of anyone who presumed to teach and perform in the extraordinary manner which had been characteristic of Jesus, especially as concerned his recent conduct in clearing the temple of all commerce. These traders and money-changers all operated by direct license from the highest rulers, and a percentage of their gains was supposed to go directly into the temple treasury. **Do not forget** that *authority* was the watchword of all Jewry. The prophets were always stirring up trouble because they so boldly presumed to teach without authority, without having been duly instructed in the rabbinic academies and subsequently regularly ordained by the Sanhedrin. Lack of this authority in pretentious public teaching was looked upon as indicating either ignorant presumption or open rebellion. At this time only the Sanhedrin could ordain an elder or teacher, and such a ceremony had to take place in the presence of at least three persons who had previously been so ordained. Such an ordination conferred the title of "rabbi" upon the teacher and also qualified him to act as a judge, "binding and loosing such matters as might be brought to him for adjudication."

Page 1897:2   *Paper 174:0.2*

This morning he greeted each of the twelve with a personal salutation. To Andrew he said: "Be not dismayed by the events just ahead. Keep a firm hold on your brethren and see that they do not find you downcast." To Peter he said: "Put not your trust in the arm of flesh nor in weapons of steel. Establish yourself on the spiritual foundations of the eternal rocks." To James he said: "Falter not because of outward appearances. Remain firm in your faith, and you shall soon know of the reality of that which you believe." To John he said: "Be gentle; love even your enemies; be tolerant. **And remember** that I have trusted you with many things." To Nathaniel he said: "Judge not by appearances; remain firm in your faith when all appears to vanish; be true to your commission as an ambassador of the kingdom." To Philip he said: "Be unmoved by the events now impending. Remain unshaken, even when you cannot see the way. Be loyal to your oath of consecration." To Matthew he said: "**Forget not** the mercy that received you into the kingdom. Let no man cheat you of your eternal reward. As you have withstood the inclinations of the mortal nature, be willing to be steadfast." To Thomas he said: "No matter how difficult it may be, just now you must walk by faith and not by sight. Doubt not that I am able to finish the work I have begun, and that I shall eventually see all of my faithful ambassadors in the kingdom beyond." To the Alpheus twins he said: "Do not allow the things which you cannot understand to crush you. Be true to the affections of your hearts and put not your trust in either great men or the changing attitude of the people. Stand by your brethren." And to Simon Zelotes he said: "Simon, you may be crushed by disappointment, but your spirit shall rise above all that may come upon you. What you have failed to learn from me, my spirit will teach you. Seek the true realities of the spirit and cease to be attracted by unreal and material shadows." And to Judas Iscariot he said: "Judas, I have loved you and have prayed that you would love your brethren. Be not weary in well doing; and I would warn you to beware the slippery paths of flattery and the poison darts of ridicule."

Page 1926:4   *Paper 177:4.11*

Judas did not realize it at this time, but he had been a subconscious critic of Jesus ever since John the Baptist was beheaded by Herod. Deep down in his heart Judas always resented the fact that Jesus did not save John. **You should not forget** that Judas had been a disciple of John before he became a follower of Jesus. And all these accumulations of human resentment and bitter disappointment which Judas had laid by in his soul in habiliments of hate were now well organized in his subconscious mind and ready to spring up to engulf him when he once dared to separate himself from the supporting influence of his brethren while at the same time exposing himself to the clever insinuations and subtle ridicule of the enemies of Jesus. Every time Judas allowed his hopes to soar high and Jesus would do or say something to dash them to pieces, there was always left in Judas's heart a scar of bitter resentment; and as these scars multiplied, presently that heart, so often wounded, lost all real affection for the one who had inflicted this distasteful experience upon a well-intentioned but cowardly and self-centered personality. Judas did not realize it, but he was a coward. Accordingly was he always inclined to assign to Jesus cowardice as the motive which led him so often to refuse to grasp for power or glory when they were apparently within his easy reach. And every mortal man knows full well how love, even when once genuine, can, through disappointment, jealousy, and long-continued resentment, be eventually turned into actual hate.

Page 1931:3   *Paper 178:1.12*

You must not seek to promulgate truth nor to establish righteousness by the power of civil governments or by the enaction of secular laws. You may always labor to persuade men's minds, but you must never dare to compel them. **You**

must not forget the great law of human fairness which I have taught you in positive form: Whatsoever you would that men should do to you, do even so to them.

Page 1932:1–2   *Paper 178:1.16–17*

And forget not: We have made no direct attack upon the persons or upon the authority of those who sit in Moses' seat; we only offered them the new light, which they have so vigorously rejected. We have assailed them only by the denunciation of their spiritual disloyalty to the very truths which they profess to teach and safeguard. We clashed with these established leaders and recognized rulers only when they threw themselves directly in the way of the preaching of the gospel of the kingdom to the sons of men. And even now, it is not we who assail them, but they who seek our destruction. **Do not forget** that you are commissioned to go forth preaching only the good news. You are not to attack the old ways; you are skillfully to put the leaven of new truth in the midst of the old beliefs. Let the Spirit of Truth do his own work. Let controversy come only when they who despise the truth force it upon you. But when the willful unbeliever attacks you, do not hesitate to stand in vigorous defense of the truth which has saved and sanctified you.

Throughout the vicissitudes of life, **remember always** to love one another. Do not strive with men, even with unbelievers. Show mercy even to those who despitefully abuse you. Show yourselves to be loyal citizens, upright artisans, praiseworthy neighbors, devoted kinsmen, understanding parents, and sincere believers in the brotherhood of the Father's kingdom. And my spirit shall be upon you, now and even to the end of the world.

Page 1934:6   *Paper 178:3.4*

"When you see this city destroyed, **forget not** that you have entered already upon the eternal life of endless service in the ever-advancing kingdom of heaven, even of the heaven of heavens. You should know that in my Father's universe and in mine are many abodes, and that there awaits the children of light the revelation of cities whose builder is God and worlds whose habit of life is righteousness and joy in the truth. I have brought the kingdom of heaven to you here on earth, but I declare that all of you who by faith enter therein and remain therein by the living service of truth, shall surely ascend to the worlds on high and sit with me in the spirit kingdom of our Father. But first must you gird yourselves and complete the work which you have begun with me. You must first pass through much tribulation and endure many sorrows—and these trials are even now upon us—and when you have finished your work on earth, you shall come to my joy, even as I have finished my Father's work on earth and am about to return to his embrace."

Page 1945:3   *Paper 180:1.6*

Keep in mind: It is loyalty, not sacrifice, that Jesus demands. The consciousness of sacrifice implies the absence of that wholehearted affection which would have made such a loving service a supreme joy. The idea of *duty* signifies that you are servant-minded and hence are missing the mighty thrill of doing your service as a friend and for a friend. The impulse of friendship transcends all convictions of duty, and the service of a friend for a friend can never be called a sacrifice. The Master has taught the apostles that they are the sons of God. He has called them brethren, and now, before he leaves, he calls them his friends.

Page 1946:6   *Paper 180:3.1*

The eleven had scarcely ceased their discussions of the discourse on the vine and the branches when the Master, indicating that he was desirous of speaking to them further and knowing that his time was short, said: "When I have left you, be not discouraged by the enmity of the world. Be not downcast even when fainthearted believers turn against you and join hands with the enemies of the kingdom. If the world shall hate you, you should recall that it hated me even before it hated you. If you were of this world, then would the world love its own, but because you are not, the world refuses to love you. You are in this world, but your lives are not to be worldlike. I have chosen you out of the world to represent the spirit of another world even to this world from which you have been chosen. **But always remember** the words I have spoken to you: The servant is not greater than his master. If they dare to persecute me, they will also persecute you. If my words offend the unbelievers, so also will your words offend the ungodly. And all of this will they do to you because they believe not in me nor in Him who sent me; so will you suffer many things for the sake of my gospel. But when you endure these tribulations, you should recall that I also suffered before you for the sake of this gospel of the heavenly kingdom.

Page 1955:6   *Paper 181:2.5*

As John Zebedee stood there in the upper chamber, the tears rolling down his cheeks, he looked into the Master's face and said: "And so I will, my Master, but how can I learn to love my brethren more?" And then answered Jesus: "You will learn to love your brethren more when you first learn to love their Father in heaven more, and after you have become truly more interested in their welfare in time and in eternity. And all such human interest is fostered by understanding sympathy, unselfish service, and unstinted forgiveness. No man should despise your youth, but I exhort you always to give due consideration to the fact that age oftentimes represents experience, and that nothing in human affairs can take the place of actual experience. Strive to live peaceably with all men, especially your friends in the brotherhood of the heavenly kingdom. And, John, **always remember,** strive not with the souls you would win for the kingdom."

Page 1956:5, 1957:1   *Paper 181:2.10–11*

Simon wanted to speak further, but Jesus raised his hand and, stopping him, went on to say: "None of my apostles are more sincere and honest at heart than you, but not one of them will be so upset and disheartened as you, after my departure. In all of your discouragement my spirit shall abide with you, and these, your brethren, will not forsake you. **Do not forget** what I have taught you regarding the relation of citizenship on earth to sonship in the Father's spiritual kingdom. Ponder well all that I have said to you about rendering to Caesar the things which are Caesar's and to God that which is God's. Dedicate your life, Simon, to showing how acceptably mortal man may fulfill my injunction concerning the simultaneous recognition of temporal duty to civil powers and spiritual service in the brotherhood of the kingdom. If you will be taught by the Spirit of Truth, never will there be conflict between the requirements of citizenship on earth and sonship in heaven unless the temporal rulers presume to require of you the homage and worship which belong only to God.

"And now, Simon, when you do finally see all of this, and after you have shaken off your depression and have gone forth proclaiming this gospel in great power, **never forget** that I was with you even through all of your season of discouragement, and that I will go on with you to the very end. You shall always be my apostle, and after you become willing to see by the eye of the spirit and more fully to yield your will to the will of the Father in heaven, then will you return to labor as my ambassador, and no one shall take away from you the authority which I have conferred upon you, because of your slowness of comprehending the truths I have taught you. And so, Simon, once more I warn you that they who fight with the sword perish with the sword, while they who labor in the spirit achieve life everlasting in the kingdom to come with joy and peace in the kingdom which now is. And when the work given into your hands is finished on earth, you, Simon, shall sit down with me in my kingdom over there. You shall really see the kingdom you have longed for, but not in this life. Continue to believe in me and in that which I have revealed to you, and you shall receive the gift of eternal life."

Page 1960:1   *Paper 181:2.20*

And then Jesus went over to Philip, who, standing up, heard this message from his Master: "Philip, you have asked me many foolish questions, but I have done my utmost to answer every one, and now would I answer the last of such questionings which have arisen in your most honest but unspiritual mind. All the time I have been coming around toward you, have you been saying to yourself, `What shall I ever do if the Master goes away and leaves us alone in the world?' O, you of little faith! And yet you have almost as much as many of your brethren. You have been a good steward, Philip. You failed us only a few times, and one of those failures we utilized to manifest the Father's glory. Your office of stewardship is about over. You must soon more fully do the work you were called to do—the preaching of this gospel of the kingdom. Philip, you have always wanted to be shown, and very soon shall you see great things. Far better that you should have seen all this by faith, but since you were sincere even in your material sightedness, you will live to see my words fulfilled. And then, when you are blessed with spiritual vision, go forth to your work, dedicating your life to the cause of leading mankind to search for God and to seek eternal realities with the eye of spiritual faith and not with the eyes of the material mind. **Remember**, Philip, you have a great mission on earth, for the world is filled with those who look at life just as you have tended to. You have a great work to do, and when it is finished in faith, you shall come to me in my kingdom, and I will take great pleasure in showing you that which eye has not seen, ear heard, nor the mortal mind conceived. In the meantime, become as a little child in the kingdom of the spirit and permit me, as the spirit of the new teacher, to lead you forward in the spiritual kingdom. And in this way will I be able to do much for you which I was not able to accomplish when I sojourned with you as a mortal of the realm. **And always remember**, Philip, he who has seen me has seen the Father."

Page 2033:1   *Paper 190:3.1*

The fifth morontia manifestation of Jesus to the recognition of mortal eyes occurred in the presence of some twenty-five women believers assembled at the home of Joseph of Arimathea, at about fifteen minutes past four o'clock on this same Sunday afternoon. Mary Magdalene had returned to Joseph's house just a few minutes before this appearance. James, Jesus' brother, had requested that nothing be said to the apostles concerning the Master's appearance at Bethany. He had not asked Mary to refrain from reporting the occurrence to her sister believers. Accordingly, after Mary had pledged all the women to secrecy, she proceeded to relate what had so recently happened while she was with Jesus' family at Bethany. And she was in the very midst of this thrilling recital when a sudden and solemn hush fell over them; they beheld in their very midst the fully visible form of the risen Jesus. He greeted them, saying: "Peace be upon you. In the fellowship of the kingdom there shall be neither Jew nor gentile, rich nor poor, free nor bond, man nor woman. You also are called to publish the good news of the liberty of mankind through the gospel of sonship with God in the kingdom of heaven. Go to all the world proclaiming this gospel and confirming believers in the faith thereof. And while you do this, **forget not** to minister to the sick and strengthen those who are fainthearted and fear-ridden. And I will be with you always, even to the ends of the earth." And when he had thus spoken, he vanished from their sight, while the women fell on their faces and worshiped in silence.

Page 2047:6   *Paper 192:2.2*

Jesus then turned toward Peter and asked, "Peter, do you love me?" Peter answered, "Lord, you know I love you with all my soul." Then said Jesus: "If you love me, Peter, feed my lambs. Do not neglect to minister to the weak, the poor, and the young. Preach the gospel without fear or favor; **remember always** that God is no respecter of persons. Serve your fellow men even as I have served you; forgive your fellow mortals even as I have forgiven you. Let experience teach you the value of meditation and the power of intelligent reflection."

Page 2049:4   *Paper 192:2.13*

Then he walked and talked with the Alpheus twins, James and Judas, and speaking to both of them, he asked, "James and Judas, do you believe in me?" And when they both answered, "Yes, Master, we do believe," he said: "I will soon leave you. You see that I have already left you in the flesh. I tarry only a short time in this form before I go to my Father. You believe in me—you are my apostles, and you always will be. Go on believing and remembering your association with me, when I am gone, and after you have, perchance, returned to the work you used to do before you came to live with me. Never allow a change in your outward work to influence your allegiance. Have faith in God to the end of your days on earth. **Never forget** that, when you are a faith son of God, all upright work of the realm is sacred. Nothing which a son of God does can be common. Do your work, therefore, from this time on, as for God. And when you are through on this world, I have other and better worlds where you shall likewise work for me. And in all of this work, on this world and on other worlds, I will work with you, and my spirit shall dwell within you."

Page 2052:4   *Paper 193:0.4*

"**I admonish you ever to remember** that your mission among men is to proclaim the gospel of the kingdom—the reality of the fatherhood of God and the truth of the sonship of man. Proclaim the whole truth of the good news, not just a part of the saving gospel. Your message is not changed by my resurrection experience. Sonship with God, by faith, is still the saving truth of the gospel of the kingdom. You are to go forth preaching the love of God and the service of man. That which the world needs most to know is: Men are the sons of God, and through faith they can actually realize, and daily experience, this ennobling truth. My bestowal should help all men to know that they are the children of God, but such knowledge will not suffice if they fail personally to faith-grasp the saving truth that they are the living spirit sons of the eternal Father. The gospel of the kingdom is concerned with the love of the Father and the service of his children on earth.

Page 2071:5   *Paper 195:1.5*

At the time Paul stood up in Athens preaching "Christ and Him Crucified," the Greeks were spiritually hungry; they were inquiring, interested, and actually looking for spiritual truth. **Never forget** that at first the Romans fought Christianity, while the Greeks embraced it, and that it was the Greeks who literally forced the Romans subsequently to accept this new religion, as then modified, as a part of Greek culture.

Page 2076:2   *Paper 195:5.11*

In confusion over man's origin, do not lose sight of his eternal destiny. **Forget not** that Jesus loved even little children, and that he forever made clear the great worth of human personality.

Page 2084:3   *Paper 195:10.3*

Ever bear in mind—God and men need each other. They are mutually necessary to the full and final attainment of eternal personality experience in the divine destiny of universe finality.

~Chapter Four~

Memorable Passages

The following passages cover all four parts of *The Urantia Book*. They have surfaced numerous times over the years and will undoubtedly remain a favorite to many current and future Urantia Book students. A few blank pages are made available to add your own. Enjoy!

Page 24:6   *Paper 1:2.5*

Those who know God have experienced the fact of his presence; such God-knowing mortals hold in their personal experience the only positive proof of the existence of the living God which one human being can offer to another. The existence of God is utterly beyond all possibility of demonstration except for the contact between the God-consciousness of the human mind and the God-presence of the Thought Adjuster that indwells the mortal intellect and is bestowed upon man as the free gift of the Universal Father.

Page 159:6   *Paper 14:5.10*

Love of adventure, curiosity, and dread of monotony—these traits inherent in evolving human nature—were not put there just to aggravate and annoy you during your short sojourn on earth, but rather to suggest to you that death is only the beginning of an endless career of adventure, an everlasting life of anticipation, an eternal voyage of discovery.

Page 281:4   *Paper 25:5.3*

Every occurrence of significance in the organized and inhabited creation is a matter of record. While events of no more than local importance find only a local recording, those of wider significance are dealt with accordingly. From the planets, systems, and constellations of Nebadon, everything of universe import is posted on Salvington; and from such universe capitals those episodes are advanced to higher recording which pertain to the affairs of the sector and supergovernments. Paradise also has a relevant summary of superuniverse and Havona data; and this historic and cumulative story of the universe of universes is in the custody of these exalted tertiary supernaphim.

Page 312:3   *Paper 28:5:16*

5. *The Joy of Existence.* By nature these beings are reflectively attuned to the superaphic harmony supervisors above and to certain of the seraphim below, but it is difficult to explain just what the members of this interesting group really do. Their principal activities are directed toward promoting reactions of joy among the various orders of the angelic hosts and the lower will creatures. The Divine Counselors, to whom they are attached, seldom use them for specific joy finding. In a more general manner and in collaboration with the reversion directors, they function as joy clearinghouses, seeking to upstep the pleasure reactions of the realms while trying to improve the humor taste, to develop a superhumor among mortals and angels. They endeavor to demonstrate that there is inherent joy in freewill existence, independent of all extraneous influences; and they are right, although they meet with great difficulty in inculcating this truth in the minds of primitive men. The higher spirit personalities and the angels are more quickly responsive to these educational efforts.

Page 364:6   *Paper 32:5.4*

To me it seems more fitting, for purposes of explanation to the mortal mind, to conceive of eternity as a cycle and the eternal purpose as an endless circle, a cycle of eternity in some way synchronized with the transient material cycles of time. As regards the sectors of time connected with, and forming a part of, the cycle of eternity, we are forced to recognize that such temporary epochs are born, live, and die just as the temporary beings of time are born, live, and die. Most human beings die because, having failed to achieve the spirit level of Adjuster fusion, the metamorphosis of death constitutes the only possible procedure whereby they may escape the fetters of time and the bonds of material creation, thereby being enabled to strike spiritual step with the progressive procession of eternity. Having survived the trial life of time and material existence, it becomes possible for you to continue on in touch with, even as a part of, eternity, swinging on forever with the worlds of space around the circle of the eternal ages.

Page 438:2   *Paper 39:5.9*

In the more advanced planetary ages these seraphim enhance man's appreciation of the truth that uncertainty is the secret of contented continuity. They help the mortal philosophers to realize that, when ignorance is essential to success, it would be a colossal blunder for the creature to know the future. They heighten man's taste for the sweetness of uncertainty, for the romance and charm of the indefinite and unknown future.

Page 1002:1   *Paper 91:8.9*

But real praying does attain reality. Even when the air currents are ascending, no bird can soar except by outstretched wings. Prayer elevates man because it is a technique of progressing by the utilization of the ascending spiritual currents of the universe.

Page 1091:8   *Paper 99:5.9*

Primitive man made little effort to put his religious convictions into words. His religion was danced out rather than thought out. Modern men have thought out many creeds and created many tests of religious faith. Future religionists must live out their religion, dedicate themselves to the wholehearted service of the brotherhood of man. It is high time that man had a religious experience so personal and so sublime that it could be realized and expressed only by "feelings that lie too deep for words."

Page 1098:2   *Paper 100:4.5*

In the mind's eye conjure up a picture of one of your primitive ancestors of cave-dwelling times—a short, misshapen, filthy, snarling hulk of a man standing, legs spread, club upraised, breathing hate and animosity as he looks fiercely just ahead. Such a picture hardly depicts the divine dignity of man. But allow us to enlarge the picture. In front of this animated human crouches a saber-toothed tiger. Behind him, a woman and two children. Immediately you recognize that such a picture stands for the beginnings of much that is fine and noble in the human race, but the man is the same in both pictures. Only, in the second sketch you are favored with a widened horizon. You therein discern the motivation of this evolving mortal. His attitude becomes praiseworthy because you understand him. If you could only fathom the motives of your associates, how much better you would understand them. If you could only know your fellows, you would eventually fall in love with them.

Page 1105:5   *Paper 101:2.1*

The fact of religion consists wholly in the religious experience of rational and average human beings. And this is the only sense in which religion can ever be regarded as scientific or even psychological. The proof that revelation is revelation is this same fact of human experience: the fact that revelation does synthesize the apparently divergent sciences of nature and the theology of religion into a consistent and logical universe philosophy, a co-ordinated and unbroken explanation of both science and religion, thus creating a harmony of mind and satisfaction of spirit which answers in human experience those questionings of the mortal mind which craves to know how the Infinite works out his will and plans in matter, with minds, and on spirit.

Page 1123:7, 1124:1   *Paper 102:5.1–2*

Although the establishment of the fact of belief is not equivalent to establishing the fact of that which is believed, nevertheless, the evolutionary progression of simple life to the status of personality does demonstrate the fact of the existence of the potential of personality to start with. And in the time universes, potential is always supreme over the actual. In the evolving cosmos the potential is what is to be, and what is to be is the unfolding of the purposive mandates of Deity.

This same purposive supremacy is shown in the evolution of mind ideation when primitive animal fear is transmuted into the constantly deepening reverence for God and into increasing awe of the universe. Primitive man had more religious fear than faith, and the supremacy of spirit potentials over mind actuals is demonstrated when this craven fear is translated into living faith in spiritual realities.

Page 1130:6   *Paper 103:2:1*

Religion is functional in the human mind and has been realized in experience prior to its appearance in human consciousness. A child has been in existence about nine months before it experiences *birth*. But the "birth" of religion is not sudden; it is rather a gradual emergence. Nevertheless, sooner or later there is a "birth day." You do not enter the kingdom of heaven unless you have been "born again"—born of the Spirit. Many spiritual births are accompanied by much anguish of spirit and marked psychological perturbations, as many physical births are characterized by a "stormy labor" and other abnormalities of "delivery." Other spiritual births are a natural and normal growth of the recognition of supreme values with an enhancement of spiritual experience, albeit no religious development occurs without conscious effort and positive and individual determinations. Religion is never a passive experience, a negative attitude. What is termed the "birth of religion" is not directly associated with so-called conversion experiences which usually characterize religious episodes occurring later in life as a result of mental conflict, emotional repression, and temperamental upheavals.

Page 1133:3   *Paper 103:4.3*

The sense of guilt (not the consciousness of sin) comes either from interrupted spiritual communion or from the lowering of one's moral ideals. Deliverance from such a predicament can only come through the realization that one's highest moral ideals are not necessarily synonymous with the will of God. Man cannot hope to live up to his highest ideals, but he can be true to his purpose of finding God and becoming more and more like him.

Page 1138:6   *Paper 103:7.8*

The truth—an understanding of cosmic relationships, universe facts, and spiritual values—can best be had through the ministry of the Spirit of Truth and can best be criticized by revelation. But revelation originates neither a science nor a religion; its function is to co-ordinate both science and religion with the truth of reality. Always, in the absence of revelation or in the failure to accept or grasp it, has mortal man resorted to his futile gesture of metaphysics, that being the only human substitute for the revelation of truth or for the mota of morontia personality.

Page 1141:4   *Paper 103:9.6*

When theology masters religion, religion dies; it becomes a doctrine instead of a life. The mission of theology is merely to facilitate the self-consciousness of personal spiritual experience. Theology constitutes the religious effort to define, clarify, expound, and justify the experiential claims of religion, which, in the last analysis, can be validated only by living faith. In the higher philosophy of the universe, wisdom, like reason, becomes allied to faith. Reason, wisdom, and faith are man's highest human attainments. Reason introduces man to the world of facts, to things; wisdom introduces him to a world of truth, to relationships; faith initiates him into a world of divinity, spiritual experience.

Page 1204:3   *Paper 110:1.5*

The Adjuster remains with you in all disaster and through every sickness which does not wholly destroy the mentality. But how unkind knowingly to defile or otherwise deliberately to pollute the physical body, which must serve as the earthly tabernacle of this marvelous gift from God. All physical poisons greatly retard the efforts of the Adjuster to exalt the material mind, while the mental poisons of fear, anger, envy, jealousy, suspicion, and intolerance likewise tremendously interfere with the spiritual progress of the evolving soul.

Page 1205:6   *Paper 110:3.2*

The success of your Adjuster in the enterprise of piloting you through the mortal life and bringing about your survival depends not so much on the theories of your beliefs as upon your decisions, determinations, and steadfast faith. All these movements of personality growth become powerful influences aiding in your advancement because they help you to co-operate with the Adjuster; they assist you in ceasing to resist. Thought Adjusters succeed or apparently fail in their terrestrial undertakings just in so far as mortals succeed or fail to co-operate with the scheme whereby they are to be advanced along the ascending path of perfection attainment. The secret of survival is wrapped up in the supreme human desire to be Godlike and in the associated willingness to do and be any and all things which are essential to the final attainment of that overmastering desire.

Page 1209:4   *Paper 110:6.4*

When the development of the intellectual nature proceeds faster than that of the spiritual, such a situation renders communication with the Thought Adjuster both difficult and dangerous. Likewise, overspiritual development tends to produce a fanatical and perverted interpretation of the spirit leadings of the divine indweller. Lack of spiritual capacity makes it very difficult to transmit to such a material intellect the spiritual truths resident in the higher superconsciousness. It is to the mind of perfect poise, housed in a body of clean habits, stabilized neural energies, and balanced chemical function—when the physical, mental, and spiritual powers are in triune harmony of development—that a maximum of light and truth can be imparted with a minimum of temporal danger or risk to the real welfare of such a being. By such a balanced growth does man ascend the circles of planetary progression one by one, from the seventh to the first.

Page 1630:4   *Paper 145:2.8*

"No more should you fear that God will punish a nation for the sin of an individual; neither will the Father in heaven punish one of his believing children for the sins of a nation, albeit the individual member of any family must often suffer the material consequences of family mistakes and group transgressions. Do you not realize that the hope of a better nation—or a better world—is bound up in the progress and enlightenment of the individual?"

Page 1640:5   *Paper 146:2.16*

15. And then Jesus said: "Be not constantly overanxious about your common needs. Be not apprehensive concerning the problems of your earthly existence, but in all these things by prayer and supplication, with the spirit of sincere thanksgiving, let your needs be spread out before your Father who is in heaven." Then he quoted from the Scriptures: "I will praise the name of God with a song and will magnify him with thanksgiving. And this will please the Lord better than the sacrifice of an ox or bullock with horns and hoofs."

Page 1729:7   *Paper 155:5.12*

And Jesus went on to say: "At Jerusalem the religious leaders have formulated the various doctrines of their traditional teachers and the prophets of other days into an established system of intellectual beliefs, a religion of authority. The appeal of all such religions is largely to the mind. And now are we about to enter upon a deadly conflict with such a religion since we will so shortly begin the bold proclamation of a new religion—a religion which is not a religion in the present-day meaning of that word, a religion that makes its chief appeal to the divine spirit of my Father which resides in the mind of man; a religion which shall derive its authority from the fruits of its acceptance that will so certainly appear in the personal experience of all who really and truly become believers in the truths of this higher spiritual communion."

Page 1780:1   *Paper 160:4.9*

And it is in this business of facing failure and adjusting to defeat that the far-reaching vision of religion exerts its supreme influence. Failure is simply an educational episode—a cultural experiment in the acquirement of wisdom—in the experience of the God-seeking man who has embarked on the eternal adventure of the exploration of a universe. To such men defeat is but a new tool for the achievement of higher levels of universe reality.

Page 1917:3   *Paper 176:3.7*

Truth is living; the Spirit of Truth is ever leading the children of light into new realms of spiritual reality and divine service. You are not given truth to crystallize into settled, safe, and honored forms. Your revelation of truth must be so enhanced by passing through your personal experience that new beauty and actual spiritual gains will be disclosed to all who behold your spiritual fruits and in consequence thereof are led to glorify the Father who is in heaven. Only those faithful servants who thus grow in the knowledge of the truth, and who thereby develop the capacity for divine appreciation of spiritual realities, can ever hope to "enter fully into the joy of their Lord." What a sorry sight for successive generations of the professed followers of Jesus to say, regarding their stewardship of divine truth: "Here, Master, is the truth you committed to us a hundred or a thousand years ago. We have lost nothing; we have faithfully preserved all you gave us; we have allowed no changes to be made in that which you taught us; here is the truth you gave us." But such a plea concerning spiritual indolence will not justify the barren steward of truth in the presence of the Master. In accordance with the truth committed to your hands will the Master of truth require a reckoning.

Page 1951:5   *Paper 180:6.4*

"This spirit will not speak of himself, but he will declare to you that which the Father has revealed to the Son, and he will even show you things to come; he will glorify me even as I have glorified my Father. This spirit comes forth from me, and he will reveal my truth to you. Everything which the Father has in this domain is now mine; wherefore did I say that this new teacher would take of that which is mine and reveal it to you.

Page 2073:8   *Paper 195:3.4*

That which gave greatest power to Christianity was the way its believers lived lives of service and even the way they died for their faith during the earlier times of drastic persecution.

Page 2088:5   *Paper 196:0.10*

Jesus brought to God, as a man of the realm, the greatest of all offerings: the consecration and dedication of his own will to the majestic service of doing the divine will. Jesus always and consistently interpreted religion wholly in terms of the Father's will. When you study the career of the Master, as concerns prayer or any other feature of the religious life, look not so much for what he taught as for what he did. Jesus never prayed as a religious duty. To him prayer was a sincere expression of spiritual attitude, a declaration of soul loyalty, a recital of personal devotion, an expression of thanksgiving, an avoidance of emotional tension, a prevention of conflict, an exaltation of intellection, an ennoblement of desire, a vindication of moral decision, an enrichment of thought, an invigoration of higher inclinations, a consecration of impulse, a clarification of viewpoint, a declaration of faith, a transcendental surrender of will, a sublime assertion of confidence, a revelation of courage, the proclamation of discovery, a confession of supreme devotion, the validation of consecration, a technique for the adjustment of difficulties, and the mighty mobilization of the combined soul powers to withstand all human tendencies toward selfishness, evil, and sin. He lived just such a life of prayerful consecration to the doing of his Father's will and ended his life triumphantly with just such a prayer. The secret of his unparalleled religious life was this consciousness of the presence of God; and he attained it by intelligent prayer and sincere worship—unbroken communion with God—and not by leadings, voices, visions, or extraordinary religious practices.

Page 2094:2   *Paper 196:3.3*

The progressive comprehension of reality is the equivalent of approaching God. The finding of God, the consciousness of identity with reality, is the equivalent of the experiencing of self-completion—self-entirety, self-totality. The experiencing of total reality is the full realization of God, the finality of the God-knowing experience.

Page 2096:8   *Paper 196:3.29*

Some men's lives are too great and noble to descend to the low level of being merely successful. The animal must adapt itself to the environment, but the religious man transcends his environment and in this way escapes the limitations of the present material world through this insight of divine love. This concept of love generates in the soul of man that superanimal effort to find truth, beauty, and goodness; and when he does find them, he is glorified in their embrace; he is consumed with the desire to live them, to do righteousness.

(Add your own favorites here.)

(Add your own favorites here.)

Chapter Five

Arithmetical Insights from a Revelatory Position

There are a number of worthwhile study aids and secondary works to *The Urantia Book*. Chapters five and six provide yet another fresh approach to stimulate your thoughts and perhaps whet your appetite for additional studies in these areas. Enjoy!

Page 14:8   *Foreword 0:11.10*

3. *The Universal Absolute*, we logically deduce, was inevitable in the Universal Father's absolute freewill act of differentiating universe realities into deified and undeified—personalizable and nonpersonalizable—values. The Universal Absolute is the Deity phenomenon indicative of the resolution of the tension created by the freewill act of thus differentiating universe reality, and functions as the associative co-ordinator of these **sum totals** of existential potentialities.

Page 79:3   *Paper 6:7.3*

The Eternal Son is truly a merciful minister, a divine spirit, a spiritual power, and a real personality. The Son is the spiritual and personal nature of God made manifest to the universes—**the sum and substance** of the First Source and Center, divested of all that which is nonpersonal, extradivine, nonspiritual, and pure potential. But it is impossible to convey to the human mind a word picture of the beauty and grandeur of the supernal personality of the Eternal Son. Everything that tends to obscure the Universal Father operates with almost equal influence to prevent the conceptual recognition of the Eternal Son. You must await your attainment of Paradise, and then you will understand why I was unable to portray the character of this absolute personality to the understanding of the finite mind.

Page 110:3   *Paper 10:2.5*

The personality of the First Source and Center is the personality of infinity **minus** the absolute personality of the Eternal Son. The personality of the Third Source and Center is the **superadditive consequence** of the union of the liberated Father-personality and the absolute Son-personality.

Page 113:3   *Paper 10:5.2*

The functions of the Paradise Trinity are **not simply the sum** of the Father's apparent endowment of divinity **plus** those specialized attributes that are unique in the personal existence of the Son and the Spirit. The Trinity association of the three Paradise Deities **results** in the evolution, eventuation, and deitization of new meanings, values, powers, and capacities for universal revelation, action, and administration. Living associations, human families, social groups, or the Paradise Trinity are not augmented by mere arithmetical summation. The group potential is always far in excess of the **simple sum** of the attributes of the component individuals.

Page 175:6   *Paper 15:8.3*

Evolving energy has substance; it has weight, although weight is always relative, depending on revolutionary velocity, mass, and antigravity. Mass in matter tends to retard velocity in energy; and the anywhere-present velocity of energy represents: the initial endowment of velocity, **minus** retardation by mass encountered in transit, **plus** the regulatory function of the living energy controllers of the superuniverse and the physical influence of near-by highly heated or heavily charged bodies.

Page 202:1   *Paper 17:4.1*

The forty-nine Reflective Image Aids were created by the Reflective Spirits, and there are just seven Aids on the headquarters of each superuniverse. The first creative act of the seven Reflective Spirits of Uversa was the production of their seven Image Aids, each Reflective Spirit creating his own Aid. The Image Aids are, in certain attributes and characteristics, perfect reproductions of their Reflective Mother Spirits; they are virtual duplications **minus** the attribute of reflectivity. They are true images and constantly function as the channel of communication between the Reflective Spirits and the superuniverse authorities. The Image Aids are not merely assistants; they are actual representations of their respective Spirit ancestors; they are images, and they are true to their name.

Page 218:3   *Paper 19:4.5*

The Censors are universe totaling personalities. When a thousand witnesses have given testimony—or a million—when the voice of wisdom has spoken and the counsel of divinity has recorded, when the testimony of ascendant perfection has been added, then the Censor functions, and there is immediately revealed an unerring and **divine totaling** of all that has transpired; and such a disclosure represents the divine conclusion, the **sum and substance** of a final and perfect decision. Therefore, when a Censor has spoken, no one else may speak, for the Censor has depicted the **true and unmistakable total** of all that has gone before. When he speaks, there is no appeal.

Page 379:5   *Paper 34:5.5*

Though the Spirit of Truth is poured out upon all flesh, this spirit of the Son is almost wholly limited in function and power by man's personal reception of that which constitutes the **sum and substance** of the mission of the bestowal Son. The Holy Spirit is partly independent of human attitude and partially conditioned by the decisions and co-operation of the will of man. Nevertheless, the ministry of the Holy Spirit becomes increasingly effective in the sanctification and spiritualization of the inner life of those mortals who the more fully obey the divine leadings.

Page 474:1   *Paper 42:4.11*

The increase of mass in matter is **equal** to the increase of energy **divided by the square** of the velocity of light. In a dynamic sense the work which resting matter can perform is **equal** to the energy expended in bringing its parts together from Paradise **minus** the resistance of the forces overcome in transit and the attraction exerted by the parts of matter on one another.

Page 539:2   *Paper 47:9.5*

You will greatly enjoy your progress through the seven dematerializing worlds; they are really demortalizing spheres. You are mostly human on the first mansion world, just a mortal being **minus** a material body, a human mind housed in a morontia form—a material body of the morontia world but not a mortal house of flesh and blood. You really pass from the mortal state to the immortal status at the time of Adjuster fusion, and by the time you have finished the Jerusem career, you will be full-fledged morontians.

Page 556:6   *Paper 48:7.6*

4. Few mortals ever dare to draw anything like the **sum** of personality credits established by the combined ministries of nature and grace. The majority of impoverished souls are truly rich, but they refuse to believe it.

Page 619:3   *Paper 54:6.6*

At first the Lucifer upheaval appeared to be an unmitigated calamity to the system and to the universe. Gradually benefits began to accrue. With the passing of twenty-five thousand years of system time (twenty thousand years of Urantia time), the Melchizedeks began to teach that the good resulting from Lucifer's folly had come to **equal** the evil incurred. The **sum** of evil had by that time become almost stationary, continuing to increase only on certain isolated worlds, while the beneficial repercussions continued to **multiply and extend** out through the universe and superuniverse, even to Havona. The Melchizedeks now teach that the good resulting from the Satania rebellion is more than a **thousand times the sum** of all the evil.

Page 647:8   *Paper 56:10.17*

Universal beauty is the recognition of the reflection of the Isle of Paradise in the material creation, while eternal truth is the special ministry of the Paradise Sons who not only bestow themselves upon the mortal races but even pour out their Spirit of Truth upon all peoples. Divine goodness is more fully shown forth in the loving ministry of the manifold personalities of the Infinite Spirit. But love, the **sum total** of these three qualities, is man's perception of God as his spirit Father.

Page 726:1   *Paper 64:6.29*

From time to time all of these different peoples experienced cultural and spiritual revivals. Mansant was a great teacher of the post-Planetary Prince days. But mention is made only of those outstanding leaders and teachers who markedly influenced and inspired a whole race. With the passing of time, many lesser teachers arose in different regions; and in the aggregate they contributed much to the **sum total** of those saving influences which prevented the total collapse of cultural civilization, especially during the long and dark ages between the Caligastia rebellion and the arrival of Adam.

Page 745:1   *Paper 66:4.11*

These mid-type creatures were of great service in carrying on the affairs of the world's headquarters. They were invisible to human beings, but the primitive sojourners at Dalamatia were taught about these unseen semispirits, and for ages they constituted the **sum total** of the spirit world to these evolving mortals.

Page 762:3   *Paper 67:8.4*

If the Lucifer rebellion has handicapped the local system and its fallen worlds, if the loss of this Son and his misled associates has temporarily hampered the progress of the constellation of Norlatiadek, then weigh the effect of the far-flung presentation of the inspiring performance of this one child of nature and his determined band of 143 comrades in standing steadfast for the higher concepts of universe management and administration in the face of such tremendous and adverse pressure exerted by his disloyal superiors. And let me assure you, this has already done more good in the universe of Nebadon and the superuniverse of Orvonton than can ever be outweighed by the **sum total** of all the evil and sorrow of the Lucifer rebellion.

Page 763:7   *Paper 68:1.4*

Primitive human beings early learned that groups are vastly greater and stronger than the mere **sum** of their individual units. One hundred men united and working in unison can move a great stone; a score of well-trained guardians of the peace can restrain an angry mob. And so society was born, not of mere association of numbers, but rather as a result of the *organization* of intelligent co-operators. But co-operation is not a natural trait of man; he learns to co-operate first through fear and then later because he discovers it is most beneficial in meeting the difficulties of time and guarding against the supposed perils of eternity.

Page 768:1   *Paper 68:5.1*

Land is the stage of society; men are the actors. And man must ever adjust his performances to conform to the land situation. The evolution of the mores is always dependent on the land-man ratio. This is true notwithstanding the difficulty of its discernment. Man's land technique, or maintenance arts, plus his standards of living, equal the **sum total** of the folkways, the mores. And the **sum** of man's adjustment to the life demands equals his cultural civilization.

Page 794:12   *Paper 70:9.7*

But this equality ideal is the child of civilization; it is not found in nature. Even culture itself demonstrates conclusively the inherent inequality of men by their very unequal capacity therefor. The sudden and nonevolutionary realization of supposed natural equality would quickly throw civilized man back to the crude usages of primitive ages. Society cannot offer equal rights to all, but it can promise to administer the varying rights of each with fairness and equity. It is the business and duty of society to provide the child of nature with a fair and peaceful opportunity to pursue self-maintenance, participate in self-perpetuation, while at the same time enjoying some measure of self-gratification, the **sum** of all three constituting human happiness.

Page 913:1   *Paper 82:0.1*

Marriage—mating—grows out of bisexuality. Marriage is man's reactional adjustment to such bisexuality, while the family life is the **sum total** resulting from all such evolutionary and adaptative adjustments. Marriage is enduring; it is not inherent in biologic evolution, but it is the basis of all social evolution and is therefore certain of continued existence in some form. Marriage has given mankind the home, and the home is the crowning glory of the whole long and arduous evolutionary struggle.

Page 953:3   *Paper 86:4.4*

Eventually the savage conceived of himself as a double—body and breath. The breath **minus** the body equaled a spirit, a ghost. While having a very definite human origin, ghosts, or spirits, were regarded as superhuman. And this belief in the existence of disembodied spirits seemed to explain the occurrence of the unusual, the extraordinary, the infrequent, and the inexplicable.

Page 953:6   *Paper 86:4.7*

Early man entertained no ideas of hell or future punishment. The savage looked upon the future life as just like this one, **minus** all ill luck. Later on, a separate destiny for good ghosts and bad ghosts—heaven and hell—was conceived. But since many primitive races believed that man entered the next life just as he left this one, they did not relish the idea of becoming old and decrepit. The aged much preferred to be killed before becoming too infirm.

Page 1005:2   *Paper 92:2.6*

Religion has at one time or another sanctioned all sorts of contrary and inconsistent behavior, has at some time approved of practically all that is now regarded as immoral or sinful. Conscience, untaught by experience and unaided by reason, never has been, and never can be, a safe and unerring guide to human conduct. Conscience is not a divine voice speaking to the human soul. It is merely the **sum total** of the moral and ethical content of the mores of any current stage of existence; it simply represents the humanly conceived ideal of reaction in any given set of circumstances.

Page 1136:1   *Paper 103:6.6*

Always must man's inner spirit depend for its expression and self-realization upon the mechanism and technique of the mind. Likewise must man's outer experience of material reality be predicated on the mind consciousness of the experiencing personality. Therefore are the spiritual and the material, the inner and the outer, human experiences always correlated with the mind function and conditioned, as to their conscious realization, by the mind activity. Man experiences matter in his mind; he experiences spiritual reality in the soul but becomes conscious of this experience in his mind. The intellect is the harmonizer and the ever-present conditioner and qualifier of the **sum total** of mortal experience. Both energy-things and spirit values are colored by their interpretation through the mind media of consciousness.

Page 1145:5   *Paper 104:2.4*

And this selfsame Paradise Trinity is a real entity—not a personality but nonetheless a true and absolute reality; not a personality but nonetheless compatible with coexistent personalities—the personalities of the Father, the Son, and the Spirit. The Trinity is a **supersummative** Deity reality eventuating out of the conjoining of the three Paradise Deities. The qualities, characteristics, and functions of the Trinity are **not the simple sum** of the attributes of the three Paradise Deities; Trinity functions are something unique, original, and not wholly predictable from an analysis of the attributes of Father, Son, and Spirit.

Page 1147:9   *Paper 104:3.10*

There is, however, one point of comparison between trinity and triunity: Both eventuate in functions that are something other than the **discernible sum** of the attributes of the component members. But while they are thus comparable from a functional standpoint, they otherwise exhibit no categorical relationship. They are roughly related as the relation of function to structure. But the function of the triunity association is not the function of the trinity structure or entity.

Page 1151:6   *Paper 104:5.3*

The Eternal Son is the absolute of spirit reality, the absolute personality. The Paradise Isle is the absolute of cosmic reality, the absolute pattern. The Conjoint Actor is the absolute of mind reality, the co-ordinate of absolute spirit reality, and the existential Deity synthesis of personality and power. This triune association eventuates the co-ordination of the **sum total** of actualized reality—spirit, cosmic, or mindal. It is unqualified in actuality.

Page 1205:4   *Paper 110:2.6*

To the extent that this identity is realized, you are mentally approaching the morontia order of existence. Morontia mind is a term signifying the substance and **sum total** of the co-operating minds of diversely material and spiritual natures. Morontia intellect, therefore, connotes a dual mind in the local universe dominated by one will. And with mortals this is a will, human in origin, which is becoming divine through man's identification of the human mind with the mindedness of God.

Page 1209:1   *Paper 110:6.1*

The **sum total** of personality realization on a material world is contained within the successive conquest of the seven psychic circles of mortal potentiality. Entrance upon the seventh circle marks the beginning of true human personality function. Completion of the first circle denotes the relative maturity of the mortal being. Though the traversal of the seven circles of cosmic growth does not equal fusion with the Adjuster, the mastery of these circles marks the attainment of those steps which are preliminary to Adjuster fusion.

Page 1230:1   *Paper 112:3.3*

2. *Intellectual (mind) death.* When the vital circuits of higher adjutant ministry are disrupted through the aberrations of intellect or because of the partial destruction of the mechanism of the brain, and if these conditions pass a certain critical point of irreparability, the indwelling Adjuster is immediately released to depart for Divinington. On the universe records a mortal personality is considered to have met with death whenever the essential mind circuits of human will-action have been destroyed. And again, this is death, irrespective of the continuing function of the living mechanism of the physical body. The body **minus** the volitional mind is no longer human, but according to the prior choosing of the human will, the soul of such an individual may survive.

Page 1279:7   *Paper 117:1.9*

The Supreme is **symmetrically** inclusive. The First Source and Center is potential in the three great Absolutes, is actual in Paradise, in the Son, and in the Spirit; but the Supreme is both actual and potential, a being of personal supremacy and of almighty power, responsive alike to creature effort and Creator purpose; self-acting upon the universe and self-reactive to the **sum total** of the universe; and at one and the same time the supreme creator and the supreme creature. The Deity of Supremacy is thus expressive of the **sum total** of the entire finite.

Page 1281:3   *Paper 117:3.1*

The cosmic reality variously designated as the Supreme Being, God the Supreme, and the Almighty Supreme, is the complex and universal synthesis of the emerging phases of all finite realities. The far-flung diversification of eternal energy, divine spirit, and universal mind attains finite culmination in the evolution of the Supreme, who is the **sum total** of all finite growth, self-realized on deity levels of finite maximum completion.

Page 1290:1   *Paper 117:6.15*

To evolutionary creatures there are seven great approaches to the Universal Father, and each of these Paradise ascensions passes through the divinity of one of the Seven Master Spirits; and each such approach is made possible by an enlargement of experience receptivity consequent upon the creature's having served in the superuniverse reflective of the nature of that Master Spirit. The **sum total** of these seven experiences constitutes the present-known limits of a creature's consciousness of the reality and actuality of God the Supreme.

Page 1318:4   *Paper 119:8.5*

The completion of these seven bestowals resulted in the liberation of Michael's supreme sovereignty and also in the creation of the possibility for the sovereignty of the Supreme in Nebadon. On none of Michael's bestowals did he reveal God the Supreme, but **the sum total** of all seven bestowals is a new Nebadon revelation of the Supreme Being.

Page 1431:2   *Paper 130:2.7*

This was a conference which lasted well into the night, in the course of which the young man requested Jesus to tell him the difference between the will of God and that human mind act of choosing which is also called will. In substance Jesus said: The will of God is the way of God, partnership with the choice of God in the face of any potential alternative. To do the will of God, therefore, is the progressive experience of becoming more and more like God, and God is the source and destiny of all that is good and beautiful and true. The will of man is the way of man, **the sum and substance** of that which the mortal chooses to be and do. Will is the deliberate choice of a self-conscious being which leads to decision-conduct based on intelligent reflection.

Page 1449:1   *Paper 131:4.6*

"The spirit of the Universe Keeper enters the soul of the simple creature. That man is wise who worships the One God. Those who strive for perfection must indeed know the Lord Supreme. He never fears who knows the blissful security of the Supreme, for the Supreme says to those who serve him, `Fear not, for I am with you.' The God of providence is our Father. God is truth. And it is the desire of God that his creatures should understand him—come fully to know the truth. Truth is eternal; it sustains the universe. Our supreme desire shall be union with the Supreme. The Great Controller is the generator of all things—all evolves from him. And this is the **sum** of duty: Let no man do to another what would be repugnant to himself; cherish no malice, smite not him who smites you, conquer anger with mercy, and vanquish hate by benevolence. And all this we should do because God is a kind friend and a gracious father who remits all our earthly offenses.

Page 1477:1   *Paper 133:5.6*

Mathematics asserts that, if one person stands for a certain unit of intellectual and moral value, ten persons would stand for ten times this value. But in dealing with human personality it would be nearer the truth to say that such a personality association is a sum equal to the square of the number of personalities concerned in the equation rather than the simple arithmetical sum. A social group of human beings in co-ordinate working harmony stands for a force far **greater than the simple sum of its parts.**

Page 1479:7   *Paper 133:7.7*

Ideas are not simply a record of sensations; ideas are sensations plus the reflective interpretations of the personal self; and the self is more than the **sum** of one's sensations. There begins to be something of an approach to unity in an evolving selfhood, and that unity is derived from the indwelling presence of a part of absolute unity which spiritually activates such a self-conscious animal-origin mind.

Page 1480:1   *Paper 133:7.9*

Neither is the human self merely the **sum** of the successive states of consciousness. Without the effective functioning of a consciousness sorter and associator there would not exist sufficient unity to warrant the designation of a selfhood. Such an ununified mind could hardly attain conscious levels of human status. If the associations of consciousness were just an accident, the minds of all men would then exhibit the uncontrolled and random associations of certain phases of mental madness.

Page 1769:9   *Paper 159:5.7*

And this is illustrative of the way Jesus, day by day, appropriated the cream of the Hebrew scriptures for the instruction of his followers and for inclusion in the teachings of the new gospel of the kingdom. Other religions had suggested the thought of the nearness of God to man, but Jesus made the care of God for man like the solicitude of a loving father for the welfare of his dependent children and then made this teaching the cornerstone of his religion. And thus did the doctrine of the fatherhood of God make imperative the practice of the brotherhood of man. The worship of God and the service of man became **the sum and substance** of his religion. Jesus took the best of the Jewish religion and translated it to a worthy setting in the new teachings of the gospel of the kingdom.

Page 1865:1   *Paper 170:5.11*

The kingdom, to the Jews, was the Israelite *community*; to the gentiles it became the Christian *church*. To Jesus the kingdom was the **sum** of those *individuals* who had confessed their faith in the fatherhood of God, thereby declaring their wholehearted dedication to the doing of the will of God, thus becoming members of the spiritual brotherhood of man.

Chapter Six

Pre-Morontia Math 101

—A Glimpse at the Master Mathematicians—

Page 28:7   *Paper 1:5.10*

The idea of the personality of the Universal Father is an enlarged and truer concept of God which has come to mankind chiefly through revelation. Reason, wisdom, and religious experience all infer and imply the personality of God, but they do not altogether validate it. Even the indwelling Thought Adjuster is prepersonal. The truth and maturity of any religion is **directly proportional** to its concept of the infinite personality of God and to its grasp of the absolute unity of Deity. The idea of a personal Deity becomes, then, the measure of religious maturity after religion has first formulated the concept of the unity of God.

Page 36:1   *Paper 2:2.3*

The reactions of a changeless God, in the execution of his eternal purpose, **may seem to vary in accordance with** the changing attitude and the shifting minds of his created intelligences; that is, they may apparently and superficially vary; but underneath the surface and beneath all outward manifestations, there is still present the changeless purpose, the everlasting plan, of the eternal God.

Page 56:7, 57:1   *Paper 4:2.3–4*

Nature is a time-space resultant of two cosmic factors: first, the immutability, perfection, and rectitude of Paradise Deity, and second, the experimental plans, executive blunders, insurrectionary errors, incompleteness of development, and imperfection of wisdom of the extra-Paradise creatures, from the highest to the lowest. Nature therefore carries a uniform, unchanging, majestic, and marvelous thread of perfection from the circle of eternity; but in each universe, on each planet, and in each individual life, this nature is modified, qualified, and perchance marred by the acts, the mistakes, and the disloyalties of the creatures of the evolutionary systems and universes; and therefore must nature ever be of a changing mood, whimsical withal, though stable underneath, and **varied in accordance** with the operating procedures of a local universe.

Nature is the perfection of Paradise **divided by** the incompletion, evil, and sin of the unfinished universes. This quotient is thus expressive of both the perfect and the partial, of both the eternal and the temporal. Continuing evolution modifies nature by augmenting the content of Paradise perfection and by diminishing the content of the evil, error, and disharmony of relative reality.

Page 135:11   *Paper 12:6.1*

The universe is nonstatic. Stability is not the result of inertia but rather the product of balanced energies, co-operative minds, co-ordinated morontias, spirit overcontrol, and personality unification. Stability is wholly and **always proportional to** divinity.

Page 141:4   *Paper 12:9.3*

Mathematics, material science, is indispensable to the intelligent discussion of the material aspects of the universe, but such knowledge is not necessarily a part of the higher realization of truth or of the personal appreciation of spiritual realities. **Not only in the realms of life but even in the world of physical energy, the sum of two or more things is very often something** *more* **than, or something** *different* **from, the predictable additive consequences of such unions.** The entire science of mathematics, the whole domain of philosophy, the highest physics or chemistry, could not predict or know that the union of two gaseous hydrogen atoms with one gaseous oxygen atom would result in a new and **qualitatively superadditive** substance—liquid water. The understanding knowledge of this one physiochemical phenomenon should have prevented the development of materialistic philosophy and mechanistic cosmology.

Page 459:6   *Paper 41:4.2*

Gaseous, liquid, and solid states are matters of atomic-molecular relationships, but density is a relationship of space and mass. Density **varies directly** with the quantity of mass in space **and inversely with** the amount of space in mass, the space between the central cores of matter and the particles which whirl around these centers as well as the space within such material particles.

Page 469:8   *Paper 42:2.8*

Passive and potential force becomes active and primordial in response to the resistance afforded by the space presence of the Primary Eventuated Master Force Organizers. Force is now emerging from the exclusive domain of the Unqualified Absolute into the realms of multiple response—response to certain primal motions initiated by the God of Action and thereupon to certain compensating motions emanating from the Universal Absolute. Primordial force is seemingly reactive to transcendental causation **in proportion to** absoluteness.

Page 477:2   *Paper 42:6.8*

If the mass of matter should be magnified until that of an electron equaled one tenth of an ounce, then were size to be **proportionately magnified**, the volume of such an electron would become as large as that of the earth. If the volume of a proton—eighteen hundred times as heavy as an electron—should be magnified to the size of the head of a pin, then, in comparison, a pin's head would attain a diameter equal to that of the earth's orbit around the sun.

Page 482:2–3   *Paper 42:11.4–5*

Motion and universe gravitation are twin facets of the impersonal time-space mechanism of the universe of universes. The levels of gravity response for spirit, mind, and matter are quite independent of time, but only true spirit levels of reality are independent of space (nonspatial). The higher mind levels of the universe—the spirit-mind levels—may also be nonspatial, but the levels of material mind, such as human mind, are responsive to the interactions of universe gravitation, losing this response only in proportion to spirit identification. Spirit-reality levels are recognized by their spirit content, and spirituality in time and space is **measured inversely** to the linear-gravity response.

Linear-gravity response is a quantitative measure of nonspirit energy. All mass—organized energy—is subject to this grasp except as motion and mind act upon it. Linear gravity is the short-range cohesive force of the macrocosmos somewhat as the forces of intra-atomic cohesion are the short-range forces of the microcosmos. Physical materialized energy, organized as so-called matter, cannot traverse space without affecting linear-gravity response. Although such gravity response is **directly proportional to** mass, it is so modified by intervening space that the final result is no more than roughly approximated when expressed as inversely according to the square of the distance. Space eventually conquers linear gravitation because of the presence therein of the antigravity influences of numerous supermaterial forces which operate to neutralize gravity action and all responses thereto.

Page 552:5   *Paper 48:6.7*

These seraphic evangels are dedicated to the proclamation of the gospel of eternal progression, the triumph of perfection attainment. On the mansion worlds they proclaim the great law of the conservation and dominance of goodness: No act of good is ever wholly lost; it may be long thwarted but never wholly annulled, and it is eternally potent **in proportion to** the divinity of its motivation.

Page 557:14   *Paper 48:7.30*

28. The argumentative defense of any proposition is **inversely proportional to** the truth contained.

Page 564:9   *Paper 49:4.7*

Human beings are all gregarious, both tribal and racial. These group segregations are inherent in their origin and constitution. Such tendencies can be modified only by advancing civilization and by gradual spiritualization. The social, economic, and governmental problems of the inhabited worlds **vary in accordance with** the age of the planets and the degree to which they have been influenced by the successive sojourns of the divine Sons.

Page 613:6   *Paper 54:1.4*

Liberty is suicidal when divorced from material justice, intellectual fairness, social forbearance, moral duty, and spiritual values. Liberty is nonexistent apart from cosmic reality, and all personality reality **is proportional to** its divinity relationships.

Page 630:1   *Paper 55:5.4*

War has become a matter of history, and there are no more armies or police forces. Government is gradually disappearing. Self-control is slowly rendering laws of human enactment obsolete. The extent of civil government and statutory regulation, in an intermediate state of advancing civilization, is **in inverse proportion to** the morality and spirituality of the citizenship.

Page 700:2   *Paper 61:6.1*

The great event of this glacial period was the evolution of primitive man. Slightly to the west of India, on land now under water and among the offspring of Asiatic migrants of the older North American lemur types, the dawn mammals *suddenly* appeared. These small animals walked mostly on their hind legs, and they possessed large brains **in proportion to** their size and in comparison with the brains of other animals. In the seventieth generation of this order of life a new and higher group of animals *suddenly* differentiated. These new mid-mammals—almost twice the size and height of their ancestors and possessing proportionately increased brain power—had only well established themselves when the Primates, the third vital mutation, *suddenly* appeared. (At this same time, a retrograde development within the mid-mammal stock gave origin to the simian ancestry; and from that day to this the human branch has gone forward by progressive evolution, while the simian tribes have remained stationary or have actually retrogressed.)

As an aside, during one of my readings, I decided to count all the italicized words in *The Urantia Book*. I counted four thousand eight hundred and twenty seven (4,827) italicized words. Care to guess what the two most italicized words are? Yep, *suddenly* is first with 58 out of 100 and what is remarkable is that *sudden* is second with 47 out of 66.

Page 769:8   *Paper 68:6.3*

Human society is controlled by a law which decrees that the population must **vary directly in accordance with** the land arts and **inversely with** a given standard of living. Throughout these early ages, even more than at present, the law of supply and demand as concerned men and land determined the estimated value of both. During the times of plentiful land—unoccupied territory—the need for men was great, and therefore the value of human life was much enhanced; hence the loss of life was more horrifying. During periods of land scarcity and associated overpopulation, human life became comparatively cheapened so that war, famine, and pestilence were regarded with less concern.

Page 789:1   *Paper 70:5.3*

In the early council of the elders there resided the potential of all governmental functions: executive, legislative, and judicial. When the council interpreted the current mores, it was a court; when establishing new modes of social usage, it was a legislature; **to the extent that** such decrees and enactments were enforced, it was the executive. The chairman of the council was one of the forerunners of the later tribal chief.

Page 901:4   *Paper 81:2.1*

The growth of culture is predicated upon the development of the tools of civilization. And the tools which man utilized in his ascent from savagery were effective **just to the extent that** they released man power for the accomplishment of higher tasks.

Page 916:5   *Paper 82:3.9*

Among later peoples, puberty was the common age of marriage, but this has advanced in **direct proportion to** the progress of civilization. Early in social evolution peculiar and celibate orders of both men and women arose; they were started and maintained by individuals more or less lacking normal sex urge.

Page 929:2   *Paper 83:7.8*

But just so long as society fails to properly educate children and youths, so long as the social order fails to provide adequate premarital training, and so long as unwise and immature youthful idealism is to be the arbiter of the entrance upon marriage, just so long will divorce remain prevalent. And in so far as the social group falls short of providing marriage preparation for youths, **to that extent** must divorce function as the social safety valve which prevents still worse situations during the ages of the rapid growth of the evolving mores.

Page 932:3   *Paper 84:1.6*

A family of some simple sort was insured by the fact that the reproductive function entails the mother-child relationship. Mother love is instinctive; it did not originate in the mores as did marriage. All mammalian mother love is the inherent endowment of the adjutant mind-spirits of the local universe and is in strength and devotion always **directly proportional to** the length of the helpless infancy of the species.

Page 936:7   *Paper 84:5.3*

The modern idea of sex equality is beautiful and worthy of an expanding civilization, but it is not found in nature. When might is right, man lords it over woman; when more justice, peace, and fairness prevail, she gradually emerges from slavery and obscurity. Woman's social position has generally **varied inversely with** the degree of militarism in any nation or age.

Page 946:7   *Paper 85:3.5*

In religion, symbolism may be either good or bad just **to the extent** that the symbol does or does not displace the original worshipful idea. And symbolism must not be confused with direct idolatry wherein the material object is directly and actually worshiped.

Page 974:4   *Paper 89:1.2*

The respect which these prohibitions commanded in the mind of the savage **exactly equaled** his fear of the powers who were supposed to enforce them. Taboos first arose because of chance experience with ill luck; later they were proposed by chiefs and shamans—fetish men who were thought to be directed by a spirit ghost, even by a god. The fear of spirit retribution is so great in the mind of a primitive that he sometimes dies of fright when he has violated a taboo, and this dramatic episode enormously strengthens the hold of the taboo on the minds of the survivors.

Page 976:5   *Paper 89:3.3*

Poverty was just a part of the ritual of the mortification of the flesh which, unfortunately, became incorporated into the writings and teachings of many religions, notably Christianity. Penance is the negative form of this ofttimes foolish ritual of renunciation. But all this taught the savage self-control, and that was a worth-while advancement in social evolution. Self-denial and self-control were two of the greatest social gains from early evolutionary religion. Self-control gave man a new philosophy of life; it taught him the art of **augmenting life's fraction by lowering the denominator** of personal demands instead of always attempting to **increase the numerator** of selfish gratification.

Page 1016:5   *Paper 93:3.3*

The symbol of the three concentric circles, which Melchizedek adopted as the insignia of his bestowal, a majority of the people interpreted as standing for the three kingdoms of men, angels, and God. And they were allowed to continue in that belief; very few of his followers ever knew that these three circles were emblematic of the infinity, eternity, and universality of the Paradise Trinity of divine maintenance and direction; even Abraham rather regarded this symbol as standing for the three Most Highs of Edentia, as he had been instructed that the three Most Highs functioned as one. **To the extent** that Melchizedek taught the Trinity concept symbolized in his insignia, he usually associated it with the three Vorondadek rulers of the constellation of Norlatiadek.

Page 1094:3   *Paper 100:1.1*

While religion produces growth of meanings and enhancement of values, evil always results when purely personal evaluations are elevated to the levels of absolutes. A child evaluates experience in accordance with the content of pleasure; maturity is **proportional to** the substitution of higher meanings for personal pleasure, even loyalties to the highest concepts of diversified life situations and cosmic relations.

Page 1095:2   *Paper 100:1.7*

Religion cannot be bestowed, received, loaned, learned, or lost. It is a personal experience which **grows proportionally to** the growing quest for final values. Cosmic growth thus attends on the accumulation of meanings and the ever-expanding elevation of values. But nobility itself is always an unconscious growth.

Page 1096:1   *Paper 100:2.4*

Spirituality becomes at once the indicator of one's nearness to God and the measure of one's usefulness to fellow beings. Spirituality enhances the ability to discover beauty in things, recognize truth in meanings, and discover goodness in values. Spiritual development is determined by capacity therefor and is **directly proportional to** the elimination of the selfish qualities of love.

Page 1099:3   *Paper 100:5.5*

But emotion alone is a false conversion; one must have faith as well as feeling. **To the extent** that such psychic mobilization is partial, and in so far as such human-loyalty motivation is incomplete, **to that extent** will the experience of conversion be a blended intellectual, emotional, and spiritual reality.

Page 1123:4   *Paper 102:4.4*

The element of error present in human religious experience is **directly proportional to** the content of materialism which contaminates the spiritual concept of the Universal Father. Man's prespirit progression in the universe consists in the experience of divesting himself of these erroneous ideas of the nature of God and of the reality of pure and true spirit. Deity is more than spirit, but the spiritual approach is the only one possible to ascending man.

Page 1133:2   *Paper 103:4.2*

When primitive man felt that his communion with God had been interrupted, he resorted to sacrifice of some kind in an effort to make atonement, to restore friendly relationship. The hunger and thirst for righteousness leads to the discovery of truth, and truth augments ideals, and this creates new problems for the individual religionists, for our ideals tend to grow by **geometrical progression**, while our ability to live up to them is enhanced only by **arithmetical progression.**

Page 1142:3   *Paper 103:9.12*

There is a reality in religious experience that is **proportional to** the spiritual content, and such a reality is transcendent to reason, science, philosophy, wisdom, and all other human achievements. The convictions of such an experience are unassailable; the logic of religious living is incontrovertible; the certainty of such knowledge is superhuman; the satisfactions are superbly divine, the courage indomitable, the devotions unquestioning, the loyalties supreme, and the destinies final—eternal, ultimate, and universal.

Page 1149:12   *Paper 104:4.15*

The fourth triunity absolutely controls the fundamental units of cosmic energy and releases them from the grasp of the Unqualified Absolute **in direct proportion** to the appearance in the experiential Deities of subabsolute capacity to control and stabilize the metamorphosing cosmos.

Page 1168:2   *Paper 106:6.3*

If we assume a cosmos-infinite—some illimitable cosmos on beyond the master universe—and if we conceive that the final developments of the Absolute Trinity will take place out on such a superultimate stage of action, then it becomes possible to conjecture that the completed function of the Trinity Absolute will achieve final expression in the creations of infinity and will consummate the absolute actualization of all potentials. The integration and association of ever-enlarging segments of reality will approach absoluteness of status **proportional to** the inclusion of all reality within the segments thus associated.

Page 1173:5   *Paper 106:9.4*

The concept of the unification of all reality, be it in this or any other universe age, is basically twofold: existential and experiential. Such a unity is in process of experiential realization in the Trinity of Trinities, but the degree of the apparent actualization of this threefold Trinity is **directly proportional to** the disappearance of the qualifications and imperfections of reality in the cosmos. But total integration of reality is unqualifiedly and eternally and existentially present in the Paradise Trinity, within which, at this very universe moment, infinite reality is absolutely unified.

Page 1174:7   *Paper 106:9.11*

Sooner or later all universe personalities begin to realize that the final quest of eternity is the endless exploration of infinity, the never-ending voyage of discovery into the absoluteness of the First Source and Center. Sooner or later we all become aware that all creature growth is **proportional to** Father identification. We arrive at the understanding that living the will of God is the eternal passport to the endless possibility of infinity itself. Mortals will sometime realize that success in the quest of the Infinite is **directly proportional to** the achievement of Fatherlikeness, and that in this universe age the realities of the Father are revealed within the qualities of divinity. And these qualities of divinity are personally appropriated by universe creatures in the experience of living divinely, and to live divinely means actually to live the will of God.

Page 1185:3   *Paper 108:1.1*

When Adjusters are dispatched for mortal service from Divinington, they are identical in the endowment of existential divinity, but they vary in experiential qualities **proportional to** previous contact in and with evolutionary creatures. We cannot explain the basis of Adjuster assignment, but we conjecture that these divine gifts are bestowed in accordance with some wise and efficient policy of eternal fitness of adaptation to the indwelt personality. We do observe that the more experienced Adjuster is often the indweller of the higher type of human mind; human inheritance must therefore be a considerable factor in determining selection and assignment.

Page 1197:3   *Paper 109:3.1*

The character of the detailed work of Mystery Monitors **varies in accordance with** the nature of their assignments, as to whether or not they are *liaison or fusion* Adjusters. Some Adjusters are merely loaned for the temporal lifetimes of their subjects; others are bestowed as personality candidates with permission for everlasting fusion if their subjects survive. There is also a slight variation in their work among the different planetary types as well as in different systems and universes. But, on the whole, their labors are remarkably uniform, more so than are the duties of any of the created orders of celestial beings.

Page 1209:3   *Paper 110:6.3*

The psychic circles are not exclusively intellectual, neither are they wholly morontial; they have to do with personality status, mind attainment, soul growth, and Adjuster attunement. The successful traversal of these levels demands the harmonious functioning of the *entire personality*, not merely of some one phase thereof. **The growth of the parts does not equal the true maturation of the whole; the parts really grow in proportion to the expansion of the entire self—the whole self—material, intellectual, and spiritual.**

Page 1210:2–3   *Paper 110:6.7–8*

It is difficult precisely to define the seven levels of human progression, for the reason that these levels are personal; they are variable for each individual and are apparently determined by the growth capacity of each human being. The conquest of these levels of cosmic evolution is reflected in three ways:

1. *Adjuster attunement.* The spiritizing mind nears the Adjuster presence **proportional to** circle attainment.

Page 1219:3   *Paper 111:3.4*

Both the human mind and the divine Adjuster are conscious of the presence and differential nature of the evolving soul—the Adjuster fully, the mind partially. The soul becomes increasingly conscious of both the mind and the Adjuster as associated identities, **proportional to** its own evolutionary growth. The soul partakes of the qualities of both the human mind and the divine spirit but persistently evolves toward augmentation of spirit control and divine dominance through the fostering of a mind function whose meanings seek to co-ordinate with true spirit value.

Page 1222:4   *Paper 111:6.5*

When man wishes to modify physical reality, be it himself or his environment, he succeeds **to the extent that** he has discovered the ways and means of controlling matter and directing energy. Unaided mind is impotent to influence anything material save its own physical mechanism, with which it is inescapably linked. But through the intelligent use of the body mechanism, mind can create other mechanisms, even energy relationships and living relationships, by the utilization of which this mind can increasingly control and even dominate its physical level in the universe.

Page 1236:1   *Paper 112:6.3*

To a certain extent, the appearance of the material body-form is responsive to the character of the personality identity; the physical body does, to a limited degree, reflect something of the inherent nature of the personality. Still more so does the morontia form. In the physical life, mortals may be outwardly beautiful though inwardly unlovely; in the morontia life, and increasingly on its higher levels, the personality form will **vary directly in accordance with** the nature of the inner person. On the spiritual level, outward form and inner nature begin to approximate complete identification, which grows more and more perfect on higher and higher spirit levels.

Page 1266:5   *Paper 115:7.4*

To the extent that the triodities are directly operative on the finite level, they impinge upon the Supreme, who is the Deity focalization and cosmic summation of the finite qualifications of the natures of the Absolute Actual and the Absolute Potential.

Page 1278:1   *Paper 117:0.1*

To the extent that we do the will of God in whatever universe station we may have our existence, in that measure the almighty potential of the Supreme becomes one step more actual. The will of God is the purpose of the First Source and Center as it is potentialized in the three Absolutes, personalized in the Eternal Son, conjoined for universe action in the Infinite Spirit, and eternalized in the everlasting patterns of Paradise. And God the Supreme is becoming the highest finite manifestation of the total will of God.

Page 1284:7   *Paper 117:4.11*

The great struggle of this universe age is between the potential and the actual—the seeking for actualization by all that is as yet unexpressed. If mortal man proceeds upon the Paradise adventure, he is following the motions of time, which flow as currents within the stream of eternity; if mortal man rejects the eternal career, he is moving counter to the stream of events in the finite universes. The mechanical creation moves on inexorably in accordance with the unfolding purpose of the Paradise Father , but the volitional creation has the choice of accepting or of rejecting the role of personality participation in the adventure of eternity. Mortal man cannot destroy the supreme values of human existence, but he can very definitely prevent the evolution of these values in his own personal experience. **To the extent that** the human self thus refuses to take part in the Paradise ascent, to just that extent is the Supreme delayed in achieving divinity expression in the grand universe.

Page 1305:5   *Paper 118:10.10*

There is a providence in the evolving universes, and it can be discovered by creatures **to just the extent that** they have attained capacity to perceive the purpose of the evolving universes. Complete capacity to discern universe purposes equals the evolutionary completion of the creature and may otherwise be expressed as the attainment of the Supreme within the limits of the present state of the incomplete universes.

Page 1641:6   *Paper 146:3.4*

And Jesus said to Thomas: "Your assurance that you have entered into the kingdom family of the Father, and that you will eternally survive with the children of the kingdom, is wholly a matter of personal experience—faith in the word of truth. Spiritual assurance is the equivalent of your personal religious experience in the eternal realities of divine truth and is **otherwise equal to** your intelligent understanding of truth realities plus your spiritual faith and **minus** your honest doubts..."

Page 1702:2   *Paper 152:3.1*

The feeding of the five thousand by supernatural energy was another of those cases where human pity plus creative power **equaled** that which happened. Now that the multitude had been fed to the full, and since Jesus' fame was then and there augmented by this stupendous wonder, the project to seize the Master and proclaim him king required no further personal direction. The idea seemed to spread through the crowd like a contagion. The reaction of the multitude to this sudden and spectacular supplying of their physical needs was profound and overwhelming. For a long time the Jews had been taught that the Messiah, the son of David, when he should come, would cause the land again to flow with milk and honey, and that the bread of life would be bestowed upon them as manna from heaven was supposed to have fallen upon their forefathers in the wilderness. And was not all of this expectation now fulfilled right before their eyes? When this hungry, undernourished multitude had finished gorging itself with the wonder-food, there was but one unanimous reaction: "Here is our king." The wonder-working deliverer of Israel had come. In the eyes of these simple-minded people the power to feed carried with it the right to rule. No wonder, then, that the multitude, when it had finished feasting, rose as one man and shouted, "Make him king!"

Page 1898:5   *Paper 174:1.5*

"Your inability or unwillingness to forgive your fellows is the measure of your immaturity, your failure to attain adult sympathy, understanding, and love. You hold grudges and nurse vengefulness **in direct proportion to** your ignorance of the inner nature and true longings of your children and your fellow beings. Love is the outworking of the divine and inner urge of life. It is founded on understanding, nurtured by unselfish service, and perfected in wisdom."

Page 2078:2   *Paper 195:6.16*

Freedom or initiative in any realm of existence is **directly proportional to** the degree of spiritual influence and cosmic-mind control; that is, in human experience, the degree of the actuality of doing "the Father's will." And so, when you once start out to find God, that is the conclusive proof that God has already found you.

The all-consuming and indomitable spiritual faith of Jesus never became fanatical, for it never attempted to run away with his well-balanced intellectual judgments concerning the **proportional values of** practical and commonplace social, economic, and moral life situations. The Son of Man was a splendidly unified human personality; he was a perfectly endowed divine being; he was also magnificently co-ordinated as a combined human and divine being functioning on earth as a single personality. Always did the Master co-ordinate the faith of the soul with the wisdom-appraisals of seasoned experience. Personal faith, spiritual hope, and moral devotion were always correlated in a matchless religious unity of harmonious association with the keen realization of the reality and sacredness of all human loyalties—personal honor, family love, religious obligation, social duty, and economic necessity.

Chapter Seven

Study Companion

—A Conceptual Road Map—

Introduction

This Study Companion is the result of thirty enthralling years of studying *The Urantia Book*. This includes seventeen readings (sixteen cover to cover and at least once in study groups around the world). The cross references reflect many Urantia Book students who have commented on these related passages over the years, as well as my own. It is my intent that these related passages throughout all parts of our beloved revelation would provide additional gems of insight that will enhance your current and future studies, either individually or in a study group setting.

It is a fact that *The Urantia Book* is not written as a novel, even though the Jesus papers on the surface appear to do just that. It has been likened to a "symphony" where themes are introduced almost anywhere in the text and reappear later (or as a pre-echo) in a different context, presenting a delicacy that seems to transfix one under a different octave. Sub themes are woven subtly inside major themes, providing a rich development of a subject that previously had no connection to the reader or student.

There are a number of ways of reading *The Urantia Book*, and probably an infinite amount of approaches when it comes to studying it. One can read alone or in small groups either sequentially or ad-hoc. Groups of longer-term readers have a habit of choosing one particular paper to read and discuss as the group decides appropriate. These sessions can get transfixed on one paragraph, which contains a subject so exquisite that supporting passages in other parts of the book are sought after. Newer readers may prefer a sequential read yet have the same need for answering a particularly vexing question that is responded to in a paper they may not have yet read.

Over the years there have been indexes developed to help the reader find passages that may be related to a particular subject. In most cases these indexes operate at the term (word) level and some go as far as short phrases. To my knowledge, none have the ability to fan out from a concept in a particular paper and draw the network of linkages to other passages that provide lucidity on a subject area. One could make the case that we are, to a certain degree, witnessing the partial unfolding of the Supreme when we are privileged to have access to other readers' insights into related passages in *The Urantia Book*.

This study companion provides a group (or an individual reader alone) the ability to start where they are in a particular paper and quickly find a large number of passages in places within the book one would hardly think contains connections. The associations in this study companion are not computer-generated linkages, but rather have been developed as experiential patterns other groups have painstakingly located and made part of humanity's knowledge base. The companion does this by relating passages to concepts.

Related passages are grouped together under a keyword, and are listed chronologically. The cross references of a given topic are indicated by page numbers and Paragraph number of the Foundation edition of *The Urantia Book* on the left margin, and Paper, Section, and Paragraph numbers on the right margin for use with the Fellowship's Uversa Press edition of *The Urantia Book*.

For example, if you or a study group is reading Paper 5, "God's Relation To The Individual," you would turn to pages 142-150 of your "Study Companion" to find several topics, or keywords, within Paper 5 that have multiple inter-related passages. If you look further within Section 6 on page 149, you will find "Personality Circuit as Divine Love" listed as one topic, and your study companion will refer you to four related passages.

Personality Circuit as Divine Love

Page 71:7   *Paper 5:6.12*

    Concerning those personalities ...

Page 640:2   *Paper 56:4.3*

    Notwithstanding that God ...

Page 647:8   *Paper 56:10.17*

    Universal beauty is ...

Page 1289:3   *Paper 117:6.10*

    All true love is ...

As you can see, these references span over 1200 pages, cover Parts I, II, and III of *The Urantia Book* to include at least three different authors (maybe four if the Mighty Messenger who co-authored Paper 56 is not the same as the Mighty Messenger who takes credit for Paper 117). So, the insights provided by these diverse authors may stimulate a discussion on "The personality circuit described as the great circuit of divine love". To facilitate focusing in on the related issues in the referenced paragraphs, certain words and sentences are in **bold**.

All the references will be listed in their entirety once and are located on the page that the first reference is noted. In the example above, Paper 5 will have the text of all four passages that have the references listed. If you happen to be reading in Paper 56 or Paper 117, the cross-reference will refer you back to Paper 5, the first reference source, to see all the related texts.

Every effort has been made to make this as user friendly as possible, so the related passages have been arranged to be viewed on the same page or opposite page, wherever possible, so as to avoid having to turn a page to view the different references.

This is by no means an exhaustive compilation of all related passages on the topics covered. These are from my margin notes over the years of personal reading, studying and from study group participation.

To the extent that this secondary work brings you closer to the truths revealed in *The Urantia Book* and brings the Epochal Revelation of Truth closer to you, to that extent I am filled with joy and blessed to be a part of this.

Study Companion

—A Conceptual Road Map—

Deity Function—Levels

Page 2:3–10   *Foreword: 1.3–10*

Deity functions on personal, prepersonal, and superpersonal levels. Total Deity is functional on the following seven levels:

1. *Static*—self-contained and self-existent Deity.
2. *Potential*—self-willed and self-purposive Deity.
3. *Associative*—self-personalized and divinely fraternal Deity.
4. *Creative*—self-distributive and divinely revealed Deity.
5. *Evolutional*—self-expansive and creature-identified Deity.
6. *Supreme*—self-experiential and creature-Creator-unifying Deity. Deity functioning on the first creature-identificational level as time-space overcontrollers of the grand universe, sometimes designated the Supremacy of Deity.
7. *Ultimate*—self-projected and time-space-transcending Deity. Deity omnipotent, omniscient, and omnipresent. **Deity functioning** on the second level of unifying divinity expression as effective overcontrollers and absonite upholders of the master universe. As compared with the ministry of the Deities to the grand universe, this absonite function in the master universe is tantamount to universal overcontrol and supersustenance, sometimes called the Ultimacy of Deity.

Page 1030:3   *Paper 94:3.3*

Brahman-Narayana was conceived as the Absolute, the infinite IT IS, the primordial creative potency of the potential cosmos, the Universal Self existing static and potential throughout all eternity. Had the philosophers of those days been able to make the next advance in deity conception, had they been able to conceive of the Brahman as associative and creative, as a personality approachable by created and evolving beings, then might such a teaching have become the most advanced portraiture of Deity on Urantia **since it would have encompassed the first five levels of total deity function and might possibly have envisioned the remaining two.**

III. The First Source and Center

I AM

From the Foreword to Paper 105—insights into prereality, the I AM, and our maximum comprehension of reality within the Supreme Being.

Page 5:19, Page 6:2   *Foreword:3.13, 15*

REALITY, as comprehended by finite beings, is partial, relative, and shadowy. The maximum Deity reality fully comprehensible by evolutionary finite creatures is embraced within the Supreme Being. **Nevertheless there are antecedent and eternal realities, superfinite realities, which are ancestral to this Supreme Deity of evolutionary time-space creatures.** In attempting to portray the origin and nature of universal reality, we are forced to employ the technique of time-space reasoning in order to reach the level of the finite mind. Therefore must many of the simultaneous events of eternity be presented as sequential transactions....

In this original transaction the theoretical I AM achieved the realization of personality by becoming the Eternal Father of the Original Son simultaneously with becoming the Eternal Source of the Isle of Paradise. Coexistent with the differentiation of the Son from the Father, and in the presence of Paradise, there appeared the person of the Infinite Spirit and the central universe of Havona. With the appearance of coexistent personal Deity, the Eternal Son and the Infinite Spirit, the Father escaped, as a personality, from otherwise inevitable diffusion throughout the potential of Total Deity. Thenceforth it is only in Trinity association with his two Deity equals that the Father fills all Deity potential, while increasingly experiential Deity is being actualized on the divinity levels of Supremacy, Ultimacy, and Absoluteness.

Page 1152:4–5, Page 1153:2   *Paper 105:1.1–2,5*

Absolute primal causation in infinity the philosophers of the universes attribute to the Universal Father functioning as the infinite, the eternal, and the absolute I AM.

There are many elements of danger attendant upon the presentation to the mortal intellect of this idea of an infinite I AM since this concept is so remote from human experiential understanding as to involve serious distortion of meanings and misconception of values. **Nevertheless, the philosophic concept of the I AM does afford finite beings some basis for an attempted approach to the partial comprehension of absolute origins and infinite destinies.** But in all our attempts to elucidate the genesis and fruition of reality, let it be made clear that this concept of the I AM is, in all personality meanings and values, synonymous with the First Person of Deity, the Universal Father of all personalities. But this postulate of the I AM is not so clearly identifiable in undeified realms of universal reality....

...To the finite mind there simply must be a beginning, and though there never was a real beginning to reality, still there are certain source relationships which reality manifests to infinity. **The prereality, primordial, eternity situation may be thought of something like this:** At some infinitely distant, hypothetical, past-eternity moment, **the I AM may be conceived as both thing and no thing,** as both cause and effect, as both volition and response. At this hypothetical eternity moment there is no differentiation throughout all infinity. Infinity is filled by the Infinite; the Infinite encompasses infinity. This is the hypothetical static moment of eternity; actuals are still contained within their potentials, and potentials have not yet appeared within the infinity of the I AM. But even in this conjectured situation we must assume the existence of the possibility of self-will.

Father-Infinite

Infinity of will, personality of infinity, volitional absolute, and additional related insights can be gleamed from the Foreword to Paper 10.

Page 6:4   *Foreword:3.17*

The Infinite is used to denote the fullness—the finality—implied by the primacy of the First Source and Center. The *theoretical* I AM is a creature-philosophic extension of the **"infinity of will,"** but the Infinite is an *actual* value-level representing the eternity-intension of the true infinity of the absolute and unfettered free will of the Universal Father. This concept is sometimes designated the Father-Infinite.

Page 74:1   *Paper 6:1.2*

The Eternal Son is the spiritual center and the divine administrator of the spiritual government of the universe of universes. The Universal Father is first a creator and then a controller; the Eternal Son is first a cocreator and then a *spiritual administrator.* "God is spirit," and the Son is a personal revelation of that spirit. The First Source and Center is the **Volitional Absolute;** the Second Source and Center is the **Personality Absolute.**

Page 109:2   *Paper 10:1.4*

For knowledge concerning the Father's personality and divine attributes we will always be dependent on the revelations of the Eternal Son, for when the conjoint act of creation was effected, when the Third Person of Deity sprang into personality existence and executed the combined concepts of his divine parents, the Father ceased to exist as the unqualified personality. With the coming into being of the Conjoint Actor and the materialization of the central core of creation, certain eternal changes took place. God gave himself as an absolute personality to his Eternal Son. Thus does the Father bestow the **"personality of infinity"** upon his only-begotten Son, while they both bestow the "conjoint personality" of their eternal union upon the Infinite Spirit.

Page 110:3   *Paper 10:2.5*

The personality of the First Source and Center is the **personality of infinity minus the absolute personality** of the Eternal Son. The personality of the Third Source and Center is the superadditive consequence of the union of the liberated Father-personality and the absolute Son-personality.

Page 111:4   *Paper 10:3.6*

We observe that the Father has divested himself of all direct manifestations of absoluteness except absolute fatherhood and absolute volition. We do not know whether volition is an inalienable attribute of the Father; we can only observe that he did *not* divest himself of volition. Such **infinity of will** must have been eternally inherent in the First Source and Center.

V. PERSONALITY REALITIES

Soul

Insights into the soul, mind, and worship—from the Foreword and three Papers.

Page 8:10   *Foreword:5.10*

4. *Soul.* The soul of man is an experiential acquirement. As a mortal creature chooses to "do the will of the Father in heaven," so the indwelling spirit becomes the father of a *new reality* in human experience. The mortal and material **mind is the mother** of this same emerging reality. The substance of this new reality is neither material nor spiritual—it is *morontial*. This is the **emerging and immortal soul** which is destined to survive mortal death and begin the Paradise ascension.

Page 66:4   *Paper 5:3.8*

The worship experience consists in the sublime attempt of the betrothed Adjuster to communicate to the divine Father the inexpressible longings and the unutterable aspirations of the human soul—the conjoint creation of the God-seeking mortal mind and the God-revealing immortal Adjuster. Worship is, therefore, the act of the material mind's assenting to the attempt of its spiritualizing self, under the guidance of the associated spirit, to communicate with God as a faith son of the Universal Father. **The mortal mind consents to worship; the immortal soul craves and initiates worship; the divine Adjuster presence conducts such worship in behalf of the mortal mind and the evolving immortal soul.** True worship, in the last analysis, becomes an experience realized on four cosmic levels: the intellectual, the morontial, the spiritual, and the personal—the consciousness of mind, soul, and spirit, and their unification in personality.

Page 551:7   *Paper 48:6.2*

You should understand that the morontia life of an ascending mortal is really initiated on the inhabited worlds at the **conception of the soul,** at that moment when the creature mind of moral status is indwelt by the spirit Adjuster. And from that moment on, the mortal soul has potential capacity for supermortal function, even for **recognition** on the higher levels of the morontia spheres of the local universe.

Page 1478:4   *Paper 133:6.5*

"The soul is the self-reflective, truth-discerning, and spirit-perceiving part of man which forever elevates the human being above the level of the animal world. Self-consciousness, in and of itself, is not the soul. Moral self-consciousness is true human self-realization and constitutes the foundation of the human soul, and the soul is that part of man which represents the potential survival value of human experience. Moral choice and spiritual attainment, the ability to know God and the urge to be like him, are the characteristics of the soul. The soul of man cannot exist apart from moral thinking and spiritual activity. A stagnant soul is a dying soul. But the soul of man is distinct from the divine spirit which dwells within the mind. The divine spirit arrives simultaneously with the **first moral activity** of the human mind, and that is the occasion of the **birth of the soul.**

VI. ENERGY AND PATTERN

Paucity of Language
Being reminded of the paucity of language—from the Foreword and Part II.

Page 9:4   *Foreword:6.2*

ENERGY we use as an all-inclusive term applied to spiritual, mindal, and material realms. *Force* is also thus broadly used. *Power* is ordinarily limited to the designation of the electronic level of material or linear-gravity-responsive matter in the grand universe. Power is also employed to designate sovereignty. We cannot follow your generally accepted definitions of force, energy, and power. There is such **paucity of language** that we must assign multiple meanings to these terms.

Page 469:1   *Paper 42:2.1*

It is indeed **difficult to find suitable words in the English language** whereby to designate and wherewith to describe the various levels of force and energy—physical, mindal, or spiritual. These narratives cannot altogether follow your accepted definitions of force, energy, and power. There is such **paucity of language** that we must use these terms in multiple meanings. In this paper, for example, the word *energy* is used to denote all phases and forms of phenomenal motion, action, and potential, while *force* is applied to the pregravity, and *power* to the postgravity, stages of energy.

Mind
The mind and how it is ministered to.

Page 9:10   *Foreword:6.8*

Mind is a phenomenon connoting the **presence-activity of** *living ministry* in addition to varied energy systems; and this is true on all levels of intelligence. In personality, mind ever intervenes between spirit and matter; **therefore is the universe illuminated by three kinds of light: material light, intellectual insight, and spirit luminosity.**

Page 103:1   *Paper 9:5.3*

The unique feature of mind is that it can be bestowed upon such a wide range of life. Through his creative and creature associates the **Third Source and Center ministers to all minds** on all spheres. He ministers to human and subhuman intellect through the adjutants of the local universes and, through the agency of the physical controllers, ministers even to the lowest nonexperiencing entities of the most primitive types of living things. And always is the direction of mind a ministry of mind-spirit or mind-energy personalities.

Page 140:7   *Paper 12:8.11*

2. *Mind.* Organized consciousness which is not wholly subject to material gravity, and which **becomes truly liberated when modified by spirit.**

Majeston—Creative Act of the Supreme Being
Wow, when the Supreme Being decides to create!

Page 11:3   *Foreword:7.7*

The Supreme Being is not a direct creator, except that he is the **father of Majeston,** but he is a synthetic co-ordinator of all creature-Creator universe activities. The Supreme Being, now actualizing in the evolutionary universes, is the Deity correlator and synthesizer of time-space divinity, of triune Paradise Deity in experiential association with the Supreme Creators of time and space. When finally actualized, this evolutionary Deity will constitute the eternal fusion of the finite and the infinite—the everlasting and indissoluble union of experiential power and spirit personality.

Page 200:3   *Paper 17:2.5*

The **creation of Majeston signalized the first supreme creative act of the Supreme Being.** This will to action was volitional in the Supreme Being, but the stupendous reaction of the Deity Absolute was not foreknown. Not since the eternity-appearance of Havona had the universe witnessed such a tremendous factualization of such a gigantic and far-flung alignment of power and co-ordination of functional spirit activities. The Deity response to the creative wills of the Supreme Being and his associates was vastly beyond their purposeful intent and greatly in excess of their conceptual forecasts.

Technique Toward Attaining Absonite Levels
Attaining the absonite for ultimate service.

Page 12:2   *Foreword:8.4*

The Creator Sons in the Deity association of God the Sevenfold provide the mechanism whereby the mortal becomes immortal and the finite attains the embrace of the infinite. The Supreme Being provides the technique for the power-personality mobilization, the divine synthesis, of *all* these manifold transactions, **thus enabling the finite to attain the absonite and, through other possible future actualizations, to attempt the attainment of the Ultimate.** The Creator Sons and their associated Divine Ministers are participants in this supreme mobilization, but the Ancients of Days and the Seven Master Spirits are probably eternally fixed as permanent administrators in the grand universe.

Page 12:4   *Foreword:9.1*

Just as the Supreme Being progressively evolves from the antecedent divinity endowment of the encompassed grand universe potential of energy and personality, so does God the Ultimate eventuate from the potentials of divinity residing in the transcended time-space domains of the master universe. The actualization of Ultimate Deity signalizes absonite unification of the first experiential Trinity and signifies unifying Deity expansion on the second level of creative self-realization. This constitutes the personality-power equivalent of the universe experiential-Deity actualization of Paradise absonite realities on the eventuating levels of transcended time-space values. The completion of such an experiential unfoldment is designed to **afford ultimate service-destiny for all time-space creatures who have attained absonite levels** through the completed realization of the Supreme Being and by the ministry of God the Sevenfold.

XI. THE THREE ABSOLUTES

Supergravity Presence
The Foreword and three Papers on the Unqualified Absolute and the supergravity presence.

Page 14:5   *Foreword:11.7*

2. *The Unqualified Absolute* is nonpersonal, extradivine, and undeified. The Unqualified Absolute is therefore devoid of personality, divinity, and all creator prerogatives. **Neither fact nor truth, experience nor revelation, philosophy nor absonity are able to penetrate the nature and character of this Absolute without universe qualification.**

Page 139:5   *Paper 12:8.2*

The bestowal of cosmic force, the domain of cosmic gravity, is the function of the Isle of Paradise. **All original force-energy proceeds from Paradise, and the matter for the making of untold universes now circulates throughout the master universe in the form of a supergravity presence which constitutes the force-charge of pervaded space.**

Page 324:2   *Paper 29:3.12*

These living power mechanisms are **not consciously related to the master universe energy overcontrol of the Unqualified Absolute,** but we surmise that their entire and almost perfect scheme of power direction **is in some unknown manner subordinated to this supergravity presence.** In any local energy situation the centers and controllers exert near-supremacy, but they are always conscious of the superenergy presence and the unrecognizable performance of the Unqualified Absolute.

Page 1155:3   Paper 105:2.10

6. *The Infinite Capacity.* I AM static-reactive. This is the endless matrix, the possibility for all future cosmic expansion. **This phase of the I AM is perhaps best conceived as the supergravity presence of the Unqualified Absolute.**

Consummator of Universe Destiny

Consummator of Universe Destiny—what a title. A little about this unrevealed order from three separate sources.

Page 16:4   Foreword:12.7

2. *The Absolute Trinity*—**the second experiential Trinity**—now in process of actualization, will consist of God the Supreme, God the Ultimate, and the **unrevealed Consummator of Universe Destiny.** This Trinity functions on both personal and superpersonal levels, even to the borders of the nonpersonal, and its unification in universality would experientialize Absolute Deity.

Page 333:12   Paper 30:1.21

VII. UNCLASSIFIED AND **UNREVEALED ORDERS.** During the present universe age it would not be possible to place all beings, personal or otherwise, within classifications pertaining to the present universe age; nor have all such categories been revealed in these narratives; **hence numerous orders have been omitted from these lists. Consider the following:**
The Consummator of Universe Destiny.
The Qualified Vicegerents of the Ultimate.
The Unqualified Supervisors of the Supreme.
The Unrevealed Creative Agencies of the Ancients of Days.
Majeston of Paradise.
The Unnamed Reflectivator Liaisons of Majeston.
The Midsonite Orders of the Local Universes.

Page 1167:2   Paper 106:5.1

The Ultimate is the apex of transcendental reality even as the Supreme is the capstone of evolutionary-experiential reality. And the actual emergence of these two experiential Deities lays the foundation for the **second experiential**

Trinity. This is the Trinity Absolute, the union of God the Supreme, God the Ultimate, and **the unrevealed Consummator of Universe Destiny**. And this Trinity has theoretical capacity to activate the Absolutes of potentiality—Deity, Universal, and Unqualified. But the completed formation of this Trinity Absolute could take place only after the completed evolution of the entire master universe, from Havona to the fourth and outermost space level.

Page 1169:2   *Paper 106:7.3*

Destiny is established by the volitional act of the Deities who constitute the Paradise Trinity; destiny is established in the vastness of the three great potentials whose absoluteness encompasses the possibilities of all future development; **destiny is probably consummated by the act of the Consummator of Universe Destiny, and this act is probably involved with the Supreme and the Ultimate in the Trinity Absolute.** Any experiential destiny can be at least partially comprehended by experiencing creatures; but a destiny which impinges on infinite existentials is hardly comprehensible. Finality destiny is an existential-experiential attainment which appears to involve the Deity Absolute. But the Deity Absolute stands in eternity relationship with the Unqualified Absolute by virtue of the Universal Absolute. And these three Absolutes, experiential in possibility, are actually existential and more, being limitless, timeless, spaceless, boundless, and measureless—truly infinite.

Page 1171:4   *Paper 106:8.8*

3. *The Absolute Trinity.* **This is the grouping of God the Supreme, God the Ultimate, and the Consummator of Universe Destiny in regard to all divinity values.** Certain other phases of this triune grouping have to do with other-than-divinity values in the expanding cosmos. But these are unifying with the divinity phases just as the power and the personality aspects of the experiential Deities are now in process of experiential synthesis.

PAPER 1: THE UNIVERSAL FATHER

Be You Perfect

Be you perfect. What does it truly mean?

Page 21:3   *Paper 1:0.3*

The enlightened worlds all recognize and worship the Universal Father, the eternal maker and infinite upholder of all creation. The will creatures of universe upon universe have embarked upon the long, long Paradise journey, the fascinating struggle of the eternal adventure of attaining God the Father. The transcendent goal of the children of time is to find the eternal God, to comprehend the divine nature, to recognize the Universal Father. **God-knowing creatures have only one supreme ambition, just one consuming desire, and that is to become, as they are in their spheres, like him as he is in his Paradise perfection of personality and in his universal sphere of righteous supremacy. From the Universal Father who inhabits eternity there has gone forth the supreme mandate, "Be you perfect, even as I am perfect."** In love and mercy the messengers of Paradise have carried this divine exhortation down through the ages and out through the universes, even to such lowly animal-origin creatures as the human races of Urantia.

Page 22:2–3   *Paper 1:0.5–6*

Urantia **mortals can hardly hope to be perfect in the infinite sense, but it is entirely possible for human beings,** starting out as they do on this planet, to attain the supernal and divine goal which the infinite God has set for mortal man; and when they do achieve this destiny, they will, in all that pertains to self-realization and mind attainment, be just as replete in their sphere of divine perfection as God himself is in his sphere of infinity and eternity. Such perfection may not be universal in the material sense, unlimited in intellectual grasp, or final in spiritual experience, but it is final and complete in all finite aspects of divinity of will, perfection of personality motivation, and God-consciousness.

This is the true meaning of that divine command, "Be you perfect, even as I am perfect," which ever urges mortal man onward and beckons him inward in that long and fascinating struggle for the attainment of higher and higher levels of spiritual values and true universe meanings. This sublime search for the God of universes is the supreme adventure of the inhabitants of all the worlds of time and space.

---

1. THE FATHER'S NAME

Gift to God
Personality and free will tied-in—1200 pages apart.

Page 22:5   *Paper 1:1.2*

The Universal Father never imposes any form of arbitrary recognition, formal worship, or slavish service upon the intelligent will creatures of the universes. The evolutionary inhabitants of the worlds of time and space must of themselves—in their own hearts—recognize, love, and voluntarily worship him. The Creator refuses to coerce or compel the submission of the spiritual free wills of his material creatures. **The affectionate dedication of the human will to the doing of the Father's will is man's choicest gift to God; in fact, such a consecration of creature will constitutes man's only possible gift of true**

value to the Paradise Father. In God, man lives, moves, and has his being; there is nothing which man can give to God except this choosing to abide by the Father's will, and such decisions, effected by the intelligent will creatures of the universes, constitute the reality of that true worship which is so satisfying to the love-dominated nature of the Creator Father.

Page 1225:2,10   *Paper 112:0.2,10*

While it would be presumptuous to attempt the definition of personality, **it may prove helpful to recount some of the things which are known about personality:...**

    8. **It can make a gift to God—dedication of the free will to the doing of the will of God.**

---

2. THE REALITY OF GOD

God Conscious—Positive Proof

Can we become God-conscious in two ways, but only positive proof of God's existence can be offered by one of these experiences?

Page 24:6   *Paper 1:2.5*

Those who know God have experienced the fact of his presence; such God-knowing mortals hold in their personal experience the only positive proof of the existence of the living God which one human being can offer to another. The existence of God is utterly beyond all possibility of demonstration except for the contact between the God-consciousness of the human mind and the God-presence of the Thought Adjuster that indwells the mortal intellect and is bestowed upon man as the free gift of the Universal Father.

Page 2052:3   *Paper 193:0.3*

[Said Jesus,] "And now you should give ear to my words lest you again make the mistake of hearing my teaching with the mind while in your hearts you fail to comprehend the meaning. From the beginning of my sojourn as one of you, I taught you that my one purpose was to reveal my Father in heaven to his children on earth. I have lived the God-revealing bestowal that you might experience the God-knowing career. I have revealed God as your Father in heaven; I have revealed you as the sons of God on earth. It is a fact that God loves you, his sons. By faith in my word this fact becomes an eternal and living truth in your hearts. **When, by living faith, you become divinely God-conscious, you are then born of the spirit as children of light and life,** even the eternal life wherewith you shall ascend the universe of universes and attain the experience of finding God the Father on Paradise.

Page 2097:2   *Paper 196:3.31*

The great challenge to modern man is to achieve better communication with the divine Monitor that dwells within the human mind. Man's greatest adventure in the flesh consists in the well-balanced and sane effort to advance the borders of self-consciousness out through the dim realms of embryonic soul-consciousness in a wholehearted **effort to reach the borderland of spirit-consciousness—contact with the divine presence. Such an experience constitutes God-consciousness, an experience mightily confirmative of the pre-existent truth of the religious experience of knowing God. Such spirit-consciousness is the equivalent of the knowledge of the actuality of sonship with God. Otherwise, the assurance of sonship is the experience of faith.**

---

6. PERSONALITY IN THE UNIVERSE

Jesus—Religious Life
The value of the Life and Teachings of the Son of Man and Son of God are pointed out early in Papers 1 and 2.

Page 30:2   *Paper 1:6.3*

Never lose sight of the antipodal viewpoints of personality as it is conceived by God and man. Man views and comprehends personality, looking from the finite to the infinite; God looks from the infinite to the finite. Man possesses the lowest type of personality; God, the highest, even supreme, ultimate, and absolute. **Therefore did the better concepts of the divine personality have patiently to await the appearance of improved ideas of human personality, especially the enhanced revelation of both human and divine personality in the Urantian bestowal life of Michael, the Creator Son.**

Page 30:7   *Paper 1:6.8*

God is spirit—spirit personality; man is also a spirit—potential spirit personality. Jesus of Nazareth attained the full realization of this potential of spirit personality in human experience; **therefore his life of achieving the Father's will becomes man's most real and ideal revelation of the personality of God.** Even though the personality of the Universal Father can be grasped only in actual religious experience, in Jesus' earth life we are inspired by the perfect demonstration of such a realization and revelation of the personality of God in a truly human experience.

Page 33:2   *Paper 2:0.2*

The nature of God can be studied in a revelation of supreme ideas, the divine character can be envisaged as a portrayal of supernal ideals, but the **most enlightening and spiritually edifying of all revelations of the divine nature is to be found in the comprehension of the religious life of Jesus of Nazareth,** both before and after his attainment of full consciousness of divinity. If the incarnated life of Michael is taken as the background of the revelation of God to man, we may attempt to put in human word symbols certain ideas and ideals concerning the divine nature which may possibly contribute to a further illumination and unification of the human concept of the nature and the character of the personality of the Universal Father.

---

7. SPIRITUAL VALUE OF THE PERSONALITY CONCEPT

Father Contact

The First Source and Center as our father and as a person—discussed in nine papers spanning 1100 pages.

Page 31:1   *Paper 1:7.1*

When Jesus talked about "the living God," he referred to a personal Deity—the Father in heaven. The concept of the personality of Deity facilitates fellowship; it favors intelligent worship; it promotes refreshing trustfulness. Interactions can be had between nonpersonal things, but not fellowship. **The fellowship relation of father and son, as between God and man, cannot be enjoyed unless both are persons. Only personalities can commune with each other, albeit this personal communion may be greatly facilitated by the presence of just such an impersonal entity as the Thought Adjuster.**

Page 62:2   *Paper 5:0.2*

God has distributed the infinity of his eternal nature throughout the existential realities of his six absolute co-ordinates, but he may, at any time, make direct personal contact with any part or phase or kind of creation through the agency of his prepersonal fragments. And the eternal God has also reserved to himself the prerogative of bestowing personality upon the divine Creators and the living creatures of the universe of universes, while he **has further reserved the prerogative of maintaining direct and parental contact with all these personal beings through the personality circuit.**

Page 64:5   *Paper 5:2.2*

God lives in every one of his spirit-born sons. **The Paradise Sons always have access to the presence of God, "the right hand of the Father," and all of his creature personalities have access to the "bosom of the Father." This refers to the personality circuit,** whenever, wherever, and however contacted, or otherwise entails personal, self-conscious contact and communion with the Universal Father, whether at the central abode or at some other designated place, as on one of the seven sacred spheres of Paradise.

Page 76:5   *Paper 6:4.5*

In his contact with personality, the Father acts in the personality circuit. In his personal and detectable contact with spiritual creation, he appears in the fragments of the totality of his Deity, and these Father fragments have a solitary, unique, and exclusive function wherever and whenever they appear in the universes. In all such situations the spirit of the Son is co-ordinate with the spiritual function of the fragmented presence of the Universal Father.

Page 100:7   *Paper 9:2.5*

The presence of the universal spirit of the Eternal Son we *know*—we can unmistakably recognize it. The presence of the Infinite Spirit, the Third Person of Deity, even mortal man may know, for material creatures can actually experience the beneficence of this divine influence which functions as the Holy Spirit of local universe bestowal upon the races of mankind. **Human beings can also in some degree become conscious of the Adjuster, the impersonal presence of the Universal Father.** These divine spirits which work for man's uplifting and spiritualization all act in unison and in perfect co-operation. They are as one in the spiritual operation of the plans of mortal ascension and perfection attainment.

Page 111:7–14   *Paper 10:3.9–16*

The First Source and Center functions outside of Havona in the phenomenal universes as follows:
1. As creator, through the Creator Sons, his grandsons.
2. As controller, through the gravity center of Paradise.
3. As spirit, through the Eternal Son.
4. As mind, through the Conjoint Creator.
5. **As a Father, he maintains parental contact with all creatures through his personality circuit.**
6. **As a person, he acts directly throughout creation by his exclusive fragments—in mortal man by the Thought Adjusters.**
7. As total Deity, he functions only in the Paradise Trinity.

Page 203:3   *Paper 17:5.5*

The Circuit Spirits are related to the native inhabitants of Havona much as the Thought Adjusters are related to the mortal creatures inhabiting the worlds of the evolutionary universes. **Like the Thought Adjusters, the Circuit Spirits are impersonal, and they consort with the perfect minds of Havona beings much as the impersonal spirits of the Universal Father indwell the finite minds of mortal men.** But the Spirits of the Circuits never become a permanent part of Havona personalities.

Page 363:6   *Paper 32:4.8*

We can see and understand the mechanism whereby the Sons enjoy intimate and complete knowledge regarding the universes of their jurisdiction; but we cannot fully comprehend the methods whereby God is so fully and personally conversant with the details of the universe of universes, although we at least can recognize the avenue whereby the Universal Father can receive information regarding, and manifest his presence to, the beings of his immense creation. **Through the personality circuit the Father is cognizant— has personal knowledge—of all the thoughts and acts of all the beings in all the systems of all the universes of all creation.** Though we cannot fully grasp this technique of God's communion with his children, we can be strengthened in the assurance that the "Lord knows his children," and that of each one of us "he takes note where we were born."

Page 1183:3   *Paper 107:7.1*

Thought Adjusters are not personalities, but they are real entities; they are truly and perfectly individualized, although they are never, while indwelling mortals, actually personalized. Thought Adjusters are not true personalities; they are *true realities*, realities of the purest order known in the universe of universes—they are the divine presence. **Though not personal, these marvelous fragments of the Father are commonly referred to as beings and sometimes, in view of the spiritual phases of their present ministry to mortals, as spirit entities.**

Page 1183:6   *Paper 107:7.4*

Why then, if Thought Adjusters possess volition, are they subservient to the mortal will? We believe it is because Adjuster volition, though absolute in nature, is prepersonal in manifestation. **Human will functions on the personality level of universe reality, and throughout the cosmos the impersonal—the nonpersonal, the subpersonal, and the prepersonal—is ever responsive to the will and acts of existent personality.**

Personal God

A personal God, a loving Father.

Page 31:3   *Paper 1:7.3*

The concept of truth might possibly be entertained apart from personality, the concept of beauty may exist without personality, but the concept of divine goodness is understandable only in relation to personality. **Only a *person* can love and be loved. Even beauty and truth would be divorced from survival hope if they were not attributes of a personal God, a loving Father.**

Page 59:2   *Paper 4:4.6*

In God the Father freewill performances are not ruled by power, nor are they guided by intellect alone; the divine personality is defined as consisting in spirit and manifesting himself to the universes as love. **Therefore, in all his personal relations with the creature personalities of the universes, the First Source and Center is always and consistently a loving Father.** God is a Father in the highest sense of the term. He is eternally motivated by the perfect idealism of divine love, and that tender nature **finds its strongest expression and greatest satisfaction in loving and being loved.**

PAPER 2: THE NATURE OF GOD

Highest Human Concept

The *Father* idea—from page 33 to the last sentence on page 2097...

Page 33:1   *Paper 2:0.1*

Inasmuch as man's highest possible concept of God is embraced within the human idea and ideal of a primal and infinite personality, it is permissible, and may prove helpful, to study certain characteristics of the divine nature which constitute the character of Deity. The nature of God can best be understood by the revelation of the Father which Michael of Nebadon unfolded in his manifold teachings and in his superb mortal life in the flesh. **The divine nature can also be better understood by man if he regards himself as a child of God and looks up to the Paradise Creator as a true spiritual Father.**

Page 1260:3   *Paper 115:1.2*

Conceptual frames of the universe are only relatively true; they are serviceable scaffolding which must eventually give way before the expansions of enlarging cosmic comprehension. The understandings of truth, beauty, and goodness, morality, ethics, duty, love, divinity, origin, existence, purpose, destiny, time, space, even Deity, are only relatively true. **God is much, much more than a Father, but the Father is man's highest concept of God;** nonetheless, the Father-Son portrayal of Creator-creature relationship will be augmented by those supermortal conceptions of Deity which will be attained in Orvonton, in Havona, and on Paradise. Man must think in a mortal universe frame, but that does not mean that he cannot envision other and higher frames within which thought can take place.

Page 2097:3   *Paper 196:3.32*

And God-consciousness is equivalent to the integration of the self with the universe, and on its highest levels of spiritual reality. Only the spirit content of any value is imperishable. Even that which is true, beautiful, and good may not perish in human experience. If man does not choose to survive, then does the surviving Adjuster conserve those realities born of love and nurtured in service. And all these things are a part of the Universal Father. The Father is living love, and this life of the Father is in his Sons. And the spirit of the Father is in his Sons' sons—mortal men. **When all is said and done, the Father idea is still the highest human concept of God.**

Jesus—Religious Life

Page 33:2   *Paper 2:0.2*

See "Jesus—Religious Life" on page 113 for full text references. (1:6.3)

Spirt of Truth and Divine Fragment

The Spirit of Truth, written word (Revelation in book form on other planets?) and Divine Fragment all play a part in ability to expand our understanding of God. Paper 72 below notes that a neighboring planet has not yet had a Magisterial bestowal son, yet look at the progress made on this one continent...

Page 33:3   *Paper 2:0.3*

In all our efforts to enlarge and spiritualize the human concept of God, we are tremendously handicapped by the limited capacity of the mortal mind. We are also seriously handicapped in the execution of our assignment by the limitations of language and by the poverty of material which can be utilized for purposes of illustration or comparison in our efforts to portray divine values and to present spiritual meanings to the finite, mortal mind of man. **All our efforts to enlarge the human concept of God would be well-nigh futile except for the fact that the mortal mind is indwelt by the bestowed Adjuster of the Universal Father and is pervaded by the Truth Spirit of the Creator Son.** Depending, therefore, on the presence of these divine spirits within the heart of man for assistance in the enlargement of the concept of God, I cheerfully undertake the execution of my mandate to attempt the further portrayal of the nature of God to the mind of man.

Page 808:3   *Paper 72:0.3*

This planet, like Urantia, was led astray by the disloyalty of its Planetary Prince in connection with the Lucifer rebellion. It received a Material Son shortly after Adam came to Urantia, and this Son also defaulted, **leaving the sphere isolated, since a Magisterial Son has never been bestowed upon its mortal races.**

Page 820:3   *Paper 72:12.5*

The pouring out of the Spirit of Truth provides the spiritual foundation for the realization of great achievements in the interests of the human race of the bestowal world. Urantia is therefore far better prepared for the more immediate realization of a planetary government with its laws, mechanisms, symbols, conventions, and language—all of which could contribute so mightily to the establishment of world-wide peace under law and could lead to the sometime dawning of a real age of spiritual striving; and such an age is the planetary threshold to the utopian ages of light and life.

Page 1106:3   *Paper 101:2.4*

There are two basic reasons for believing in a God who fosters human survival:
1. Human experience, personal assurance, the somehow registered hope and trust initiated by the indwelling Thought Adjuster.
2. The revelation of truth, whether by direct personal ministry of the Spirit of Truth, by the world bestowal of divine Sons, **or through the revelations of the written word.**

Page 2092:2   *Paper 196:2.4*

But the greatest mistake was made in that, while the human Jesus was recognized as *having* a religion, the divine Jesus (Christ) almost overnight became a religion. Paul's Christianity made sure of the adoration of the divine Christ, but it almost wholly lost sight of the struggling and valiant human Jesus of Galilee, who, by the valor of his personal religious faith and the heroism of his indwelling Adjuster, ascended from the lowly levels of humanity to become one with divinity, thus becoming the new and living way whereby all mortals may so ascend from humanity to divinity. **Mortals in all stages of spirituality and on all worlds** may find in the personal life of Jesus that which will strengthen and inspire them as they progress from the lowest spirit levels up to the highest divine values, from the beginning to the end of all personal religious experience.

2. THE FATHER'S ETERNAL PERFECTION

Personality and Personality Circuit
Seven papers on the bestowal of personality and the personality circuit.

Page 36:3–4   *Paper 2:2.5–6*

God's primal perfection consists not in an assumed righteousness but rather in the inherent perfection of the goodness of his divine nature. He is final, complete, and perfect. There is no thing lacking in the beauty and perfection of his righteous character. And the whole scheme of living existences on the worlds of space is centered in the divine purpose of elevating all will creatures to the high destiny of the **experience of sharing the Father's Paradise perfection. God is neither self-centered nor self-contained; he never ceases to bestow himself upon all self-conscious creatures of the vast universe of universes.**

God is eternally and infinitely perfect; he cannot personally know imperfection as his own experience, but he does share the consciousness of all the experience of imperfectness of all the struggling creatures of the evolutionary universes of all the Paradise Creator Sons. **The personal and liberating touch of the God of perfection overshadows the hearts and encircuits the natures of all those mortal creatures who have ascended to the universe level of moral discernment. In this manner, as well as through the contacts of the divine presence,** the Universal Father actually participates in the experience *with* immaturity and imperfection in the evolving career of every moral being of the entire universe.

Page 71:1   *Paper 5:6.6*

Capacity for divine personality is inherent in the prepersonal Adjuster; capacity for human personality is potential in the cosmic-mind endowment of the human being. **But the experiential personality of mortal man is not observable as an active and functional reality until after the material life vehicle of the mortal creature has been touched by the liberating divinity of** the Universal Father, being thus launched upon the seas of experience as a self-conscious and a (relatively) self-determinative and self-creative personality. The material self is truly and *unqualifiedly personal*.

Page 76:5   *Paper 6:4.5*

In his contact with personality, the Father acts in the personality circuit. In his personal and detectable contact with spiritual creation, he appears in the fragments of the totality of his Deity, and these Father fragments have a solitary, unique, and exclusive function wherever and whenever they appear in the universes. In all such situations the spirit of the Son is co-ordinate with the spiritual function of the fragmented presence of the Universal Father.

Page 104:2   *Paper 9:6.4*

Selfhood of personality dignity, human or divine, immortal or potentially immortal, does not however originate in either spirit, mind, or matter; it is the bestowal of the Universal Father. Neither is the interaction of spirit, mind, and material gravity a prerequisite to the appearance of personality gravity. The Father's circuit may embrace a mind-material being who is unresponsive to spirit gravity, or it may include a mind-spirit being who is unresponsive to material gravity. The operation of personality gravity is always a volitional act of the Universal Father.

Page 111:12–13   *Paper 10:3.14–15*

5. **As a Father, he maintains parental contact with all creatures through his personality circuit.**
6. **As a person,** he acts *directly* throughout creation by his exclusive fragments—in mortal man by the Thought Adjusters.

Page 195:2   *Paper 16:8.8*

The Urantia type of human personality may be viewed as functioning in a physical mechanism consisting of the planetary modification of the Nebadon type of organism belonging to the electrochemical order of life activation and endowed with the Nebadon order of the Orvonton series of the cosmic mind of parental reproductive pattern. The **bestowal of the divine gift of personality** upon such a mind-endowed mortal mechanism confers the dignity of cosmic citizenship and **enables such a mortal creature forthwith to become reactive** to the constitutive recognition of the three basic mind realities of the cosmos:

1. The mathematical or logical recognition of the uniformity of physical causation.
2. The reasoned recognition of the obligation of moral conduct.
3. The faith-grasp of the fellowship worship of Deity, associated with the loving service of humanity.

Page 363:6   *Paper 32:4.8*

We can see and understand the mechanism whereby the Sons enjoy intimate and complete knowledge regarding the universes of their jurisdiction; but we cannot fully comprehend the methods whereby God is so fully and personally conversant with the details of the universe of universes, although we at **least can recognize the avenue whereby the Universal Father can receive information regarding, and manifest his presence to, the beings of his immense creation. Through the personality circuit the Father is cognizant— has personal knowledge—of all the thoughts and acts of all the beings in all the systems of all the universes of all creation.** Though we cannot fully grasp this technique of God's communion with his children, we can be strengthened in the assurance that the "Lord knows his children," and that of each one of us "he takes note where we were born."

---

5. THE LOVE OF GOD

"Behold, what manner of love..."
Divine Counselor and Mighty Messenger echo the same sentiment...

Page 39:3   *Paper 2:5.4*

God is divinely kind to sinners. When rebels return to righteousness, they are mercifully received, "for our God will abundantly pardon." "I am he who blots out your transgressions for my own sake, and I will not remember your sins." **"Behold what manner of love the Father has bestowed upon us that we should be called the sons of God."**

Page 448:1   *Paper 40:6.2*

It is a solemn and supernal fact that such lowly and material creatures as Urantia human beings are the sons of God, faith children of the Highest. **"Behold, what manner of love the Father has bestowed upon us that we should be called the sons of God."** "As many as received him, to them gave he the power to recognize that they are the sons of God." While "it does not yet appear what you shall be," even now "you are the faith sons of God"; "for you have not received the spirit of bondage again to fear, but you have received the spirit of sonship, whereby you cry, `our Father.'" Spoke the prophet of old in the name of the eternal God: "Even to them will I give in my house a place and a name better than sons; I will give them an everlasting name, one that shall not be cut off." "And because you are sons, God has sent forth the spirit of his Son into your hearts."

Atonement Doctrine

Parts I and IV on the ransom concept of the atonement doctrine.

Page 41:3　　　　　　　　　　　　　　　　　　　*Paper 2:6.5*

Righteousness implies that God is the source of the moral law of the universe. Truth exhibits God as a revealer, as a teacher. But love gives and craves affection, seeks understanding fellowship such as exists between parent and child. Righteousness may be the divine thought, but love is a father's attitude. The **erroneous supposition that** the righteousness of God was irreconcilable with the selfless love of the heavenly Father, presupposed absence of unity in the nature of Deity and **led directly to the elaboration of the atonement doctrine,** which is a philosophic assault upon both the unity and the free-willness of God.

Page 60:5–6　　　　　　　　　　　　　　　　　　*Paper 4:5.6–7*

The bestowal of a Paradise Son on your world was inherent in the situation of closing a planetary age; it was inescapable, and it was not made necessary for the purpose of winning the favor of God. This bestowal also happened to be the final personal act of a Creator Son in the long adventure of earning the experiential sovereignty of his universe. What a travesty upon the infinite character of God! this teaching that his fatherly heart in all its austere coldness and hardness was so untouched by the misfortunes and sorrows of his creatures that his tender mercies were not forthcoming until he saw his blameless Son bleeding and dying upon the cross of Calvary!

But the inhabitants of Urantia are to find deliverance from these ancient errors and pagan superstitions respecting the nature of the Universal Father. The revelation of the truth about God is appearing, and the human race is destined to know the Universal Father in all that beauty of character and loveliness of attributes so magnificently portrayed by the Creator Son who sojourned on Urantia as the Son of Man and the Son of God.

Page 2016:6　　　　　　　　　　　　　　　　　　*Paper 188:4.1*

Although Jesus did not die this death on the cross to atone for the racial guilt of mortal man nor to provide some sort of effective approach to an otherwise offended and unforgiving God; even though the Son of Man did not offer himself as a sacrifice to appease the wrath of God and to open the way for sinful man to obtain salvation; **notwithstanding that these ideas of atonement and propitiation are erroneous, nonetheless, there are** significances attached to this death of Jesus on the cross which should not be overlooked. It is a fact that Urantia has become known among other neighboring inhabited planets as the "World of the Cross."

Page 2016:8–10   *Paper 188:4.3–5*

Mortal man was never the property of the archdeceivers. **Jesus did not die to ransom man from the clutch of the apostate rulers and fallen princes of the spheres.** The Father in heaven never conceived of such crass injustice as damning a mortal soul because of the evil-doing of his ancestors. Neither was the Master's death on the cross a sacrifice which consisted in an effort to pay God a debt which the race of mankind had come to owe him.

Before Jesus lived on earth, you might possibly have been justified in believing in such a God, but not since the Master lived and died among your fellow mortals. Moses taught the dignity and justice of a Creator God; but Jesus portrayed the love and mercy of a heavenly Father.

The animal nature—the tendency toward evil-doing—may be hereditary, but sin is not transmitted from parent to child. Sin is the act of conscious and deliberate rebellion against the Father's will and the Sons' laws by an individual will creature.

Page 2017:3   *Paper 188:4.8*

When once you grasp the idea of God as a true and loving Father, the only concept which Jesus ever taught, you must forthwith, in all consistency, utterly abandon all those primitive notions about God as an offended monarch, a stern and all-powerful ruler whose chief delight is to detect his subjects in wrongdoing and to see that they are adequately punished, unless some being almost equal to himself should volunteer to suffer for them, to die as a substitute and in their stead. **The whole idea of ransom and atonement is incompatible with the concept of God as it was taught and exemplified by Jesus of Nazareth.** The infinite love of God is not secondary to anything in the divine nature.

---

7. DIVINE TRUTH AND BEAUTY

Happiness
Happiness from Paper 2 and Part IV.

Page 42:7   *Paper 2:7.6*

Intellectual self-consciousness can discover the beauty of truth, its spiritual quality, not only by the philosophic consistency of its concepts, but more certainly and surely by the unerring response of the ever-present Spirit of Truth. **Happiness ensues from the recognition of truth because it can be** *acted out;* **it can be lived.** Disappointment and sorrow attend upon error because, not being a reality, it cannot be realized in experience. Divine truth is best known by its *spiritual flavor.*

Page 1480:4   *Paper 133:7.12*

The human mind does not well stand the conflict of double allegiance. It is a severe strain on the soul to undergo the experience of an effort to serve both good and evil. **The supremely happy and efficiently unified mind is the one wholly dedicated to the doing of the will of the Father in heaven.** Unresolved conflicts destroy unity and may terminate in mind disruption. But the survival character of a soul is not fostered by attempting to secure peace of mind at any price, by the surrender of noble aspirations, and by the compromise of spiritual ideals; rather is such peace attained by the stalwart assertion of the triumph of that which is true, and this victory is achieved in the overcoming of evil with the potent force of good.

Page 1674:3–4,6   *Paper 149:5.1–2,4*

When Jesus was visiting the group of evangelists working under the supervision of Simon Zelotes, during their evening conference Simon asked the Master: **"Why are some persons so much more happy and contented than others? Is contentment a matter of religious experience?"** Among other things, Jesus said in answer to Simon's question:

"Simon, some persons are naturally more happy than others. Much, very much, depends upon the willingness of man to be led and directed by the Father's spirit which lives within him. Have you not read in the Scriptures the words of the wise man, 'The spirit of man is the candle of the Lord, searching all the inward parts'? And also that such spirit-led mortals say: 'The lines are fallen to me in pleasant places; yes, I have a goodly heritage.' 'A little that a righteous man has is better than the riches of many wicked,' for 'a good man shall be satisfied from within himself.' 'A merry heart makes a cheerful countenance and is a continual feast. Better is a little with the reverence of the Lord than great treasure and trouble therewith. Better is a dinner of herbs where love is than a fatted ox and hatred therewith. Better is a little with righteousness than great revenues without rectitude.' 'A merry heart does good like a medicine.' 'Better is a handful with composure than a superabundance with sorrow and vexation of spirit....'

"Seek not, then, for false peace and transient joy but rather for the assurance of faith and the sureties of divine sonship which yield composure, contentment, and supreme joy in the spirit."

Education—Purpose
Purpose of education—from Paper 2 and two thousand pages later...

Page 43:5   *Paper 2:7.12*

Truth is coherent, beauty attractive, goodness stabilizing. And when these values of that which is real are co-ordinated in personality experience, the result is a high order of love conditioned by wisdom and qualified by loyalty. **The real purpose of all universe education is to effect the better co-ordination of the isolated child of the worlds with the larger realities of his expanding experience.** Reality is finite on the human level, infinite and eternal on the higher and divine levels.

Page 2086:3   *Paper 195:10.17*

Even secular education could help in this great spiritual renaissance if it would pay more attention to the work of teaching youth how to engage in life planning and character progression. The purpose of all education should be to foster and further the supreme purpose of life, the development of a majestic and well-balanced personality. There is great need for the teaching of moral discipline in the place of so much self-gratification. Upon such a foundation religion may contribute its spiritual incentive to the enlargement and enrichment of mortal life, even to the security and enhancement of life eternal.

PAPER 3: THE ATTRIBUTES OF GOD

---

1. GOD'S EVERYWHERENESS

Most Highs
The Most Highs in dealing with quarantined planets and how they rule in the kingdoms of men.

Page 46:2   *Paper 3:1.10*

Concerning God's presence in a planet, system, constellation, or a universe, the degree of such presence in any creational unit is a measure of the degree of the evolving presence of the Supreme Being: It is determined by the en masse recognition of God and loyalty to him on the part of the vast universe organization, running down to the systems and planets themselves. **Therefore it is sometimes with the hope of conserving and safeguarding these phases of God's precious presence that, when some planets (or even systems) have plunged far into spiritual darkness, they are in a certain sense**

quarantined, or partially isolated from intercourse with the larger units of creation. And all this, as it operates on Urantia, is a spiritually defensive reaction of the majority of the worlds to save themselves, as far as possible, from suffering the isolating consequences of the alienating acts of a headstrong, wicked, and rebellious minority.

Page 363:2   *Paper 32:4.4*

The Creator Son rules supreme in all matters of ethical associations, the relations of any division of creatures to any other class of creatures or of two or more individuals within any given group; but such a plan does not mean that the Universal Father may not in his own way intervene and do aught that pleases the divine mind with any *individual creature* throughout all creation, as pertains to that individual's present status or future prospects and as concerns the Father's eternal plan and infinite purpose.

Page 390:3   *Paper 35:5.6*

On those worlds segregated in spiritual darkness, those spheres which have, through rebellion and default, suffered planetary isolation, an observer Vorondadek is usually present pending the restoration of normal status. In certain emergencies this Most High observer could exercise absolute and arbitrary authority over every celestial being assigned to that planet. It is of record on Salvington that the Vorondadeks have sometimes exercised such authority as Most High regents of such planets. And this has also been true even of inhabited worlds that were untouched by rebellion.

Page 491:12–13   *Paper 43:5.16–17*

Ever since the Lucifer rebellion the Edentia Fathers have exercised a special care over Urantia and the other isolated worlds of Satania. Long ago the prophet recognized the controlling hand of the Constellation Fathers in the affairs of nations. "When the Most High divided to the nations their inheritance, when he separated the sons of Adam, he set the bounds of the people."

Every quarantined or isolated world has a Vorondadek Son acting as an observer. He does not participate in planetary administration except when ordered by the Constellation Father to intervene in the affairs of the nations. **Actually it is this Most High observer who "rules in the kingdoms of men."** Urantia is one of the isolated worlds of Norlatiadek, and a Vorondadek observer has been stationed on the planet ever since the Caligastia betrayal. When Machiventa Melchizedek ministered in semimaterial form on Urantia, he paid respectful homage to the Most High observer then on duty, as it is written, "And Melchizedek, king of Salem, was the priest of the Most High." Melchizedek revealed the relations of this Most High observer to Abraham when he said, "And blessed be the Most High, who has delivered your enemies into your hand."

Page 495:6   *Paper 43:9.4*

Ascending mortals on Edentia are chiefly occupied with the assignments on the seventy progressive univitatia worlds. They also serve in varied capacities on Edentia itself, mainly in conjunction with the constellation program concerned with group, racial, national, and planetary welfare. **The Most Highs are not so much engaged in fostering individual advancement on the inhabited worlds; they rule in the kingdoms of men rather than in the hearts of individuals.**

Page 865:6   *Paper 77:8.13*

Their chief work today is that of unperceived personal-liaison associates of those men and women who constitute the planetary reserve corps of destiny. It was the work of this secondary group, ably seconded by certain of the primary corps, that brought about the co-ordination of personalities and circumstances on Urantia which finally induced the planetary celestial supervisors to initiate those petitions that resulted in the granting of the mandates making possible the series of revelations of which this presentation is a part. But it should be made clear that the midway creatures are not involved in the sordid performances taking place under the general designation of "spiritualism." The midwayers at present on Urantia, all of whom are of honorable standing, are not connected with the phenomena of so-called "mediumship"; and they do not, ordinarily, permit humans to witness their sometimes necessary physical activities or other contacts with the material world, as they are perceived by human senses.

Page 908:8   *Paper 81:6.19*

While very little progress has been made on Urantia toward developing an international language, much has been accomplished by the establishment of international commercial exchange. **And all these international relations should be fostered,** whether they involve language, trade, art, science, competitive play, or religion.

Page 1255:3,7   *Paper 114:6.4,8*

The twelve corps of the master seraphim of planetary supervision are functional on Urantia as follows:
4. *The angels of nation life.* These are the "angels of the trumpets," directors of the political performances of Urantia national life. The group now functioning in the **overcontrol of international relations** is the fourth corps to serve on the planet. It is particularly through the ministry of this seraphic division that **"the Most Highs rule in the kingdoms of men."**

Page 1257:7, 1258:1   *Paper 114:7.4,6*

Each division of planetary celestial service is entitled to a liaison corps of these mortals of destiny standing. The average inhabited world employs seventy separate corps of destiny, which are intimately connected with the superhuman current conduct of world affairs. **On Urantia there are twelve reserve corps of destiny, one for each of the planetary groups of seraphic supervision.**

On many worlds the better adapted secondary midway creatures are able to attain varying degrees of contact with the Thought Adjusters of certain favorably constituted mortals through the skillful penetration of the minds of the latters' indwelling. (And it was by just such a fortuitous combination of cosmic adjustments that these revelations were materialized in the English language on Urantia.) **Such potential contact mortals of the evolutionary worlds are mobilized in the numerous reserve corps, and it is, to a certain extent, through these small groups of forward-looking personalities that spiritual civilization is advanced and the Most Highs are able to rule in the kingdoms of men.** The men and women of these reserve corps of destiny thus have various degrees of contact with their Adjusters through the intervening ministry of the midway creatures; but these same mortals are little known to their fellows except in those rare social emergencies and spiritual exigencies wherein these reserve personalities function for the prevention of the breakdown of evolutionary culture or the extinction of the light of living truth. On Urantia these reservists of destiny have seldom been emblazoned on the pages of human history. ·

Page 1258:6   *Paper 114:7.11*

Urantia mortals should not allow the comparative spiritual isolation of their world from certain of the local universe circuits to produce a feeling of cosmic desertion or planetary orphanage. **There is operative on the planet a very definite and effective superhuman supervision of world affairs and human destinies.**

Ideal Government—Struggle For
Mankind's struggle for an ideal government.

Page 46:2   *Paper 3:1.10*

Concerning God's presence in a planet, system, constellation, or a universe, the degree of such presence in any creational unit is a measure of the degree of the evolving presence of the Supreme Being: It is determined by the en masse recognition of God and loyalty to him on the part of the vast universe organization, running down to the systems and **planets themselves.** Therefore it is sometimes with the hope of conserving and safeguarding these phases of God's precious presence that, when some planets (or even systems) have plunged far into spiritual darkness, **they are in a certain sense quarantined,** or partially isolated from intercourse with the larger units of creation. And all this, **as it operates on Urantia,** is a spiritually defensive reaction of the majority of the worlds to save themselves, as far as possible, from suffering the isolating consequences of the alienating acts of a headstrong, wicked, and rebellious minority.

Page 578:1   *Paper 50:6.1*

The isolation of Urantia renders it impossible to undertake the presentation of many details of the life and environment of your Satania neighbors. In these presentations we are limited by the planetary quarantine and by the system isolation. We must be guided by these restrictions in all our efforts to enlighten Urantia mortals, but in so far as is permissible, you have been instructed in the progress of an average evolutionary world, and you are able to compare such a world's career with the present state of Urantia.

Page 799:1   *Paper 70:12.8*

Mankind's struggle to perfect government on Urantia has to do with perfecting channels of administration, with adapting them to ever-changing current needs, with improving power distribution within government, and then with selecting such administrative leaders as are truly wise. While there is a divine and ideal form of government, **such cannot be revealed but must be slowly and laboriously discovered by the men and women of each planet throughout the universes of time and space.**

God—No Respecter of Person or Planet

Page 46:4   *Paper 3:1.12*

The fluctuations of the Father's presence are not due to the changeableness of God. The Father does not retire in seclusion because he has been slighted; his affections are not alienated because of the creature's wrongdoing. Rather, having been endowed with the power of choice (concerning himself), his children, in the exercise of that choice, directly determine the degree and limitations of the Father's divine influence in their own hearts and souls. **The Father has freely bestowed himself upon us without limit and without favor. He is no respecter of persons, planets, systems, or universes.** In the sectors of time he confers differential honor only on the Paradise personalities of God the Sevenfold, the co-ordinate creators of the finite universes.

Page 56:5   *Paper 4:2.1*

Nature is in a limited sense the physical habit of God. The conduct, or action, of God is qualified and provisionally modified by the experimental plans and the evolutionary patterns of a local universe, a constellation, a system, or a planet. God acts in accordance with a well-defined, unchanging, immutable law throughout the wide-spreading master universe; **but he modifies the patterns of his action so as to contribute to the co-ordinate and balanced conduct of each universe, constellation, system, planet, and personality in accordance with the local objects, aims, and plans of the finite projects of evolutionary unfolding.**

---

2. GOD'S INFINITE POWER

Energy—Undiscovered Nonspiritual
As yet, undiscovered nonspiritual energy.

Page 47:1   *Paper 3:2.3*

Of all the divine attributes, his omnipotence, especially as it prevails in the material universe, is the best understood. Viewed as an unspiritual phenomenon, God is energy. This declaration of physical fact is predicated on the incomprehensible truth that the First Source and Center is the primal cause of the universal physical phenomena of all space. From this divine activity all physical energy and other material manifestations are derived. Light, that is, light without heat, is another of the nonspiritual manifestations of the Deities. **And there is still another form of nonspiritual energy which is virtually unknown on Urantia; it is as yet unrecognized.**

Page 467:5   *Paper 42:1.3*

There is innate in matter and present in universal space a form of energy not known on Urantia. When this discovery is finally made, then will physicists feel that they have solved, almost at least, the mystery of matter. And so will they have approached one step nearer the Creator; so will they have mastered one more phase of the divine technique; but in no sense will they have found God, neither will they have established the existence of matter or the operation of natural laws apart from the cosmic technique of Paradise and the motivating purpose of the Universal Father.

Page 468:1   *Paper 42:1.4*

Subsequent to even still greater progress and further discoveries, after Urantia has advanced immeasurably in comparison with present knowledge, **though you should gain control of the energy revolutions of the electrical**

units of matter to the extent of modifying their physical manifestations—even after all such possible progress, forever will scientists be powerless to create one atom of matter or to originate **one flash of energy or ever to add to matter that which we call life.**

Page 472:12   *Paper 42:4.1*

Light, heat, electricity, magnetism, chemism, energy, and matter are—in origin, nature, and destiny—one and the same thing, **together with other material realities as yet undiscovered on Urantia.**

Thought Adjusters—God Adjusts with Mind of Imperfection
The Universal Father *adjusts* with our thoughts.

Page 47:3   *Paper 3:2.5*

The omnipotence of the Father pertains to the everywhere dominance of the absolute level, whereon the three energies, material, mindal, and spiritual, are indistinguishable in close proximity to him—the Source of all things. Creature mind, being neither Paradise monota nor Paradise spirit, is not directly responsive to the Universal Father. **God *adjusts* with the mind of imperfection—with Urantia mortals through the Thought Adjusters.**

Page 447:2   *Paper 40:5.14*

Series three—mortals of Adjuster-fusion potential. **All Father-fused mortals are of animal origin, just like the Urantia races.** They embrace mortals of the one-brained, two-brained, and three-brained types of Adjuster-fusion potential. Urantians are of the intermediate or two-brained type, being in many ways humanly superior to the one-brained groups but definitely limited in comparison with the three-brained orders. These three types of physical-brain endowment are not factors in Adjuster bestowal, in seraphic service, or in any other phase of spirit ministry. The intellectual and spiritual differential between the three brain types characterizes individuals who are otherwise quite alike in mind endowment and spiritual potential, being greatest in the temporal life and tending to diminish as the mansion worlds are traversed one by one. From the system headquarters on, the progression of these three types is the same, and their eventual Paradise destiny is identical.

God—Foreknowledge

Foreknowledge, full allowance for all finite choice, from potentials to actuals and the potentializing of existing actuals. Is this the manner in which the Original, I AM is able to possess foreknowledge of ALL our well nigh infinite choices, and through the Supreme respond in a Fatherly way to whatever choice is made?

Page 47:4   *Paper 3:2.6*

The Universal Father is not a transient force, a shifting power, or a fluctuating energy. The power and wisdom of the Father are wholly adequate to cope with any and all universe exigencies. **As the emergencies of human experience arise, he has foreseen them all,** and therefore he does not react to the affairs of the universe in a detached way but rather in accordance with the dictates of eternal wisdom and in consonance with the mandates of infinite judgment. Regardless of appearances, the power of God is not functioning in the universe as a blind force.

Page 49:3   *Paper 3:3.4*

We are **not wholly certain as to whether or not God chooses to foreknow events of sin.** But even if God should foreknow the freewill acts of his children, such foreknowledge does not in the least abrogate their freedom. One thing is certain: God is never subjected to surprise.

Page 136:3   *Paper 12:6.5*

The universe is highly predictable only in the quantitative or gravity-measurement sense; even the primal physical forces are not responsive to linear gravity, nor are the higher mind meanings and true spirit values of ultimate universe realities. Qualitatively, the universe is not highly predictable as regards new associations of forces, either physical, mindal, or spiritual, although many such combinations of energies or forces become partially predictable when subjected to critical observation. **When matter, mind, and spirit are unified by creature personality, we are unable fully to predict the decisions of such a freewill being.**

Page 315:4   *Paper 28:6.11*

These time evaluators are also the secret of prophecy; they portray the element of time which will be required in the completion of any undertaking, and they are just as dependable as indicators as are the frandalanks and chronoldeks of other living orders. The **Gods foresee, hence foreknow;** but the ascendant authorities of the universes of time must consult the Imports of Time to be able to forecast events of the future.

Page 1185:2   *Paper 108:0.2*

Nothing in the entire universe can substitute for the fact of experience on nonexistential levels. **The infinite God is, as always, replete and complete, infinitely inclusive of all things except evil and creature experience.** God cannot do wrong; he is infallible. God cannot experientially know what he has never personally experienced; God's preknowledge is existential. Therefore does the spirit of the Father descend from Paradise to participate with finite mortals in every bona fide experience of the ascending career; it is only by such a method that the existential God could become in truth and in fact man's experiential Father. The infinity of the eternal God encompasses the potential for finite experience, which indeed becomes actual in the ministry of the Adjuster fragments that actually share the life vicissitude experiences of human beings.

Page 1262:3   *Paper 115:3.6*

1. *The Original.* The unqualified concept of the First Source and Center, that source manifestation of the I AM **from which all reality takes origin.**

Page 1262:9   *Paper 115:3.12*

From a creature's viewpoint, actuality is substance, potentiality is capacity. Actuality exists centermost and expands therefrom into peripheral infinity; potentiality comes inward from the infinity periphery and converges at the center of all things. **Originality is that which first causes and then balances the dual motions of the cycle of reality metamorphosis from potentials to actuals and the potentializing of existing actuals.**

Page 1300:5   *Paper 118:7.1*

The function of Creator will and creature will, in the grand universe, operates within the limits, and in accordance with the possibilities, established by the Master Architects. This foreordination of these maximum limits does not, however, in the least abridge the sovereignty of creature will within these boundaries. **Neither does ultimate foreknowledge—full allowance for all finite choice—constitute an abrogation of finite volition.** A mature and farseeing human being might be able to forecast the decision of some younger associate most accurately, but this foreknowledge takes nothing away from the freedom and genuineness of the decision itself. The Gods have wisely limited the range of the action of immature will, but it is true will, nonetheless, within these defined limits.

3. GOD'S UNIVERSAL KNOWLEDGE

God and Son Knows All

Paper 3 states that the Universal Father is the "only personality..."—Paper 6 states that like the Father, "Son knows all," *but* Paper 6 also says, "With the exceptions noted," several paragraphs later. So, the statement in Paper 3 is qualified in Paper 6.

Page 49:1   *Paper 3:3.2*

The Universal Father is the only personality in all the universe who does actually know the number of the stars and planets of space. All the worlds of every universe are constantly within the consciousness of God. He also says: "I have surely seen the affliction of my people, I have heard their cry, and I know their sorrows." For "the Lord looks from heaven; he beholds all the sons of men; from the place of his habitation he looks upon all the inhabitants of the earth." Every creature child may truly say: "He knows the way I take, and when he has tried me, I shall come forth as gold." "God knows our downsittings and our uprisings; he understands our thoughts afar off and is acquainted with all our ways." "All things are naked and open to the eyes of him with whom we have to do." And it should be a real comfort to every human being to understand that "he knows your frame; he remembers that you are dust." Jesus, speaking of the living God, said, "Your Father knows what you have need of even before you ask him."

Page 76:7   *Paper 6:4.7*

The Original Son is universally and spiritually self-conscious. In wisdom the Son is the full equal of the Father. **In the realms of knowledge, omniscience, we cannot distinguish between the First and Second Sources;** like the Father, the Son knows all; he is never surprised by any universe event; he comprehends the end from the beginning.

Page 77:3   *Paper 6:4.10*

It is needless further to expatiate on the attributes of the Eternal Son. **With the exceptions noted,** it is only necessary to study the spiritual attributes of God the Father to understand and correctly evaluate the attributes of God the Son.

God—Foreknowledge

Page 49:3 *Paper 3:3.4*

See "God—Foreknowledge" on page 133 for full text references. (3:2.6)

---

5. THE FATHER'S SUPREME RULE

Predestination
Predestination with regard to persons and planets.

Page 51:2 *Paper 3:5.3*

In the affairs of men's hearts the Universal Father may not always have his way; but in the conduct and destiny of a planet the divine plan prevails; the eternal purpose of wisdom and love triumphs.

Page 1204:5 *Paper 110:2.1*

When Thought Adjusters indwell human minds, they bring with them the model careers, the ideal lives, as determined and foreordained by themselves and the Personalized Adjusters of Divinington, which have been certified by the Personalized Adjuster of Urantia. Thus they begin work with a definite and predetermined plan for the intellectual and spiritual development of their human subjects, but it is not incumbent upon any human being to accept this plan. **You are all subjects of predestination, but it is not foreordained that you must accept this divine predestination; you are at full liberty to reject any part or all of the Thought Adjusters' program.** It is their mission to effect such mind changes and to make such spiritual adjustments as you may willingly and intelligently authorize, to the end that they may gain more influence over the personality directionization; but under no circumstances do these divine Monitors ever take advantage of you or in any way arbitrarily influence you in your choices and decisions. **The Adjusters respect your sovereignty of personality;** *they are always subservient to your will.*

Inevitabilities
Papers 3 and 154 on inevitabilities, and why.

Page 51:4-13 *Paper 3:5.5-14*

The uncertainties of life and the vicissitudes of existence do not in any manner contradict the concept of the universal sovereignty of God. **All evolutionary creature life is beset by certain** *inevitabilities.* Consider the following:

1. Is *courage*—strength of character—desirable? Then must man be reared in an environment which necessitates grappling with hardships and reacting to disappointments.

2. Is *altruism*—service of one's fellows—desirable? Then must life experience provide for encountering situations of social inequality.

3. Is *hope*—the grandeur of trust—desirable? Then human existence must constantly be confronted with insecurities and recurrent uncertainties.

4. Is *faith*—the supreme assertion of human thought—desirable? Then must the mind of man find itself in that troublesome predicament where it ever knows less than it can believe.

5. Is the *love of truth* and the willingness to go wherever it leads, desirable? Then must man grow up in a world where error is present and falsehood always possible.

6. Is *idealism*—the approaching concept of the divine—desirable? Then must man struggle in an environment of relative goodness and beauty, surroundings stimulative of the irrepressible reach for better things.

7. Is *loyalty*—devotion to highest duty—desirable? Then must man carry on amid the possibilities of betrayal and desertion. The valor of devotion to duty consists in the implied danger of default.

8. Is *unselfishness*—the spirit of self-forgetfulness—desirable? Then must mortal man live face to face with the incessant clamoring of an inescapable self for recognition and honor. Man could not dynamically choose the divine life if there were no self-life to forsake. Man could never lay saving hold on righteousness if there were no potential evil to exalt and differentiate the good by contrast.

9. Is *pleasure*—the satisfaction of happiness—desirable? Then must man live in a world where the alternative of pain and the likelihood of suffering are ever-present experiential possibilities.

Page 1719:1   *Paper 154:2.5*

Universe difficulties must be met and planetary obstacles must be encountered as a part of the experience training provided for the growth and development, the progressive perfection, of the evolving souls of mortal creatures. The spiritualization of the human soul requires intimate experience with the educational solving of a wide range of real universe problems. **The animal nature and the lower forms of will creatures do not progress favorably in environmental ease.** Problematic situations, coupled with exertion stimuli, conspire to produce those activities of mind, soul, and spirit which contribute mightily to the achievement of worthy goals of mortal progression and to the attainment of higher levels of spirit destiny.

6. THE FATHER'S PRIMACY

Spiritual Wisdom—Divine Destinies to Human Origins Perspective

Why was the Central and Superuniverses revealed in Part I first? Five Papers shed light on this.

Page 53:1   *Paper 3:6.3*

All religious philosophy, sooner or later, arrives at the concept of unified universe rule, of one God. Universe causes cannot be lower than universe effects. The source of the streams of universe life and of the cosmic mind must be above the levels of their manifestation. The human mind cannot be consistently explained in terms of the lower orders of existence. Man's mind can be truly comprehended only by recognizing the reality of higher orders of thought and purposive will. Man as a moral being is inexplicable unless the reality of the Universal Father is acknowledged.

Page 164:3   *Paper 15:0.3*

Of the vast body of knowledge concerning the superuniverses, I can hope to tell you little, but there is operative throughout these realms a technique of intelligent control for both physical and spiritual forces, and the universal gravity presences there function in majestic power and perfect harmony. It is important first to gain an adequate idea of the physical constitution and material organization of the superuniverse domains, for then you will be the better prepared to grasp the significance of the marvelous organization provided for their spiritual government and for the intellectual advancement of the will creatures who dwell on the myriads of inhabited planets scattered hither and yon throughout these seven superuniverses.

Page 215:2   *Paper 19:1.5*

For example: The human mind would ordinarily crave to approach the cosmic philosophy portrayed in these revelations by proceeding from the simple and the finite to the complex and the infinite, from human origins to divine destinies. But that path does not lead to *spiritual wisdom.* Such a procedure is the easiest path to a certain form of *genetic knowledge,* but at best it can only reveal man's origin; it reveals little or nothing about his divine destiny.

Page 1097:5   *Paper 100:4.1*

Religious living is devoted living, and devoted living is creative living, original and spontaneous. New religious insights arise out of conflicts which initiate the choosing of new and better reaction habits in the place of older and inferior reaction patterns. **New meanings only emerge amid conflict; and conflict persists only in the face of refusal to espouse the higher values connoted in superior meanings.**

Page 1162:1   *Paper 106:0.1*

It is not enough that the ascending mortal should know something of the relations of Deity to the genesis and manifestations of cosmic reality; he should also comprehend something of the relationships existing between himself and the numerous levels of existential and experiential realities, of potential and actual realities. Man's terrestrial orientation, his cosmic insight, and his spiritual directionization are all enhanced by a better comprehension of universe realities and their techniques of interassociation, integration, and unification.

PAPER 4: GOD'S RELATION TO THE UNIVERSE

---

2. GOD AND NATURE

God—No Respecter of Person or Planet

Page 56:5   *Paper 4:2.1*

See "God—No Respecter of Person or Planet" on page 130 for full text references. (3:1.12)

Revelation—Substitute for Morontia Mota

Papers 4 and 102 remind us that revelation is substitute for morontia mota.

Page 57:4   *Paper 4:2.7*

The apparent defects of the natural world are not indicative of any such corresponding defects in the character of God. Rather are such observed imperfections merely the inevitable stop-moments in the exhibition of the ever-moving reel of infinity picturization. **It is these very defect-interruptions of perfection-continuity which make it possible for the finite mind of material man to catch a fleeting glimpse of divine reality in time and space.** The material manifestations of divinity appear defective to the evolutionary mind of man only because mortal man persists in viewing the phenomena of nature through natural eyes, **human vision unaided by morontia mota or by revelation, its compensatory substitute on the worlds of time.**

Page 1122:1   *Paper 102:3.5*

Science, knowledge, leads to *fact* consciousness; religion, experience, leads to *value* consciousness; philosophy, wisdom, leads to *co-ordinate* consciousness; **revelation (the substitute for morontia mota) leads to the consciousness of *true reality;* while the co-ordination of the consciousness of fact, value, and true reality constitutes awareness of personality reality, maximum of being, together with the belief in the possibility of the survival of that very personality.**

---

3. God's Unchanging Character

Goodness and Evil
Goodness and evil during our ascension career.

Page 58:4   *Paper 4:3.6*

The infinite goodness of the Father is beyond the comprehension of the finite mind of time; **hence must there always be afforded a contrast with comparative evil (not sin) for the effective exhibition of all phases of relative goodness.** Perfection of divine goodness can be discerned by mortal imperfection of insight only because it stands in contrastive association with relative imperfection in the relationships of time and matter in the motions of space.

Page 342:4   *Paper 30:4.14*

Before departing from their native local universes for the superuniverse receiving worlds, the mortals of time are **recipients of spirit confirmation from the Creator Son and the local universe Mother Spirit.** From this point on, the status of the ascending mortal is forever settled. Superuniverse wards have never been known to go astray. Ascending seraphim are also advanced in angelic standing at the time of their departure from the local universes.

Page 1458:2–5   *Paper 132:2.5–8*

Goodness is always growing toward new levels of the increasing liberty of moral self-realization and spiritual personality attainment—the discovery of, and identification with, the indwelling Adjuster. **An experience is good when** it heightens the appreciation of beauty, augments the moral will, **enhances the discernment of truth,** enlarges the capacity to love and serve one's fellows, exalts the spiritual ideals, and unifies the supreme human motives of time with the eternal plans of the indwelling Adjuster, all of which lead directly to an increased desire to do the Father's will, thereby fostering the divine passion to find God and to be more like him.

As you ascend the universe scale of creature development, you will find increasing goodness and diminishing evil in perfect accordance with your **capacity for goodness-experience and truth-discernment.** The ability to entertain error or experience evil will **not be fully lost until the ascending human soul achieves final spirit levels.**

Goodness is living, relative, always progressing, invariably a personal experience, and everlastingly correlated with the discernment of truth and beauty. Goodness is found in the recognition of the positive truth-values of the spiritual level, which must, in human experience, be contrasted with the negative counterpart—the shadows of potential evil.

Until you attain Paradise levels, goodness will always be more of a quest than a possession, more of a goal than an experience of attainment. But even as you hunger and thirst for righteousness, you experience increasing satisfaction in the partial attainment of goodness. **The presence of goodness and evil in the world is in itself positive proof of the existence and reality of man's moral will,** the personality, which thus identifies these values and is also able to choose between them.

---

4. The Realization of God

Personal God

Page 59:2   *Paper 4:4.6*

See "Personal God" on page 117 for full text references. (1:7.3)

---

5. Erroneous Ideas of God

Atonement Doctrine

Page 60:5–6   *Paper 4:5.6–7*

See "Atonement Doctrine" on page 123 for full text references. (2:6.5)

PAPER 5: GOD'S RELATION TO THE INDIVIDUAL

Communion with Indwelling Presence
Attempted communion discussed twice within couple of pages of each other.

Page 62:1   *Paper 5:0.1*

If the finite mind of man is unable to comprehend how so great and so majestic a God as the Universal Father can descend from his eternal abode in infinite perfection to fraternize with the individual human creature, then must such a finite intellect rest assurance of divine fellowship upon the truth of the fact that an actual fragment of the living God resides within the intellect of every normal-minded and morally conscious Urantia mortal. The indwelling Thought Adjusters are a part of the eternal Deity of the Paradise Father. **Man does not have to go farther than his own inner experience of the soul's contemplation of this spiritual-reality presence to find God and attempt communion with him.**

Page 64:6   *Paper 5:2.3*

The divine presence cannot, however, be discovered anywhere in nature or even in the lives of God-knowing mortals so fully and so certainly as in your attempted communion with the indwelling Mystery Monitor, the Paradise Thought Adjuster. What a mistake to dream of God far off in the skies when the spirit of the Universal Father lives within your own mind!

Father Contact

Page 62:2   *Paper 5:0.2*

See "Father Contact" on page 114 for full text references. (1:7.1)

1. THE APPROACH TO GOD

Spiritual Endowment—Uniform and Unique
Pages 63 and 69 point out on three occasions how we may differ in a number of endowments, but our spiritual endowment is uniform.

Page 63:2–3   *Paper 5:1.4–5*

The mortals of the realms of time and space may differ greatly in innate abilities and intellectual endowment, they may enjoy environments exceptionally favorable to social advancement and moral progress, or they may suffer from the lack of almost every human aid to culture and supposed advancement in the arts of civilization; **but the possibilities for spiritual progress in the ascension career are equal to all;** increasing levels of spiritual insight and cosmic meanings are attained quite independently of all such sociomoral differentials of the diversified material environments on the evolutionary worlds.

However Urantia mortals may differ in their intellectual, social, economic, and even moral opportunities and endowments, **forget not that their spiritual endowment is uniform and unique.** They all enjoy the same divine presence of the gift from the Father, and they are all equally privileged to seek intimate personal communion with this indwelling spirit of divine origin, while they may all equally choose to accept the uniform spiritual leading of these Mystery Monitors.

Page 69:8   *Paper 5:5.13*

Eternal survival of personality is wholly dependent on the choosing of the mortal mind, whose decisions determine the survival potential of the immortal soul. When the mind believes God and the soul knows God, and when, with the fostering Adjuster, they all *desire* God, then is survival assured. **Limitations of intellect, curtailment of education, deprivation of culture, impoverishment of social status, even inferiority of the human standards of morality resulting from the unfortunate lack of educational, cultural, and social advantages, cannot invalidate the presence of the divine spirit in such unfortunate and humanly handicapped but believing individuals.** The indwelling of the Mystery Monitor constitutes the inception and insures the possibility of the potential of growth and survival of the immortal soul.

Irrevocable Choice
Irrevocable choice in Papers 5 and 112—soul death and what our Mystery Monitor and man offer each other.

Page 64:2   *Paper 5:1.11*

Mortal man may draw near God and may repeatedly forsake the divine will so long as the power of choice remains. Man's final doom is not sealed until he has lost the power to choose the Father's will. There is never a closure of the Father's heart to the need and the petition of his children. Only do his offspring close their hearts forever to the Father's drawing power when they finally and forever **lose the desire** to do his divine will—to know him and to be like him. **Likewise is man's eternal destiny assured when Adjuster fusion proclaims to the universe that such an ascender has made the final and irrevocable choice to live the Father's will.**

Page 1182:4   *Paper 107:6.2*

The Adjuster is man's eternity possibility; man is the Adjuster's personality possibility. Your individual Adjusters work to spiritize you in the hope of eternalizing your temporal identity. The Adjusters are saturated with the beautiful and self-bestowing love of the Father of spirits. They truly and divinely love you; they are the prisoners of spirit hope confined within the minds of men. They long for the divinity attainment of your mortal minds that their loneliness may end, that they may be delivered with you from the limitations of material investiture and the habiliments of time.

Page 1229:9   *Paper 112:3.2*

1. *Spiritual (soul) death.* If and when mortal man has **finally rejected survival,** when he has been pronounced spiritually insolvent, morontially bankrupt, in the conjoint opinion of the Adjuster and the surviving seraphim, when such co-ordinate advice has been recorded on Uversa, and after the Censors and their reflective associates have verified these findings, thereupon do the rulers of Orvonton order the immediate release of the indwelling Monitor. But this release of the Adjuster in no way affects the duties of the personal or group seraphim concerned with that Adjuster-abandoned individual. This kind of death is final in its significance irrespective of the temporary continuation of the living energies of the physical and mind mechanisms. From the cosmic standpoint the mortal is already dead; the continuing life merely indicates the persistence of the material momentum of cosmic energies.

Page 1237:7   *Paper 112:7.5*

Fusion with the Adjuster never occurs until the mandates of the superuniverse have pronounced that the human nature has made a final and irrevocable choice for the eternal career. This is the at-onement authorization, which, when issued, constitutes the clearance authority for the fused personality eventually to leave the confines of the local universe to proceed sometime to the headquarters of the superuniverse, from which point the pilgrim of time will, in the distant future, enseconaphim for the long flight to the central universe of Havona and the Deity adventure.

---

2. The Presence of God

Father Contact

Page 64:5   *Paper 5:2.2*

See "Father Contact" on page 114 for full text references. (1:7.1)

Communion with Indwelling Presence

Page 64:6   *Paper 5:2.3*

See "Communion with Indwelling Presence" on page 142 for full text references. (5:0.1)

3. TRUE WORSHIP

Ministering Spirits from the Conjoint Actor— Relation with Humans

The nuts and bolts on the help we get from the children of the Third Source and Center, and how it becomes personally meaningful to us—from Parts I and III.

Page 66:1   *Paper 5:3.5*

When you deal with the practical affairs of your daily life, you are in the hands of the spirit personalities having origin in the Third Source and Center; you are co-operating with the agencies of the Conjoint Actor. And so it is: You worship God; pray to, and commune with, the Son; and work out the details of your earthly sojourn in connection with the intelligences of the Infinite Spirit operating on your world and throughout your universe.

Page 425:1   *Paper 38:9.9*

The gap between the material and spiritual worlds is perfectly bridged by the serial association of mortal man, secondary midwayer, primary midwayer, morontia cherubim, mid-phase cherubim, and seraphim. **In the personal experience of an individual mortal these diverse levels are undoubtedly more or less unified and made personally meaningful by the unobserved and mysterious operations of the divine Thought Adjuster.**

Page 1244:2   *Paper 113:3.1*

One of the most important things a destiny guardian does for her mortal subject is to effect a personal co-ordination of the numerous impersonal spirit influences which indwell, surround, and impinge upon the mind and soul of the evolving material creature. Human beings are personalities, and it is exceedingly difficult for nonpersonal spirits and prepersonal entities to make direct contact with such highly material and discretely personal minds. In the ministry of the guarding angel all of **these influences are more or less unified and made more nearly appreciable by the expanding moral nature of the evolving human personality.**

Soul

Page 66:4   *Paper 5:3.8*

> See "Soul" on page 104 for full text references. (0:5.10)

5. THE CONSCIOUSNESS OF GOD

Spiritual Endowment—Uniform and Unique

Page 69:8   *Paper 5:5.13*

> "Spiritual Endowment—Uniform and Unique" on page 142 (5:1.4)

6. THE GOD OF PERSONALITY

Personality—Unsolved Mystery
An unsolved mystery designed by the Universal Father.

Page 70:3   *Paper 5:6.2*

Personality is one of the unsolved mysteries of the universes. We are able to form adequate concepts of the factors entering into the make-up of various orders and levels of personality, but we do not fully comprehend the real nature of the personality itself. We clearly perceive the numerous factors which, when put together, constitute the vehicle for human personality, but we do not fully comprehend the nature and significance of such a finite personality.

Page 236:1–3   *Paper 21:2.3–5*

The departure of a Michael Son on this occasion forever liberates his creator prerogatives from the Paradise Sources and Centers, subject only to certain limitations inherent in the pre-existence of these Sources and Centers and to certain other antecedent powers and presences. **Among these limitations to the otherwise all-powerful creator prerogatives of a local universe Father are the following:**

1. *Energy-matter* is dominated by the Infinite Spirit. Before any new forms of things, great or small, may be created, before any new transformations of energy-matter may be attempted, a Creator Son must secure the consent and working co-operation of the Infinite Spirit.

2. *Creature designs and types* are controlled by the Eternal Son. Before a Creator Son may engage in the creation of any new type of being, any new design of creature, he must secure the consent of the Eternal and Original Mother Son.
3. *Personality* is designed and bestowed by the Universal Father.

Personality and Personality Circuit

Page 71:1   *Paper 5:6.6*

See "Personality and Personality Circuit" on page 120 for full text references. (2:2.5–6)

Human Will—Sovereignty
The sovereignty of our mortal will confirmed in Parts I and III.

Page 71:3   *Paper 5:6.8*

Having thus provided for the growth of the immortal soul and having liberated man's inner self from the fetters of absolute dependence on antecedent causation, the Father stands aside. Now, man having thus been liberated from the fetters of causation response, at least as pertains to eternal destiny, and provision having been made for the growth of the immortal self, the soul, it remains for man himself to will the creation or to inhibit the creation of this surviving and eternal self which is his for the choosing. No other being, force, creator, or agency in all the wide universe of universes can interfere to any degree with the absolute sovereignty of the mortal free will, as it operates within the realms of choice, regarding the eternal destiny of the personality of the choosing mortal. **As pertains to eternal survival, God has decreed the sovereignty of the material and mortal will, and that decree is absolute.**

Page 1245:7   *Paper 113:5.1*

Angels do not invade the sanctity of the human mind; they do not manipulate the will of mortals; neither do they directly contact with the indwelling Adjusters. The guardian of destiny influences you in every possible manner consistent with the dignity of your personality; **under no circumstances do these angels interfere with the free action of the human will.** Neither angels nor any other order of universe personality have power or authority to curtail or abridge the prerogatives of human choosing.

Personality

A little about personality in these three papers.

Page 71:5   *Paper 5:6.10*

The **personality circuit** of the universe of universes is centered in the person of the Universal Father, and the **Paradise Father is personally conscious of, and in personal touch with, all personalities** of all levels of self-conscious existence. And **this personality consciousness of all creation exists independently of the mission of the Thought Adjusters.**

Page 1220:5   *Paper 111:4.6*

Snow crystals are always hexagonal in form, but no two are ever alike. Children conform to types, but no two are exactly alike, even in the case of twins. **Personality follows types but is always unique.**

Page 1225:2–16   *Paper 112:0.2–16*

While it would be presumptuous to attempt the definition of personality, it may prove helpful to recount some of the things which are known about personality:

1. Personality is that quality in reality which is bestowed by the Universal Father himself or by the Conjoint Actor, acting for the Father.
2. It may be bestowed upon any living energy system which includes mind or spirit.
3. **It is not wholly subject to the fetters of antecedent causation. It is relatively creative or cocreative.**
4. When bestowed upon evolutionary material creatures, it causes spirit to strive for the mastery of energy-matter through the mediation of mind.
5. Personality, while devoid of identity, can unify the identity of any living energy system.
6. It discloses only qualitative response to the personality circuit in contradistinction to the three energies which show both qualitative and quantitative response to gravity.
7. Personality is changeless in the presence of change.
8. **It can make a gift to God—dedication of the free will to the doing of the will of God.**
9. It is characterized by morality—awareness of relativity of relationship with other persons. It discerns conduct levels and choosingly discriminates between them.

10. Personality is unique, absolutely unique: It is unique in time and space; it is unique in eternity and on Paradise; it is unique when bestowed—there are no duplicates; it is unique during every moment of existence; it is unique in relation to God—he is no respecter of persons, but neither does he add them together, for they are nonaddable—they are associable but nontotalable.

11. Personality responds directly to other-personality presence.

12. It is one thing which can be added to spirit, thus illustrating the primacy of the Father in relation to the Son. (Mind does not have to be added to spirit.)

13. Personality may survive mortal death with identity in the surviving soul. The Adjuster and the personality are changeless; the relationship between them (in the soul) is nothing but change, continuing evolution; and if this change (growth) ceased, the soul would cease.

14. Personality is uniquely conscious of time, and this is something other than the time perception of mind or spirit.

Personality Circuit as Divine Love

The personality circuit described as the great circuit of divine love. Three papers over a thousand pages apart talk about this great circuit of love.

Page 71:7   *Paper 5:6.12*

Concerning those personalities who are not Adjuster indwelt: The attribute of choice-liberty is also bestowed by the Universal Father, **and such persons are likewise embraced in the great circuit of divine love, the personality circuit of the Universal Father.** God provides for the sovereign choice of all true personalities. No personal creature can be coerced into the eternal adventure; the portal of eternity opens only in response to the freewill choice of the freewill sons of the God of free will.

Page 640:2   *Paper 56:4.3*

Notwithstanding that God is manifest from the domains of the Sevenfold up through supremacy and ultimacy to God the Absolute, **the personality circuit, centering on Paradise and in the person of God the Father,** provides for the complete and perfect unification of all these diverse expressions of divine personality so far as concerns all creature personalities on all levels of intelligent existence and in all the realms of the perfect, perfected, and perfecting universes.

Page 647:8   *Paper 56:10.17*

Universal beauty is the recognition of the reflection of the Isle of Paradise in the material creation, while eternal truth is the special ministry of the Paradise Sons who not only bestow themselves upon the mortal races but even pour out their Spirit of Truth upon all peoples. Divine goodness is more fully shown forth in the loving ministry of the manifold personalities of the Infinite Spirit. **But love, the sum total of these three qualities, is man's perception of God as his spirit Father.**

Page 1289:3   *Paper 117:6.10*

All true love is from God, and man receives the divine affection as he himself bestows this love upon his fellows. Love is dynamic. It can never be captured; it is alive, free, thrilling, and always moving. Man can never take the love of the Father and imprison it within his heart. The Father's love can become real to mortal man only by passing through that man's personality as he in turn bestows this love upon his fellows. **The great circuit of love is from the Father,** through sons to brothers, and hence to the Supreme. The love of the Father appears in the mortal personality by the ministry of the indwelling Adjuster. Such a God-knowing son reveals this love to his universe brethren, and this fraternal affection is the essence of the love of the Supreme.

PAPER 6: THE ETERNAL SON

---

1. IDENTITY OF THE ETERNAL SON

Father-Infinite

Page 74:1   *Paper 6:1.2*

See "Father-Infinite" on page 103 for full text references. (0:3.17)

---

3. MINISTRY OF THE FATHER'S LOVE

Love of the Creator Son

There is a difference between the love of the Father and the love of the Son. Six papers on this, and which one our Creator Son more resembles.

Page 75:10   *Paper 6:3.5*

God is love, the Son is mercy. Mercy is applied love, the Father's love in action in the person of his Eternal Son. The love of this universal Son is likewise universal. **As love is comprehended on a sex planet, the love of God is**

more comparable to the love of a father, while the love of the Eternal Son is more like the affection of a mother. Crude, indeed, are such illustrations, but I employ them in the hope of conveying to the human mind the thought that **there is a difference,** not in divine content but in quality and technique of expression, between the love of the Father and the love of the Son.

Page 94:4   *Paper 8:4.2*

God is love, the Son is mercy, the Spirit is ministry—the ministry of divine love and endless mercy to all intelligent creation. The Spirit is the personification of the Father's love and the Son's mercy; in him are they eternally united for universal service. **The Spirit is** *love applied* to the creature creation, the combined love of the Father and the Son.

Page 367:3   *Paper 33:2.1*

Observation of Creator Sons discloses that some resemble more the Father, some the Son, while others are a blend of both their infinite parents. **Our Creator Son very definitely manifests traits and attributes which more resemble the Eternal Son.**

---

4. ATTRIBUTES OF THE ETERNAL SON

Father Contact

Page 76:5   *Paper 6:4.5*

See "Father Contact" on page 114 for full text references. (1:7.1)

Personality and Personality Circuit

Page 76:5   *Paper 6:4.5*

See "Personality and Personality Circuit" on page 120 for full text references. (2:2.5–6)

God and Son Knows All

Page 76:7   *Paper 6:4.7*

Page 77:3   *Paper 6:4.10*

See "God and Son Knows All" on page 135 for full text references. (3:3.2)

PAPER 7: RELATION OF THE ETERNAL SON TO THE UNIVERSE

---

1. THE SPIRIT-GRAVITY CIRCUIT

Relativity
Keeping the concept of relativity in perspective.

Page 82:1   *Paper 7:1.2*

This gravity control of spiritual things operates independently of time and space; therefore is spirit energy undiminished in transmission. Spirit gravity never suffers time delays, nor does it undergo space diminution. It does not decrease in accordance with the square of the distance of its transmission; the circuits of pure spirit power are not retarded by the mass of the material creation. And this transcendence of time and space by pure spirit energies is inherent in the absoluteness of the Son; it is not due to the interposition of the antigravity forces of the Third Source and Center.

Page 474:1   *Paper 42:4.11*

The increase of mass in matter is equal to the increase of energy divided by the square of the velocity of light. **In a dynamic sense the work which resting matter can perform is equal to the energy expended in bringing its parts together from Paradise minus the resistance of the forces overcome in transit and the attraction exerted by the parts of matter on one another.**

Page 1436:1   *Paper 130:4.15*

All static, dead, concepts are potentially evil. The finite shadow of relative and living truth is continually moving. Static concepts invariably retard science, politics, society, and religion. Static concepts may represent a certain knowledge, but they are deficient in wisdom and devoid of truth. **But do not permit the concept of relativity so to mislead you that you fail to recognize the co-ordination of the universe under the guidance of the cosmic mind, and its stabilized control by the energy and spirit of the Supreme.**

Page 2078:8   *Paper 195:7.5*

The realities and values of spiritual progress are not a "psychologic projection"—a mere glorified daydream of the material mind. Such things are the spiritual forecasts of the indwelling Adjuster, the spirit of God living in the mind of man. **And let not your dabblings with the faintly glimpsed findings of "relativity" disturb your concepts of the eternity and infinity of**

God. And in all your solicitation concerning the necessity for *self-expression* do not make the mistake of failing to provide for *Adjuster-expression*, the manifestation of your real and better self.

---

4. THE DIVINE PERFECTION PLANS

Bestowal Plan of Paradise Sons

Bestowal Plan of Paradise Sons—one subject to will of the Father, the other to the will of the Eternal Son.

Page 85:6   *Paper 7:4.5*

2. *The Bestowal Plan.* The next universal plan is the great Father-revelation enterprise of the Eternal Son and his co-ordinate Sons. **This is the proposal of the Eternal Son and consists of his bestowal of the Sons of God upon the evolutionary creations, there to personalize and factualize, to incarnate and make real, the love of the Father and the mercy of the Son to the creatures of all universes.** Inherent in the bestowal plan, and as a provisional feature of this ministration of love, the Paradise Sons act as rehabilitators of that which misguided creature will has placed in spiritual jeopardy. Whenever and wherever there occurs a delay in the functioning of the attainment plan, if rebellion, perchance, should mar or complicate this enterprise, then do the emergency provisions of the bestowal plan become active forthwith. The Paradise Sons stand pledged and ready to function as retrievers, to go into the very realms of rebellion and there restore the spiritual status of the spheres. And such a heroic service a co-ordinate Creator Son did perform on Urantia in connection with his experiential bestowal career of sovereignty acquirement.

Page 229:4   *Paper 20:6.5*

The mortal-bestowal careers of the Michaels and the Avonals, while comparable in most respects, are not identical in all: Never does a Magisterial Son proclaim, "Whosoever has seen the Son has seen the Father," as did your Creator Son when on Urantia and in the flesh. But a bestowed Avonal does declare, "Whosoever has seen me has seen the Eternal Son of God." The Magisterial Sons are not of immediate descent from the Universal Father, nor do they incarnate subject to the Father's will; **always do they bestow themselves as Paradise *Sons* subject to the will of the Eternal Son of Paradise.**

5. The Spirit of Bestowal

Bestowals of the Second Source and Center and the Original Michael

The bestowals of the Second Source and center and the Original Michael.

Page 86:6   *Paper 7:5.5*

Long, long ago the **Eternal Son bestowed himself upon each of the circuits of the central creation** for the enlightenment and advancement of all the inhabitants and pilgrims of Havona, including the ascending pilgrims of time. **On none of these seven bestowals did he function as either an ascender or a Havoner.** He existed as himself. His experience was unique; it was not *with* or *as* a human or other pilgrim but in some way associative in the superpersonal sense.

Page 87:2   *Paper 7:5.8*

Whatever our difficulty in comprehending the bestowals of the Second Person of Deity, we do comprehend the **Havona bestowal of a Son of the Eternal Son,** who literally passed through the circuits of the central universe and actually shared those experiences which constitute an ascender's preparation for Deity attainment. This was **the original Michael, the first-born Creator Son, and he passed through the life experiences of the ascending pilgrims from circuit to circuit, personally journeying a stage of each circle with them in the days of Grandfanda,** the first of all mortals to attain Havona.

Page 234:4   *Paper 21:0.4*

The original or first-born Michael has never experienced incarnation as a material being, but seven times he passed through the experience of spiritual creature ascent on the seven circuits of Havona, advancing from the outer spheres to the innermost circuit of the central creation. The order of Michael knows the grand universe from one end to the other; there is no essential experience of any of the children of time and space in which the Michaels have not personally participated; they are in fact partakers not only of the divine nature but also of your nature, meaning all natures, from the highest to the lowest.

Page 270:2   *Paper 24:6.4*

The number of Graduate Guides is beyond the power of human minds to grasp, and they continue to appear. Their origin is something of a mystery. They have not existed from eternity; they mysteriously appear as they are needed. **There is no record of a Graduate Guide in all the realms of the central universe until that far-distant day when the first mortal pilgrim of all time made his way to the outer belt of the central creation.** The instant he arrived on the pilot world of the outer circuit, he was met with friendly greetings by Malvorian, the first of the Graduate Guides and now the chief of their supreme council and the director of their vast educational organization.

Page 270:6   *Paper 24:6.8*

The name of this pilgrim discoverer of Havona is *Grandfanda,* **and he hailed from planet 341 of system 84 in constellation 62 of local universe 1,131 situated in superuniverse number one.** His arrival was the signal for the establishment of the broadcast service of the universe of universes. Theretofore only the broadcasts of the superuniverses and the local universes had been in operation, but the announcement of the arrival of Grandfanda at the portals of Havona signalized the inauguration of the "space reports of glory," so named because the initial universe broadcast reported the Havona arrival of the first of the evolutionary beings to attain entrance upon the goal of ascendant existence.

---

6. *THE PARADISE SONS OF GOD*

Magisterial Sons—Judges of Survival

Magisterial Sons have earned the right to judge the dead but do not execute such judgement.

Page 88:3   *Paper 7:6.5*

Much as the Creator Sons are personalized by the Father and the Son, so are the *Magisterial Sons* personalized by the Son and the Spirit. These are the Sons who, in the experiences of creature incarnation, **earn the right to serve as the judges of survival** in the creations of time and space.

Page 210:1   *Paper 18:3.7*

In power, scope of authority, and extent of jurisdiction the **Ancients of Days** are the most powerful and mighty of any of the direct rulers of the time-space creations. In all the vast universe of universes **they alone are invested with the high powers of final executive judgment concerning the eternal extinction of will creatures.** And all three Ancients of Days must participate in the final decrees of the supreme tribunal of a superuniverse.

Page 226:2   *Paper 20:3.2*

When they sit in judgment on the destinies of an age, the **Avonals** decree the fate of the evolutionary races, but **though they may render judgments extinguishing the identity of personal creatures, they do not execute such sentences.** Verdicts of this nature are executed by none but the authorities of a superuniverse.

Page 231:6   *Paper 20:9.2*

Trinity Teacher Sons have nothing to do with terminating planetary dispensations. **They neither judge the dead nor translate the living,** but on each planetary mission they are **accompanied by a Magisterial Son who performs these services.** Teacher Sons are wholly concerned with the initiation of a spiritual age, with the dawn of the era of spiritual realities on an evolutionary planet. They make real the spiritual counterparts of material knowledge and temporal wisdom.

Page 396:2   *Paper 36:1.1*

Though the Life Carriers belong to the family of divine sonship, they are a peculiar and distinct type of universe Sons, being the only group of intelligent life in a local universe in whose creation the rulers of a superuniverse participate. The Life Carriers are the offspring of three pre-existent personalities: the Creator Son, the Universe Mother Spirit, and, by designation, one of the three Ancients of Days presiding over the destinies of the superuniverse concerned. **These Ancients of Days, who alone can decree the extinction of intelligent life, participate in the creation of the Life Carriers, who are intrusted with establishing physical life on the evolving worlds.**

PAPER 8: THE INFINITE SPIRIT

---

3. RELATION OF THE SPIRIT TO THE FATHER AND THE SON

Creator Son and Creative Spirit

Paper 8 below portrays a Creator Son of the Eternal Son; the other two papers elaborate on the Father's participation.

Page 93:6   *Paper 8:3.4*

A Creator Son of the Eternal Son and a Creative Spirit of the Infinite Spirit created you and your universe; and while the Father in faithfulness upholds that which they have organized, it devolves upon this Universe Son and this Universe Spirit to foster and sustain their work as well as to minister to the creatures of their own making.

Page 235:2   *Paper 21:1.3*

The divine natures of these Creator Sons are, in principle, derived equally from the attributes of both Paradise parents. All partake of the fullness of the divine nature of the Universal Father and of the creative prerogatives of the Eternal Son, but as we observe the practical outworking of the Michael functions in the universes, we discern apparent differences. Some Creator Sons appear to be more like God the Father; others more like God the Son. For example: The trend of administration in the universe of Nebadon suggests that its Creator and ruling Son is one whose nature and character more resemble that of the Eternal Mother Son. It should be further stated that some universes are presided over by Paradise Michaels who appear equally to resemble God the Father and God the Son. And these observations are in no sense implied criticisms; they are simply a recording of fact.

Page 374:1   *Paper 34:0.1*

When a Creator Son is personalized by the Universal Father and the Eternal Son, then does the Infinite Spirit individualize a new and unique representation of himself to accompany this Creator Son to the realms of space, there to be his companion, first, in physical organization and, later, in creation and ministry to the creatures of the newly projected universe.

4. THE SPIRIT OF DIVINE MINISTRY

Love of the Creator Son

Page 94:4   *Paper 8:4.2*

> See "Love of the Creator Son" on page 150 for full text references. (6:3.5)

PAPER 9: RELATION OF THE INFINITE SPIRIT TO THE UNIVERSE

2. THE OMNIPRESENT SPIRIT

Father Contact

Page 100:7   *Paper 9:2.5*

> See "Father Contact" on page 114 for full text references. (1:7.1)

4. THE ABSOLUTE MIND

Pattern and Space
An idea pattern doesn't contain space, but does it occupy space?

Page 102:4   *Paper 9:4.4*

Infinite mind ignores time, ultimate mind transcends time, **cosmic mind is conditioned by time.** And so with space: The Infinite Mind is independent of space, but as descent is made from the infinite to the adjutant levels of mind, **intellect must increasingly reckon with the fact and limitations of space.**

Page 1297:8   *Paper 118:3.7*

All patterns of reality occupy space on the material levels, but spirit patterns only exist in relation to space; they do not occupy or displace space, neither do they contain it. But to us the master riddle of space pertains to the pattern of an idea. When we enter the mind domain, we encounter many a puzzle. **Does the pattern—the reality—of an idea occupy space? We really do not know,** albeit we are sure that an idea pattern does not contain space. But it would hardly be safe to postulate that the immaterial is always nonspatial.

5. THE MINISTRY OF MIND

Mind

Page 103:1   *Paper 9:5.3*

See "Mind" on page 106 for full text references. (0:6.8)

Impersonal Entities, Mind, and Mortal Will

Impersonal entities and the impersonal mind circuit in conjunction with mortal will.

Page 103:2   *Paper 9:5.4*

Since the Third Person of Deity is the source of mind, it is quite natural that the evolutionary will creatures find it easier to form comprehensible concepts of the Infinite Spirit than they do of either the Eternal Son or the Universal Father. The reality of the Conjoint Creator is disclosed imperfectly in the very existence of human mind. **The Conjoint Creator is the ancestor of the cosmic mind, and the mind of man is an individualized circuit, an impersonal portion, of that cosmic mind as it is bestowed in a local universe by a Creative Daughter of the Third Source and Center.**

Page 203:3   *Paper 17:5.5*

The Circuit Spirits are related to the native inhabitants of Havona much as the Thought Adjusters are related to the mortal creatures inhabiting the worlds of the evolutionary universes. **Like the Thought Adjusters, the Circuit Spirits are impersonal, and they consort with the perfect minds of Havona beings much as the impersonal spirits of the Universal Father indwell the finite minds of mortal men.** But the Spirits of the Circuits never become a permanent part of Havona personalities.

Page 1183:6   *Paper 107:7.4*

Why then, if Thought Adjusters possess volition, **are they subservient to the mortal will?** We believe it is because Adjuster volition, though absolute in nature, is prepersonal in manifestation. **Human will functions on the personality level of universe reality, and throughout the cosmos the impersonal**—the nonpersonal, the subpersonal, and the prepersonal—is ever responsive to the will and acts of existent personality.

Page 1203:2   *Paper 110:0.2*

As far as I am conversant with the affairs of a universe, I regard the love and devotion of a Thought Adjuster as the most truly divine affection in all creation. The love of the Sons in their ministry to the races is superb, but the devotion of an Adjuster to the individual is touchingly sublime, divinely Fatherlike. The Paradise Father has apparently reserved this form of personal contact with his individual creatures as an exclusive Creator prerogative. **And there is nothing in all the universe of universes exactly comparable to the marvelous ministry of these impersonal entities that so fascinatingly indwell the children of the evolutionary planets.**

---

6. THE MIND-GRAVITY CIRCUIT

Personality and Personality Circuit

Page 104:2   *Paper 9:6.4*

See "Personality and Personality Circuit" on page 120 for full text references. (2:2.5–6)

---

7. UNIVERSE REFLECTIVITY

Reflectivity

Is Paper 13 referring to Paper 9? An example of one paper's author directing us to a previous paper.

Page 105:1   *Paper 9:7.1*

The Conjoint Actor is able to co-ordinate all levels of universe actuality in such manner as to make possible the simultaneous recognition of the mental, the material, and the spiritual. This is the phenomenon of *universe reflectivity,* that unique and inexplicable power to see, hear, sense, and know all things as they transpire throughout a superuniverse, and to focalize, by reflectivity, all this information and knowledge at any desired point. The action of reflectivity is shown in perfection on each of the headquarters worlds of the seven superuniverses. It is also operative throughout all sectors of the superuniverses and within the boundaries of the local universes. Reflectivity finally focalizes on Paradise.

Page 105:5   *Paper 9:7.5*

Reflectivity appears to be omniscience within the limits of the experiential finite and may represent the emergence of the presence-consciousness of the Supreme Being. **If this assumption is true,** then the utilization of reflectivity in any of its phases is equivalent to partial contact with the consciousness of the Supreme.

Page 145:5   *Paper 13:1.10*

The secrets of Spiritington involve the impenetrable mysteries of reflectivity. We tell you of the vast and universal phenomenon of reflectivity, more particularly as it is operative on the headquarters worlds of the seven superuniverses, but we never fully explain this phenomenon, for we do not fully understand it. Much, very much, we do comprehend, but many basic details are still mysterious to us. Reflectivity is a secret of God the Spirit. **You have been instructed concerning reflectivity functions in relation to the ascension scheme of mortal survival, and it does so operate, but reflectivity is also an indispensable feature of the normal working of numerous other phases of universe occupation.** This endowment of the Infinite Spirit is also utilized in channels other than those of intelligence gathering and information dissemination. And there are other secrets of Spiritington.

---

8. PERSONALITIES OF THE INFINITE SPIRIT

Subjective Self-consciousness
Subjective experience expressed objectively—Parts I and IV.

Page 106:4   *Paper 9:8.6*

The Third Source and Center is represented in the grand universe by a vast array of ministering spirits, messengers, teachers, adjudicators, helpers, and advisers, together with supervisors of certain circuits of physical, morontial, and spiritual nature. Not all of these beings are personalities in the strict meaning of the term. Personality of the finite-creature variety is characterized by:

1. **Subjective self-consciousness.**
2. Objective response to the Father's personality circuit.

Page 1431:5   *Paper 130:2.10*

It was on this same day that we first heard that momentous truth which, stated in modern terms, would signify: **"Will is that manifestation of the human mind which enables the subjective consciousness to express itself** objectively and to experience the phenomenon of aspiring to be Godlike." And it is in this same sense that every reflective and spiritually minded human being can become *creative*.

Page 2095:5   *Paper 196:3.18*

The exquisite and transcendent experience of loving and being loved is not just a psychic illusion because it is so purely subjective. The one truly divine and objective reality that is associated with mortal beings, the Thought Adjuster, functions to human observation apparently as an exclusively subjective phenomenon. **Man's contact with the highest objective reality, God, is only through the purely subjective experience of knowing him, of worshiping him, of realizing sonship with him.**

PAPER 10: THE PARADISE TRINITY

---

1. SELF-DISTRIBUTION OF THE FIRST SOURCE AND CENTER

Father-Infinite

Page 109:2   *Paper 10:1.4*

See "Father-Infinite" on page 103 for full text references. (0:3.17)

---

2. DEITY PERSONALIZATION

Father-Infinite

Page 110:3   *Paper 10:2.5*

See "Father-Infinite" on page 103 for full text references. (0:3.17)

---

3. THE THREE PERSONS OF DEITY

Father-Infinite

Page 111:4   *Paper 10:3.6*

See "Father-Infinite" on page 103 for full text references. (0:3.17)

Father Contact

Page 111:7–14   *Paper 10:3.9–16*

See "Father Contact" on page 114 for full text references. (1:7.1)

Personality and Personality Circuit

Page 111:12–13   *Paper 10:3.14–15*

See "Personality and Personality Circuit" on page 120 for full text
references. (2:2.5–6)

5. Functions of the Trinity

Seven Triunities
The seven triunities in Papers 10 and 104.

Page 113:8   *Paper 10:5.7*

The Trinity Infinite involves the co-ordinate action of all triunity
relationships of the First Source and Center—undeified as well as deified—
and hence is very difficult for personalities to grasp. **In the contemplation of
the Trinity as infinite, do not ignore the seven triunities;** thereby certain
difficulties of understanding may be avoided, and certain paradoxes may be
partially resolved.

Page 1147:11, 1148:1–2   *Paper 104:4.1–3*

In attempting the description of seven triunities, attention is directed to
the fact that the Universal Father is the primal member of each. He is, was,
and ever will be: the First Universal Father-Source, Absolute Center, Primal
Cause, Universal Controller, Limitless Energizer, Original Unity,
Unqualified Upholder, First Person of Deity, Primal Cosmic Pattern, and
Essence of Infinity. The Universal Father is the personal cause of the
Absolutes; he is the absolute of Absolutes.

The nature and meaning of the seven triunities may be suggested as:
The First Triunity—the personal-purposive triunity. This is the grouping of
the three Deity personalities:
1. The Universal Father.
2. The Eternal Son.
3. The Infinite Spirit.

PAPER 12: THE UNIVERSE OF UNIVERSES

---

6. UNIVERSAL OVERCONTROL

God—Foreknowledge

Page 136:3   *Paper 12:6.5*

See "God—Foreknowledge" on page 133 for full text references.
(3:2.6)

---

8. MATTER, MIND, AND SPIRIT

Spirit Beings Live on Physical Spheres
Spirit beings live on worlds just as real as ours.

Page 139:4   *Paper 12:8.1*

"God is spirit," but Paradise is not. The material universe is always the
arena wherein take place all spiritual activities; **spirit beings and spirit
ascenders live and work on physical spheres of material reality.**

Page 154:3   *Paper 14:2.1*

**Spirit beings do not dwell in nebulous space; they do not inhabit ethereal
worlds; they are domiciled on actual spheres of a material nature, worlds just
as real as those on which mortals live.** The Havona worlds are actual and
literal, albeit their literal substance differs from the material organization of
the planets of the seven superuniverses.

Supergravity Presence

Page 139:5   *Paper 12:8.2*

See "Supergravity Presence" on page 108 for full text references.
(0:11.7)

Mind

Page 140:7   *Paper 12:8.11*

See "Mind" on page 106 for full text references. (0:6.8)

1. THE SEVEN SACRED WORLDS OF THE FATHER

Reflectivity

Page 145:5   Paper 13:1.10

> See "Reflectivity" on page 160 for full text references. (9:7.1)

PAPER 14: THE CENTRAL AND DIVINE UNIVERSE

2. CONSTITUTION OF HAVONA

Spirit Beings Live on Physical Spheres

Page 154:3   Paper 14:2.1

> See "Spirit Beings Live on Physical Spheres" on page 164 for full text references. (12:8.1)

4. CREATURES OF THE CENTRAL UNIVERSE

Spiritual Economy
Spiritual economy and the economy of spiritual growth—Parts I and II.

Page 157:3   Paper 14:4.4

The natives of Havona live on the billion spheres of the central universe in the same sense that other orders of permanent citizenship dwell on their respective spheres of nativity. **As the material order of sonship carries on the material, intellectual, and spiritual economy of a billion local systems in a superuniverse,** so, in a larger sense, do the Havona natives live and function on the billion worlds of the central universe. You might possibly regard these Havoners as material creatures in the sense that the word "material" could be expanded to describe the physical realities of the divine universe.

Page 855:4   Paper 77:1.2

It is well always to bear in mind that the successive bestowals of the Sons of God on an evolving planet produce marked changes in the spiritual economy of the realm and sometimes so modify the workings of the interassociation of spiritual and material agencies on a planet as to create situations indeed difficult of understanding. The status of the one hundred corporeal members of Prince Caligastia's staff illustrates just such a unique interassociation: As ascendant morontia citizens of Jerusem they were supermaterial creatures without reproductive prerogatives. As descendant planetary ministers on Urantia they were material sex creatures capable of procreating material offspring (as some of them later did). What we cannot satisfactorily explain is how these one hundred could function in the parental role on a supermaterial level, but that is exactly what happened. A supermaterial (nonsexual) liaison of a male and a female member of the corporeal staff resulted in the appearance of the first-born of the primary midwayers.

Page 1095:3   *Paper 100:1.8*

Religious habits of thinking and acting are contributory to the economy of spiritual growth. One can develop religious predispositions toward favorable reaction to spiritual stimuli, a sort of conditioned spiritual reflex. Habits which favor religious growth embrace cultivated sensitivity to divine values, recognition of religious living in others, reflective meditation on cosmic meanings, worshipful problem solving, sharing one's spiritual life with one's fellows, avoidance of selfishness, refusal to presume on divine mercy, living as in the presence of God. The factors of religious growth may be intentional, but the growth itself is unvaryingly unconscious.

Page 1196:9   *Paper 109:2.7*

...a self-acting Adjuster is one who:

6. **Has served in a time of crisis in the experience of some human being who was the material complement of a spirit personality intrusted with the enactment of some cosmic achievement essential to the spiritual economy of the planet.**

Relationship Between the Supreme and Superuniverse Citizens

Relationship between the Supreme and Superuniverse Citizens is analogous to the relationship between Havona natives and the Paradise Trinity.

Page 157:5   *Paper 14:4.6*

As the worship of the faith sons of the evolutionary worlds ministers to the satisfaction of the Universal Father's love, so the exalted adoration of the Havona creatures satiates the perfect ideals of divine beauty and truth. As mortal man strives to do the will of God, these beings of the central universe live to gratify the ideals of the Paradise Trinity. In their very nature they *are* the will of God. Man rejoices in the goodness of God, Havoners exult in the divine beauty, while you both enjoy the ministry of the liberty of living truth.

Page 1292:10   *Paper 117:7.13*

It is possible that the Supreme may then be personally resident on Uversa, the headquarters of Orvonton, from which he will direct the administration of the time creations, but this is really only a conjecture. Certainly, though, the personality of the Supreme Being will be definitely contactable at some specific locality, although the ubiquity of his Deity presence will probably continue to permeate the universe of universes. **What the relation of the superuniverse citizens of that age will be to the Supreme we do not know, but it may be something like the present relationship between the Havona natives and the Paradise Trinity.**

---

5. LIFE IN HAVONA

Death
Why we die and what we have to look forward to.

Page 159:6   *Paper 14:5.10*

Love of adventure, curiosity, and dread of monotony—these traits inherent in evolving human nature—were not put there just to aggravate and annoy you during your short sojourn on earth, but rather to suggest to you that **death is only the beginning** of an endless career of adventure, an everlasting life of anticipation, an eternal voyage of discovery.

Page 364:6   *Paper 32:5.4*

To me it seems more fitting, for purposes of explanation to the mortal mind, to conceive of eternity as a cycle and the eternal purpose as an endless circle, a cycle of eternity in some way synchronized with the transient material cycles of time. As regards the sectors of time connected with, and forming a part of, the cycle of eternity, we are forced to recognize that such temporary epochs are born, live, and die just as the temporary beings of time are born, live, and die. **Most human beings die because,** having failed to achieve the spirit level of Adjuster fusion, the metamorphosis of death constitutes the only possible procedure whereby they may escape the fetters of time and the bonds of material creation, thereby being enabled to strike spiritual step with the progressive procession of eternity. Having survived the trial life of time and material existence, it becomes possible for you to continue on in touch with, even as a part of, eternity, swinging on forever with the worlds of space around the circle of the eternal ages.

PAPER 15: THE SEVEN SUPERUNIVERSES

Spiritual Wisdom—Divine Destinies to Human Origins Perspective

Page 164:3   *Paper 15:0.3*

See "Spiritual Wisdom—Divine Destinies to Human Origins Perspective" on page 138 for full text references. (3:6.3)

2. ORGANIZATION OF THE SUPERUNIVERSES

Solar System— in Contradistinction to System
Parts I and II on solar systems, including our own.

Page 166:1–2   *Paper 15:2.2–3*

There are seven superuniverses in the grand universe, and they are constituted approximately as follows:

1. *The System.* The basic unit of the supergovernment consists of about one thousand inhabited or inhabitable worlds. Blazing suns, cold worlds, planets too near the hot suns, and other spheres not suitable for creature habitation are not included in this group. **These one thousand worlds adapted to support life are called a system, but in the younger systems only a comparatively small number of these worlds may be inhabited.** Each inhabited planet is presided over by a Planetary Prince, and each local system has an architectural sphere as its headquarters and is ruled by a System Sovereign.

Page 457:1   *Paper 41:2.2*

Satania itself is composed of over seven thousand astronomical groups, or physical systems, few of which had an origin similar to that of your solar system. The astronomic center of Satania is an enormous dark island of space which, with its attendant spheres, is situated not far from the headquarters of the system government.

Page 466:1   *Paper 41:10.2*

The majority of solar systems, however, had an origin entirely different from yours, and this is true even of those which were produced by gravity-tidal technique. **But no matter what technique of world building obtains, gravity always produces the solar system type of creation; that is, a central sun or dark island with planets, satellites, subsatellites, and meteors.**

Minor Sector
About the Minor Sector—over 500 pages apart with different authors.

Page 166:5   *Paper 15:2.6*

4. *The Minor Sector.* One hundred local universes (about 1,000,000,000 inhabitable planets) constitute a minor sector of the superuniverse government; it has a wonderful headquarters world, wherefrom its rulers, the Recents of Days, administer the affairs of the minor sector. There are three Recents of Days, Supreme Trinity Personalities, on each minor sector headquarters.

Page 635:4   *Paper 55:11.2*

The minor sector age. As far as observations can penetrate, **the fifth or minor sector stage of stabilization has exclusively to do with physical status and with the co-ordinate settling of the one hundred associated local universes in the established circuits of the superuniverse.** Apparently none but the power centers and their associates are concerned in these realignments of the material creation.

3. THE SUPERUNIVERSE OF ORVONTON

Orvonton
A little about our superuniverse and how it got its name.

Page 167:4   *Paper 15:3.1*

Practically all of the starry realms visible to the naked eye on Urantia belong to the seventh section of the grand universe, the superuniverse of Orvonton. **The vast Milky Way starry system represents the central nucleus of Orvonton,** being largely beyond the borders of your local universe. This great aggregation of suns, dark islands of space, double stars, globular clusters, star clouds, spiral and other nebulae, together with myriads of individual planets, forms a watchlike, elongated-circular grouping of about one seventh of the inhabited evolutionary universes.

Page 198:5   *Paper 17:1.5*

Each of the executives and the facilities of his sphere are devoted to the efficient administration of a single superuniverse. Supreme Executive Number One, functioning on executive sphere number one, is wholly occupied with the affairs of superuniverse number one, and so on to Supreme Executive Number Seven, working from the seventh Paradise satellite of the Spirit and devoting his energies to the management of the seventh superuniverse. **The name of this seventh sphere is Orvonton, for the Paradise satellites of the Spirit have the same names as their related superuniverses; in fact, the superuniverses were named after them.**

---

6. THE SPHERES OF SPACE   •

Planets Suited to Harbor Life
Venus, Mars, and Urantia—from Parts I and II.

Page 173:5   *Paper 15:6.10*

In your superuniverse not one cool planet in forty is habitable by beings of your order. And, of course, the superheated suns and the frigid outlying worlds are unfit to harbor higher life. **In your solar system only three planets are at present suited to harbor life.** Urantia, in size, density, and location, is in many respects ideal for human habitation.

Page 561:12   *Paper 49:2.6*

Beings such as the Urantia races are classified as mid-breathers; you represent the average or typical breathing order of mortal existence. **If intelligent creatures should exist on a planet with an atmosphere similar to that of your near neighbor, Venus, they would belong to the superbreather group,** while those inhabiting a planet with an **atmosphere as thin as that of your outer neighbor, Mars, would be denominated subbreathers.**

---

8. ENERGY CONTROL AND REGULATION

Power Centers
Power Centers, some personal, intelligently functional with no will—covered over 100 pages apart.

Page 175:5   *Paper 15:8.2*

Further regulative functions are performed by the superuniverse power centers and physical controllers, living and semiliving intelligent entities constituted for this express purpose. **These power centers and controllers are difficult of understanding; the lower orders are not volitional, they do not possess will, they do not choose, their functions are very intelligent but apparently automatic and inherent in their highly specialized organization.** The power centers and physical controllers of the superuniverses assume direction and partial control of the thirty energy systems which comprise the gravita domain. The physical-energy circuits administered by the power centers of Uversa require a little over 968 million years to complete the encirclement of the superuniverse.

Page 323:6   *Paper 29:3.8*

It is utterly beyond my ability to explain the manner in which these living beings encompass the manipulation and regulation of the master circuits of universe energy. **To undertake to inform you further concerning the size and function of these gigantic and almost perfectly efficient power centers, would only add to your confusion and consternation.** They are both living and "personal," but they are beyond your comprehension.

Page 325:5   *Paper 29:4.5*

Not all of these orders are persons in the sense of possessing individual powers of choice. Especially do the last four seem to be wholly automatic and mechanical in response to the impulses of their superiors and in reaction to existing energy conditions. But though such response appears wholly mechanistic, it is not; they may seem to be automatons, but all of them disclose the differential function of intelligence.

---

10. RULERS OF THE SUPERUNIVERSES

Qualified Vicegerents of the Ultimate

The Qualified Vicegerents of the Ultimate and a couple of their buddies covered in four papers in Parts I and III.

Page 179:1   *Paper 15:10.7*

The Reflective Image Aids **also** function as the **representatives** of numerous groups of beings who are influential in the superuniverse governments, but who are not, at present, for various reasons, fully active in their individual capacities. Embraced within this group are: the evolving superuniverse personality manifestation of the **Supreme Being,** the **Unqualified Supervisors of the Supreme,** the **Qualified Vicegerents of the Ultimate,** the **unnamed liaison reflectivators of Majeston,** and the **superpersonal** spirit representatives of the Eternal Son.

Page 333:12   *Paper 30:1.21*

VII. *UNCLASSIFIED AND UNREVEALED ORDERS.* During the present universe age it would not be possible to place all beings, personal or otherwise, within classifications pertaining to the present universe age; nor have all such categories been revealed in these narratives; hence numerous orders have been omitted from these lists. Consider the following:

The Consummator of Universe Destiny.
The Qualified Vicegerents of the Ultimate.
The Unqualified Supervisors of the Supreme.
The Unrevealed Creative Agencies of the Ancients of Days.
Majeston of Paradise.
The Unnamed Reflectivator Liaisons of Majeston.
The Midsonite Orders of the Local Universes.

Page 1166:6   *Paper 106:4.3*

What changes will be inaugurated by the full emergence of the Ultimate we do not know. But as the Supreme is now spiritually and personally present in Havona, so also is the Ultimate there present but in the absonite and superpersonal sense. **And you have been informed of the existence of the Qualified Vicegerents of the Ultimate, though you have not been informed of their present whereabouts or function.**

Page 1291:7–8   *Paper 117:7.3–4*

1. **The Unqualified Supervisors of the Supreme** could hardly be deitized at any stage prior to his completed evolution, and **yet these same supervisors even now qualifiedly exercise the sovereignty of supremacy concerning** the universes settled in light and life.

2. The Supreme could hardly function in the Trinity Ultimate until he had attained complete actuality of universe status, and yet the Trinity Ultimate is even now a qualified reality, **and you have been informed of the existence of the Qualified Vicegerents of the Ultimate.**

Communication Between Superuniverses—Paradise Clearinghouse

Superuniverses ordinarily communicate through Paradise clearinghouse. Paper 15 sets the stage for reinforcement of cosmology for other parts, in this case Paper 39 in Part II.

Page 179:3   *Paper 15:10.9*

The superuniverses do not maintain any sort of ambassadorial representation; they are completely isolated from each other. They know of mutual affairs only through the Paradise clearinghouse maintained by the

Seven Master Spirits. Their rulers work in the councils of divine wisdom for the welfare of their own superuniverses regardless of what may be transpiring in other sections of the universal creation. This isolation of the superuniverses will persist until such time as their co-ordination is achieved by the more complete factualization of the personality-sovereignty of the evolving experiential Supreme Being.

Page 429:7   *Paper 39:2.4*

The intelligence corps of the various local universes can and do intercommunicate but only within a given superuniverse. There is a differential of energy which effectively segregates the business and transactions of the various supergovernments. **One superuniverse can ordinarily communicate with another superuniverse only through the provisions and facilities of the Paradise clearinghouse.**

---

14. PURPOSES OF THE SEVEN SUPERUNIVERSES

Major Sector—Splandon

Splandon and a little about our ascension career with the other nine Major Sectors.

Page 182:6   *Paper 15:14.7*

The minor sector of Ensa consists of one hundred local universes and has a capital called Uminor the third. This minor sector is number three in the major sector of Splandon. **Splandon consists of one hundred minor sectors and has a headquarters world called Umajor the fifth.** It is the fifth major sector of the superuniverse of Orvonton, the seventh segment of the grand universe. Thus you can locate your planet in the scheme of the organization and administration of the universe of universes.

Page 211:5   *Paper 18:4.9*

Although you are entered only upon the registry of the major sector of Splandon, which embraces the local universe of your origin, **you will have to pass through every one of the ten major divisions of our superuniverse.** You will see all thirty of the Orvonton Perfections of Days before you reach Uversa.

PAPER 16: THE SEVEN MASTER SPIRITS

8. URANTIA PERSONALITY

Personality and Personality Circuit

Page 195:2   *Paper 16:8.8*

See "Personality and Personality Circuit" on page 120 for full text
references. (2:2.5–6)

9. REALITY OF HUMAN CONSCIOUSNESS

Cosmic Mind
The cosmic mind in conjunction with Spirit of Truth and Thought Adjuster—
Parts I, III, and IV.

Page 195:7–8   *Paper 16:9.1–2*

The cosmic-mind-endowed, Adjuster-indwelt, personal creature
possesses innate recognition-realization of energy reality, mind reality, and
spirit reality. The will creature is thus **equipped to discern the fact, the law,
and the love of God.** Aside from these three inalienables of human
consciousness, all human experience is really subjective except that intuitive
realization of validity attaches to the *unification* of these three universe reality
responses of cosmic recognition.

**The God-discerning mortal is able to sense the unification value of these
three cosmic qualities in the evolution of the surviving soul,** man's supreme
undertaking in the physical tabernacle where the moral mind collaborates
with the indwelling divine spirit to dualize the immortal soul. From its earliest
inception the soul is *real*; it has cosmic survival qualities.

Page 230:1   *Paper 20:6.8*

When the mortal incarnation is finished, the Avonal of service proceeds
to Paradise, is accepted by the Universal Father, returns to the local universe
of assignment, and is acknowledged by the Creator Son. **Thereupon the
bestowal Avonal and the Creator Son send their conjoint Spirit of Truth to
function in the hearts of the mortal races dwelling on the bestowal world.** In
the presovereignty ages of a local universe, this is the joint spirit of both Sons,
implemented by the Creative Spirit. It differs somewhat from the Spirit of
Truth which characterizes the local universe ages following a Michael's
seventh bestowal.

Page 757:2   *Paper 67:3.9*

Caligastia, with a maximum of intelligence and a vast experience in universe affairs, went astray—embraced sin. Amadon, with a minimum of intelligence and utterly devoid of universe experience, remained steadfast in the service of the universe and in loyalty to his associate. Van utilized both mind and spirit in a magnificent and effective combination of intellectual determination and spiritual insight, thereby achieving an experiential level of personality realization of the highest attainable order. **Mind and spirit, when fully united, are potential for the creation of superhuman values, even morontia realities.**

Page 1108:1   *Paper 101:3.2*

Faith-insight, or spiritual intuition, is the endowment of the cosmic mind in association with the Thought Adjuster, which is the Father's gift to man. Spiritual reason, soul intelligence, is the endowment of the Holy Spirit, the Creative Spirit's gift to man. Spiritual philosophy, the wisdom of spirit realities, is the endowment of the Spirit of Truth, the combined gift of the bestowal Sons to the children of men. And the co-ordination and interassociation of these spirit endowments constitute man a spirit personality in potential destiny.

Page 1949:5   *Paper 180:5.3*

Intelligence grows out of a material existence which is illuminated by the **presence of the cosmic mind.** Wisdom comprises the consciousness of knowledge **elevated to new levels of meaning** and activated by the presence of the universe endowment of the adjutant of wisdom. Truth is a spiritual reality value experienced only by spirit-endowed beings who function upon supermaterial levels of universe consciousness, and who, after the realization of truth, permit its spirit of activation to live and reign within their souls.

PAPER 17: THE SEVEN SUPREME SPIRIT GROUPS

---

1. THE SEVEN SUPREME EXECUTIVES

Orvonton

Page 198:5   *Paper 17:1.5*

See "Orvonton" on page 169 for full text references. (15:3.1)

Majeston

Majeston—almost a thousand pages apart.

Page 199:6   *Paper 17:2.2*

This momentous transaction, occurring in the dawn of time, represents the initial effort of the Supreme Creator Personalities, represented by the Master Spirits, to function as cocreators with the Paradise Trinity. This union of the creative power of the Supreme Creators with the creative potentials of the Trinity is the very source of the actuality of the Supreme Being. Therefore, when the cycle of reflective creation had run its course, when each of the Seven Master Spirits had found perfect creative synchrony with the Paradise Trinity, when the forty-ninth Reflective Spirit had personalized, then a new and far-reaching reaction occurred in the Deity Absolute which imparted new personality prerogatives to the Supreme Being and culminated in the personalization of Majeston, the reflectivity chief and Paradise center of all the work of the forty-nine Reflective Spirits and their associates throughout the universe of universes.

Page 1172:5   *Paper 106:8.17*

The first level consists of three Trinities; the second level exists as the personality association of experiential-evolved, experiential-eventuated, and experiential-existential Deity personalities. **And regardless of any conceptual difficulty in understanding the complete Trinity of Trinities, the personal association of these three Deities on the second level has become manifest to our own universe age in the phenomenon of the deitization of Majeston, who was actualized on this second level by the Deity Absolute, acting through the Ultimate and in response to the initial creative mandate of the Supreme Being.**

Majeston—Creative Act of the Supreme Being

Page 200:3   *Paper 17:2.5*

See "Majeston—Creative Act of the Supreme Being" on page 107 for full text references. (0:7.7)

3. THE REFLECTIVE SPIRITS

Light and Life—Jubilee Event
Jubilee event as described in Part I and near the end of Part II.

Page 201:9   *Paper 17:3.11*

On the headquarters of each superuniverse the reflective organization acts as a segregated unit; but on certain special occasions, under the direction of Majeston, all seven may and do act in universal unison, **as in the event of the jubilee occasioned by the settling of an entire local universe in light and life** and at the times of the millennial greetings of the Seven Supreme Executives.

Page 634:1   *Paper 55:10.1*

When a universe becomes settled in light and life, it soon swings into the established superuniverse circuits, and the Ancients of Days proclaim the establishment of the *supreme council of unlimited authority.* This new governing body consists of the one hundred Faithfuls of Days, presided over by the Union of Days, and the first act of this supreme council is to acknowledge the continued sovereignty of the Master Creator Son.

5. THE SEVEN SPIRITS OF THE CIRCUITS

Father Contact

Page 203:3   *Paper 17:5.5*

See "Father Contact" on page 114 for full text references. (1:7.1)

Impersonal Entities, Mind, and Mortal Will

Page 203:3   *Paper 17:5.5*

See "Impersonal Entities, Mind, and Mortal Will" on page 159 for full text references. (9:5.4)

PAPER 18: THE SUPREME TRINITY PERSONALITIES

---

3. THE ANCIENTS OF DAYS

Magisterial Sons—Judges of Survival

Page 210:1   *Paper 18:3.7*

See "Magisterial Sons—Judges of Survival" on page 155 for full text references. (7:6.5)

---

4. THE PERFECTIONS OF DAYS

Major Sector—Splandon

Page 211:5   *Paper 18:4.9*

See "Major Sector—Splandon" on page 173 for full text references. (15:14.7)

---

6. THE UNION OF DAYS

Order of "Days"
A little about Nebadon's ambassador and his order of "Days"—Parts I and IV.

Page 212:3   *Paper 18:6.1*

The Trinity personalities of the order of "Days" do not function in an administrative capacity below the level of the superuniverse governments. In the evolving local universes they act only as counselors and advisers. The Unions of Days are a group of liaison personalities accredited by the Paradise Trinity to the dual rulers of the local universes. Each organized and inhabited local universe has assigned to it one of these Paradise counselors, **who acts as the representative of the Trinity, and in some respects, of the Universal Father, to the local creation.**

Page 213:1   *Paper 18:6.6*

When a local universe is settled in light and life, its glorified beings associate freely with the Union of Days, who then functions in an enlarged capacity in such a realm of evolutionary perfection. **But he is still primarily a Trinity ambassador and Paradise counselor.**

Page 1324:3   *Paper 120:0.6*

Having determined the time of his final bestowal and having selected the planet whereon this extraordinary event would take place, Michael held the usual prebestowal conference with Gabriel and **then presented himself before his elder brother and Paradise counselor, Immanuel.** All powers of universe administration which had not previously been conferred upon Gabriel, Michael now assigned to the custody of Immanuel. And just before Michael's departure for the Urantia incarnation, Immanuel, in accepting the custody of the universe during the time of the Urantia bestowal, proceeded to impart the bestowal counsel which would serve as the incarnation guide for Michael when he would presently grow up on Urantia as a mortal of the realm.

PAPER 19: THE CO-ORDINATE TRINITY-ORIGIN BEINGS

---

1. THE TRINITY TEACHER SONS

Spiritual Wisdom—Divine Destinies to Human Origins Perspective

Page 215:2   *Paper 19:1.5*

See "Spiritual Wisdom—Divine Destinies to Human Origins Perspective" on page 138 for full text references. (3:6.3)

---

3. THE DIVINE COUNSELORS

God and Man—Combined Being
Paradise bestowal Sons as God and man.

Page 217:6   *Paper 19:3.7*

Seven Divine Counselors in liaison with a trinitized evolutionary trio—a Mighty Messenger, One High in Authority, and One without Name and Number—represent the nearest superuniverse approach to the union of the human viewpoint and the divine attitude on near-paradisiacal levels of spiritual meanings and reality values. Such close approximation of the united cosmic attitudes of the creature and the Creator **is only surpassed in the Paradise bestowal Sons, who are, in every phase of personality experience, God and man.**

Page 1331:2　　　　　　　　　　　　　　　　　　　　Paper 120:4.2

But make no mistake; Christ Michael, while truly a dual-origin being, was not a double personality. He was not God in association *with* man but, rather, God *incarnate* in man. And he was always just that combined being. The only progressive factor in such a nonunderstandable relationship **was the progressive self-conscious realization and recognition (by the human mind) of this fact of being God and man.**

---

5. *INSPIRED TRINITY SPIRITS*

Inspired Spirits and Our Divine Fragment

Inspired Spirits and mysterious operations of our Divine Fragment—in Parts I and II.

Page 220:1–3　　　　　　　　　　　　　　　　　　　Paper 19:5.6–8

I may relate a further interesting fact: When a Solitary Messenger is on a planet whose inhabitants are indwelt by Thought Adjusters, as on Urantia, he is aware of a qualitative excitation in his detection-sensitivity to spirit presence. In such instances there is no quantitative excitation, only a qualitative agitation. When on a planet to which Adjusters do not come, contact with the natives does not produce any such reaction. **This suggests that Thought Adjusters are in some manner related to, or are connected with, the Inspired Spirits of the Paradise Trinity. In some way they may possibly be associated in certain phases of their work; but we do not really know.** They both originate near the center and source of all things, but they are not the same order of being. Thought Adjusters spring from the Father alone; Inspired Spirits are the offspring of the Paradise Trinity.

The Inspired Spirits do not apparently belong to the evolutionary scheme of the individual planets or universes, and yet they seem to be almost everywhere. Even as I am engaged in the formulation of this statement, my associated Solitary Messenger's personal sensitivity to the presence of this order of Spirit indicates that there is with us at this very moment, not over twenty-five feet away, a Spirit of the Inspired order and of the third volume of power presence. The third volume of power presence suggests to us the probability that three Inspired Spirits are functioning in liaison.

Of more than twelve orders of beings associated with me at this time, the Solitary Messenger is the only one aware of the presence of these mysterious entities of the Trinity. And further, while we are thus apprised of the nearness of these divine Spirits, we are all equally ignorant of their mission. **We really do not know whether they are merely interested observers of our doings, or whether they are, in some manner unknown to us, actually contributing to the success of our undertaking.**

Page 425:1   *Paper 38:9.9*

The gap between the material and spiritual worlds is perfectly bridged by the serial association of mortal man, secondary midwayer, primary midwayer, morontia cherubim, mid-phase cherubim, and seraphim. In the personal experience of an individual mortal these diverse levels are undoubtedly **more or less unified and made personally meaningful by the unobserved and mysterious operations of the divine Thought Adjuster.**

Conscious, Unconscious, and Superconscious— Technique
Consciously unconscious and superconscious techniques.

Page 220:4   *Paper 19:5.9*

We know that the **Trinity Teacher Sons** are devoted to the *conscious* enlightenment of universe creatures. I have arrived at the settled conclusion that the Inspired Trinity Spirits, **by** *superconscious* **techniques,** are also functioning as teachers of the realms. I am persuaded that there is a **vast body of essential spiritual knowledge,** truth indispensable to high spiritual attainment, which **cannot be consciously received; self-consciousness would effectively jeopardize the certainty of reception.** If we are right in this concept, and my entire order of being shares it, it may be the mission of these Inspired Spirits to overcome this difficulty, to bridge this gap in the universal scheme of moral enlightenment and spiritual advancement. We think that **these two types** of Trinity-origin teachers effect some kind of liaison in their activities, but we do not really know.

Page 430:7   *Paper 39:2.11*

The process of being enseraphimed is not unlike the experience of death or sleep except that there is an automatic time element in the transit slumber. **You are consciously unconscious during seraphic rest.** But the Thought Adjuster is wholly and fully conscious, in fact, exceptionally efficient since you are unable to oppose, resist, or otherwise hinder creative and transforming work.

6. HAVONA NATIVES

Havona Natives

The Havona natives and their supreme adventure.

Page 221:3   *Paper 19:6.1*

The Havona natives are the direct creation of the Paradise Trinity, and their number is beyond the concept of your circumscribed minds. **Neither is it possible for Urantians to conceive of the inherent endowments of such divinely perfect creatures as these Trinity-origin races of the eternal universe.** You can never truly envisage these glorious creatures; you must await your arrival in Havona, when you can greet them as spirit comrades.

Page 346:5   *Paper 31:1.4*

The Havona recruits follow the company of their assignment; wherever the group goes, they go. And you should see their enthusiasm in the new work of the finaliters. The possibility of attaining the Corps of the Finality is one of the superb thrills of Havona; **the possibility of becoming a finaliter is one of the supreme adventures of these perfect races.**

PAPER 20: THE PARADISE SONS OF GOD

---

3. JUDICIAL ACTIONS

Magisterial Sons—Judges of Survival

Page 226:2   *Paper 20:3.2*

See "Magisterial Sons—Judges of Survival" on page 155 for full text references. (7:6.5)

---

4. MAGISTERIAL MISSIONS

Avonal Sons—Magisterial Visitations

Twelve Melchizedeks and a Paradise Avonal make a visit—in Parts I and II.

Page 226:5–6   *Paper 20:4.1–2*

Prior to the planetary appearance of a bestowal Son, an inhabited world is usually visited by a Paradise Avonal on a magisterial mission. **If it is an initial magisterial visitation, the Avonal is always incarnated as a material being. He** appears on the planet of assignment as a full-fledged male of the mortal races, a being fully visible to, and in physical contact with, the mortal creatures of his day and generation. Throughout a magisterial incarnation the connection of the Avonal Son with the local and the universal spiritual forces is complete and unbroken.

A planet may experience many magisterial visitations both before and after the appearance of a bestowal Son. It may be visited many times by the same or other Avonals, acting as dispensational adjudicators, but such technical missions of judgment are neither bestowal nor magisterial, and the Avonals are never incarnated at such times. Even when a planet is blessed with repeated magisterial missions, the Avonals do not always submit to mortal incarnation; and **when they do serve in the likeness of mortal flesh, they always appear as adult beings of the realm;** they are not born of woman.

Page 386:4   *Paper 35:2.6*

When a Creator Son enters upon the bestowal career on an evolutionary world, he goes alone; but when one of his Paradise brothers, an Avonal Son, enters upon a bestowal, he is accompanied by the Melchizedek supporters, **twelve in number,** who so efficiently contribute to the success of the bestowal mission. They also support the Paradise Avonals **on magisterial missions to the inhabited worlds, and in these assignments the Melchizedeks are visible to mortal eyes if the Avonal Son is also thus manifest.**

---

6. THE MORTAL-BESTOWAL CAREERS

Incarnational Bestowals of Paradise Sons—Mysterious and Miraculous
One of the few miracles—a Paradise Son becoming enmothered—a thousand pages apart.

Page 228:5   *Paper 20:6.1*

The method whereby a Paradise Son becomes ready for mortal incarnation as a bestowal Son, becomes enmothered on the bestowal planet, is a universal mystery; and any effort to detect the working of this Sonarington technique is doomed to meet with certain failure. Let the sublime knowledge of the mortal life of Jesus of Nazareth sink into your souls, **but waste no thought in useless speculation as to how this mysterious incarnation of Michael of Nebadon was effected.** Let us all rejoice in the knowledge and assurance that such achievements are possible to the divine nature and waste no time on futile conjectures about the technique employed by divine wisdom to effect such phenomena.

Page 1331:5   *Paper 120:4.5*

Urantia mortals have varying concepts of the miraculous, but to us who live as citizens of the local universe there are few miracles, and of these by far the most intriguing are the incarnational bestowals of the Paradise Sons. The appearance in and on your world, by apparently natural processes, of a divine Son, we regard as a miracle—the operation of universal laws beyond our understanding. Jesus of Nazareth was a miraculous person.

Bestowals and Offspring

Page 229:1   *Paper 20:6.2*

On a mortal-bestowal mission a Paradise Son is always born of woman and grows up as a male child of the realm, as Jesus did on Urantia. These Sons of supreme service all pass from infancy through youth to manhood just as does a human being. In every respect they become like the mortals of the race into which they are born. They make petitions to the Father as do the children of the realms in which they serve. **From a material viewpoint, these human-divine Sons live ordinary lives with just one exception: They do not beget offspring on the worlds of their sojourn;** that is a universal restriction imposed on all orders of the Paradise bestowal Sons.

Page 1015:6   *Paper 93:2.6*

Though Machiventa lived after the manner of the men of the realm, he never married, **nor could he have left offspring on earth.** His physical body, while resembling that of the human male, was in reality on the order of those especially constructed bodies used by the one hundred materialized members of Prince Caligastia's staff except that it did not carry the life plasm of any human race. Nor was there available on Urantia the tree of life. Had Machiventa remained for any long period on earth, his physical mechanism would have gradually deteriorated; as it was, he terminated his bestowal mission in ninety-four years long before his material body had begun to disintegrate.

Page 1017:1   *Paper 93:3.7*

Melchizedek taught that at some future time another Son of God would come in the flesh as he had come, but that he would be born of a woman; and that is why numerous later teachers held that Jesus was a priest, or minister, "forever after the order of Melchizedek."

Page 1316:6   *Paper 119:7.3*

While we believed that this would be the method, we never knew, until the time of the event itself, that Michael would appear on earth as a helpless infant of the realm. Theretofore had he always appeared as a fully developed individual of the personality group of the bestowal selection, and it was a thrilling announcement which was broadcast from Salvington telling that the babe of Bethlehem had been born on Urantia.

Page 1330:3   *Paper 120:3.8*

"7. While you will live the normal and average social life of the planet, being a normal individual of the male sex, you will probably not enter the marriage relation, which relation would be wholly honorable and consistent with your bestowal; but I must remind you that one of the incarnation mandates of Sonarington **forbids the leaving of human offspring behind on any planet by a bestowal Son of Paradise origin.**

Page 1404:7   *Paper 127:6.8*

It was during this year that Mary had a long talk with Jesus about marriage. She frankly asked him if he would get married if he were free from his family responsibilities. Jesus explained to her that, since immediate duty forbade his marriage, he had given the subject little thought. He expressed himself as doubting that he would ever enter the marriage state; he said that all such things must await "my hour," the time when "my Father's work must begin." **Having settled already in his mind that he was not to become the father of children in the flesh, he gave very little thought to the subject of human marriage.**

Bestowal Plan of Paradise Sons

Page 229:4   *Paper 20:6.5*

See "Bestowal Plan of Paradise Sons" on page 153 for full text references. (7:4.5)

Bestowals—Resurrection on Third Day
Third day versus third period.

Page 229:5   *Paper 20:6.6*

When the bestowal Sons, Creator or Magisterial, enter the portals of death, they reappear on the third day. But you should not entertain the idea that they always meet with the tragic end encountered by the Creator Son who sojourned on your world nineteen hundred years ago. The extraordinary and unusually cruel experience through which Jesus of Nazareth passed has caused Urantia to become locally known as "the world of the cross." It is not necessary that such inhuman treatment be accorded a Son of God, and the vast majority of planets have afforded them a more considerate reception, allowing them to finish their mortal careers, terminate the age, adjudicate the sleeping survivors, and inaugurate a new dispensation, without imposing a violent death. A bestowal Son must encounter death, must pass through the whole of the actual experience of mortals of the realms, but it is not a requirement of the divine plan that this death be either violent or unusual.

Page 2021:3   *Paper 189:1.4*

As far as we can judge, no creature of this universe nor any personality from another universe had anything to do with this morontia resurrection of Jesus of Nazareth. **On Friday he laid down his life as a mortal of the realm; on Sunday morning he took it up again as a morontia being of the system of Satania in Norlatiadek.** There is much about the resurrection of Jesus which we do not understand. But we know that it occurred as we have stated and at about the time indicated. We can also record that all known phenomena associated with this mortal transit, or morontia resurrection, occurred right there in Joseph's new tomb, where the mortal material remains of Jesus lay wrapped in burial cloths.

Cosmic Mind

Page 230:1   *Paper 20:6.8*

See "Cosmic Mind" on page 174 (16:9.1–2)

Spirit of Truth—Bestowal
Phenomena related to the special visitation—from Papers 20 to 194

Page 230:2   *Paper 20:6.9*

Upon the completion of a Creator Son's final bestowal the Spirit of Truth previously sent into all Avonal-bestowal worlds of that local universe changes in nature, becoming more literally the spirit of the sovereign Michael. **This phenomenon takes place concurrently with the liberation of the Spirit of Truth for service on the Michael-mortal-bestowal planet.** Thereafter, each world honored by a Magisterial bestowal will receive the same spirit Comforter from the sevenfold Creator Son, in association with that Magisterial Son, which it would have received had the local universe Sovereign personally incarnated as its bestowal Son.

Page 2059:1   *Paper 194:0.1*

About one o'clock, as the one hundred and twenty believers were engaged in prayer, they all became aware of a strange presence in the room. **At the same time these disciples all became conscious of a new and profound sense of spiritual joy, security, and confidence.** This new consciousness of spiritual strength was immediately followed by a strong urge to go out and publicly proclaim the gospel of the kingdom and the good news that Jesus had risen from the dead.

Page 2065:2   *Paper 194:3.14*

Before the teachings of Jesus which culminated in Pentecost, women had little or no spiritual standing in the tenets of the older religions. After Pentecost, in the brotherhood of the kingdom woman stood before God on an equality with man. **Among the one hundred and twenty who received this special visitation of the spirit were many of the women disciples, and they shared these blessings equally with the men believers.** No longer can man presume to monopolize the ministry of religious service. The Pharisee might go on thanking God that he was "not born a woman, a leper, or a gentile," but among the followers of Jesus woman has been forever set free from all religious discriminations based on sex. Pentecost obliterated all religious discrimination founded on racial distinction, cultural differences, social caste, or sex prejudice. No wonder these believers in the new religion would cry out, "Where the spirit of the Lord is, there is liberty."

Page 2065:5   *Paper 194:3.17*

Pentecost was the call to spiritual unity among gospel believers. When the spirit descended on the disciples at Jerusalem, **the same thing happened in Philadelphia, Alexandria, and at all other places where true believers dwelt.** It was literally true that "there was but one heart and soul among the multitude of the believers." The religion of Jesus is the most powerful unifying influence the world has ever known.

8. LOCAL UNIVERSE MINISTRY OF THE DAYNALS

Lucifer Rebellion—Spiritual Isolation

Insights on our spiritual isolation—three papers and several hundred pages apart.

Page 231:4   *Paper 20:8.4*

In all universes all the Sons of God are beholden to these ever-faithful and universally efficient Trinity Teacher Sons. They are the exalted teachers of all spirit personalities, even the tried and true teachers of the Sons of God themselves. But of the endless details of the duties and functions of the Teacher Sons I can hardly instruct you. The vast domain of Daynal-sonship activities **will be better understood on Urantia when you are more advanced in intelligence, and after the spiritual isolation of your planet has been terminated.**

Page 609:7   *Paper 53:8.4*

The Son of Man was confident of success, and he knew that his triumph on your world would forever settle the status of his agelong enemies, not only in Satania but also in the other two systems where sin had entered. There was survival for mortals and security for angels when your Master, in reply to the Lucifer proposals, calmly and with divine assurance replied, "Get you behind me, Satan." That was, in principle, the real end of the Lucifer rebellion. True, the Uversa tribunals have not yet rendered the executive decision regarding the appeal of Gabriel praying for the destruction of the rebels, but such a decree will, no doubt, be forthcoming in the fullness of time since **the first step in the hearing of this case has already been taken.**

Page 611:3   *Paper 53:9.4*

Satan could come to Urantia because you had no Son of standing in residence—neither Planetary Prince nor Material Son. Machiventa Melchizedek has since been proclaimed vicegerent Planetary Prince of Urantia, and the opening of the case of Gabriel *vs.* Lucifer has signalized the inauguration of temporary planetary regimes on all the isolated worlds. It is true that Satan did periodically visit Caligastia and others of the fallen princes **right up to the time of the presentation of these revelations, when there occurred the first hearing of Gabriel's plea for the annihilation of the archrebels.** Satan is now unqualifiedly detained on the Jerusem prison worlds.

Page 616:7   *Paper 54:4.8*

But these mercy delays are not interminable. Notwithstanding the long delay (as time is reckoned on Urantia) in adjudicating the Lucifer rebellion, **we may record that, during the time of effecting this revelation, the first hearing in the pending case of Gabriel *vs.* Lucifer was held on Uversa,** and soon thereafter there issued the mandate of the Ancients of Days directing that Satan be henceforth confined to the prison world with Lucifer. This ends the ability of Satan to pay further visits to any of the fallen worlds of Satania. Justice in a mercy-dominated universe may be slow, but it is certain.

---

9. PLANETARY SERVICE OF THE DAYNALS

Magisterial Sons—Judges of Survival

Page 231:6   *Paper 20:9.2*

See "Magisterial Sons—Judges of Survival" on page 155 for full text references. (7:6.5)

Evening Stars—Liaison Between Mortals and Teacher Sons

Activities of Evening Stars confirmed 17 papers later—Parts I and II with two different authors.

Page 232:1   *Paper 20:9.3*

The Teacher Sons usually remain on their visitation planets for one thousand years of planetary time. One Teacher Son presides over the planetary millennial reign and is assisted by seventy associates of his order. The Daynals do not incarnate or otherwise so materialize themselves as to be visible to mortal beings; **therefore is contact with the world of visitation maintained through the activities of the Brilliant Evening Stars,** local universe personalities who are associated with the Trinity Teacher Sons.

Page 408:1   *Paper 37:2.9*

Similar pairs of these superangels are assigned to the planetary corps of Trinity Teacher Sons that functions to establish the postbestowal or dawning spiritual age of an inhabited world. On such assignments the **Evening Stars serve as liaisons between the mortals of the realm and the invisible corps of Teacher Sons.**

PAPER 21: THE PARADISE CREATOR SONS

Bestowals of the Second Source and Center and the Original Michael

Page 234:4   *Paper 21:0.4*

See "Bestowals of the Second Source and Center and the Original Michael" on page 154 for full text references. (7:5.5)

1. ORIGIN AND NATURE OF CREATOR SONS

Creator Son and Creative Spirit

Page 235:2   *Paper 21:1.3*

See "Creator Son and Creative Spirit" on page 157 for full text references. (8:3.4)

2. THE CREATORS OF LOCAL UNIVERSES

Personality—Unsolved Mystery

Page 236:1–3   *Paper 21:2.3–5*

See "Personality—Unsolved Mystery" on page 146 for full text references. (5:6.2)

Unqualified Absolute—Space Presence
Space presence of the Unqualified Absolute and concept periphery of Master Universe.

Page 237:2   *Paper 21:2.12*

The personal presence of a Creator Son in his local universe is not necessary to the smooth running of an established material creation. Such Sons may journey to Paradise, and still their universes swing on through space. They may lay down their lines of power to incarnate as the children of time; still their realms whirl on about their respective centers. No material organization is independent of the absolute-gravity grasp of Paradise **or of the cosmic overcontrol inherent in the space presence of the Unqualified Absolute.**

Page 1156:4   *Paper 105:3.7*

6. *The Unqualified Absolute.* Static, reactive, and abeyant; the unrevealed cosmic infinity of the I AM; totality of nondeified reality and finality of all nonpersonal potential. Space limits the function of the Unqualified, but the presence of the Unqualified is without limit, infinite. **There is a concept periphery to the master universe, but the presence of the Unqualified is limitless; even eternity cannot exhaust the boundless quiescence of this nondeity Absolute.**

Page 1298:6   *Paper 118:4.6*

2. *Eventuation of universe capacities.* This involves the transformation of undifferentiated potentials into segregated and defined plans. This is the act of the Ultimacy of Deity and of the manifold agencies of the transcendental level. Such acts are in perfect anticipation of the future needs of the entire master universe. It is in connection with the segregation of potentials that the Architects of the Master Universe exist as the veritable embodiments of the Deity concept of the universes. Their plans appear to be ultimately space limited in extent by the **concept periphery of the master universe,** but as *plans* they are not otherwise conditioned by time or space.

---

3. LOCAL UNIVERSE SOVEREIGNTY

Creator Sons—Sovereignty

If Creator Sons assumed unearned supreme sovereignty, it would be for personal reasons.—Parts I and IV, over thousand pages apart.

Page 237:4   *Paper 21:3.2*

A Creator Son could assert full sovereignty over his personal creation at any time, but he wisely chooses not to. If, prior to passing through the creature bestowals, he assumed an unearned supreme sovereignty, **the Paradise personalities resident in his local universe would withdraw.** But this has never happened throughout all the creations of time and space.

Page 1326:3   *Paper 120:1.6*

"As in each of your previous bestowals, I would remind you that I am recipient of your universe jurisdiction as brother-trustee. I exercise all authority and wield all power in your name. I function as would our Paradise Father and in accordance with your explicit request that I thus act in your stead. And such being the fact, all this delegated authority is yours again to exercise at any moment you may see fit to requisition its return. Your bestowal is, throughout, wholly voluntary. As a mortal incarnate in the realm you are without celestial endowments, but all your relinquished power may be had at any time you may choose to reinvest yourself with universe authority. **If you should choose to reinstate yourself in power and authority, remember, it will be wholly for *personal* reasons since I am the living and supreme pledge whose presence and promise guarantee the safe administration of your universe in accordance with your Father's will. Rebellion, such as has three times occurred in Nebadon, cannot occur during your absence from Salvington on this bestowal.** For the period of the Urantia bestowal the Ancients of Days have decreed that rebellion in Nebadon shall be invested with the automatic seed of its own annihilation.

---

4. The Michael Bestowals

Creator Sons—Selection of Bestowal Race
Selected from among the higher orders of mortal races—covered in Parts I and IV.

Page 239:4   *Paper 21:4.3*

Though these seven bestowals vary in the different sectors and universes, they always embrace the mortal-bestowal adventure. In the **final bestowal a Creator Son appears as a member of one of the higher mortal races on some inhabited world,** usually as a member of that racial group which contains the largest hereditary legacy of the Adamic stock which has previously been imported to upstep the physical status of the animal-origin peoples. Only once in his sevenfold career as a bestowal Son is a Paradise Michael born of woman as you have the record of the babe of Bethlehem. Only once does he live and die as a member of the lowest order of evolutionary will creatures.

Page 1344:2   *Paper 122:0.2*

After a study of the special report on the status of segregated worlds prepared by the Melchizedeks, in counsel with Gabriel, Michael finally chose Urantia as the planet whereon to enact his final bestowal. Subsequent to this decision Gabriel made a personal visit to Urantia, and, **as a result of his study of human groups and his survey of the spiritual, intellectual, racial, and geographic features of the world and its peoples, he decided that the Hebrews possessed those relative advantages which warranted their selection as the bestowal race.** Upon Michael's approval of this decision, Gabriel appointed and dispatched to Urantia **the Family Commission of Twelve—selected from among the higher orders of universe personalities—which was intrusted with the task of making an investigation of Jewish family life.** When this commission ended its labors, Gabriel was present on Urantia and received the report nominating three prospective unions as being, in the opinion of the commission, equally favorable as bestowal families for Michael's projected incarnation.

Creator Son Final Bestowal—"It is Finished"

Jesus could literally say "It is finished"—Paper 21 gives fuller meaning to Paper 187

Page 240:1   *Paper 21:4.5*

It is of record that the divine Son of last appearance on your planet was a Paradise Creator Son who had completed six phases of his bestowal career; consequently, when he gave up the conscious grasp of the incarnated life on Urantia, he could, and did, truly say, "It is finished"—it was literally finished. His death on Urantia completed his bestowal career; it was the last step in fulfilling the sacred oath of a Paradise Creator Son. And when this experience has been acquired, such Sons are supreme universe sovereigns; no longer do they rule as vicegerents of the Father but in their own right and name as "King of Kings and Lord of Lords." With certain stated exceptions these sevenfold bestowal Sons are unqualifiedly supreme in the universes of their abode. Concerning his local universe, "all power in heaven and on earth" was relegated to this triumphant and enthroned Master Son.

Page 2011:2   *Paper 187:5.6*

Jesus died royally—as he had lived. He freely admitted his kingship and remained master of the situation throughout the tragic day. He went willingly to his ignominious death, after he had provided for the safety of his chosen apostles. He wisely restrained Peter's trouble-making violence and provided that John might be near him right up to the end of his mortal existence. He revealed his true nature to the murderous Sanhedrin and reminded Pilate of the source of his sovereign authority as a Son of God. He started out to Golgotha bearing his own crossbeam **and finished up his loving bestowal by handing over his spirit of mortal acquirement to the Paradise Father. After such a life—and at such a death—the Master could truly say, "It is finished."**

PAPER 22: THE TRINITIZED SONS OF GOD

6. THE TRINITIZED AMBASSADORS

Spirit- and Son-fused Mortals
Papers 22 and 40, authored by Might Messengers, in sync with each other—from Parts I and II.

Page 248:6   Paper 22:6.1

Trinitized Ambassadors are the second order of the Trinitized Sons of Selection and like their associates, the Custodians, are recruited from two types of ascendant creatures. Not all ascending mortals are Adjuster or Father fused; some are Spirit fused, some are Son fused. **Certain of these reach Havona and attain Paradise.** From among these Paradise ascenders, candidates are selected for the Trinity embrace, and from time to time they are trinitized in classes of seven thousand. They are then commissioned in the superuniverses as Trinitized Ambassadors of the Ancients of Days. Almost one-half billion are registered on Uversa.

Page 452:2   Paper 40:10.2

Spirit-fused mortals are of the local universe; **they do not, ordinarily, ascend beyond the confines of their native realm,** beyond the boundaries of the space range of the spirit that pervades them. Son-fused ascenders likewise rise to the source of spirit endowment, for much as the Truth Spirit of a Creator Son focalizes in the associated Divine Minister, so is his "fusion spirit" implemented in the Reflective Spirits of the higher universes. Such spirit relationship between the local and the superuniverse levels of God the Sevenfold may be difficult of explanation but not of discernment, being unmistakably revealed in those children of the Reflective Spirits—the secoraphic Voices of the Creator Sons. The Thought Adjuster, hailing from the Father on Paradise, never stops until the mortal son stands face to face with the eternal God.

PAPER 24: HIGHER PERSONALITIES OF THE INFINITE SPIRIT

2. THE CENSUS DIRECTORS

Personality—When Is it Bestowed?

This is for those who like to discuss the time of personality bestowal. Six papers in Parts I, II, and III are submitted here for clarification on this issue.

Page 267:6   *Paper 24:2.8*

Census Directors register the existence of a new will creature when the first act of will is performed; they indicate the death of a will creature when the last act of will takes place. The partial emergence of will observed in the reactions of certain of the higher animals does not belong to the domain of the Census Directors. They keep count of nothing but bona fide will creatures, and **they are responsive to nothing but** *will function.* Exactly how they register the function of will, we do not know.

Page 313:3   *Paper 28:5.21*

I assure you that all these transactions of the spirit world are real, that they take place in accordance with established usages and in harmony with the immutable laws of the universal domains. **The beings of every newly created order, immediately upon receiving the breath of life, are instantly reflected on high; a living portrayal of the creature nature and potential is flashed to the superuniverse headquarters.** Thus, by means of the discerners, are the Censors made fully cognizant of exactly "what manner of spirit" has been born on the worlds of space.

Page 409:4   *Paper 37:3.7*

The Worlds of the Archangels. The seventh group of the encircling Salvington worlds, with their associated satellites, is assigned to the archangels. Sphere number one and all of its six tributary satellites are **occupied by the personality record keepers.** This enormous corps of **recorders busy themselves with keeping straight the record of each mortal of time from the moment of birth up through the universe career** until such an individual either leaves Salvington for the superuniverse regime or is "blotted out of recorded existence" by the mandate of the Ancients of Days.

Page 413:7   *Paper 37:8.4*

The Nebadon *Census Director*, Salsatia, maintains headquarters within the Gabriel sector of Salvington. **He is automatically cognizant of the birth and death of will and currently registers the exact number of will creatures functioning in the local universe. He works in close association with the personality recorders domiciled on the record worlds of the archangels.**

Page 570:1   *Paper 49:6.12*

Children who die when too young to have Thought Adjusters are repersonalized on the finaliter world of the local systems concomitant with the arrival of either parent on the mansion worlds. **A child acquires physical entity at mortal birth,** but in the matter of survival all Adjusterless children are reckoned as still attached to their parents.

Page 1130:6   *Paper 103:2.1*

Religion is functional in the human mind and has been realized in experience prior to its appearance in human consciousness. **A child has been in existence about nine months before it experiences** *birth*. But the "birth" of religion is not sudden; it is rather a gradual emergence. Nevertheless, sooner or later there is a "birth day." You do not enter the kingdom of heaven unless you have been "born again"—born of the Spirit. Many spiritual births are accompanied by much anguish of spirit and marked psychological perturbations, as many physical births are characterized by a "stormy labor" and other abnormalities of "delivery." Other spiritual births are a natural and normal growth of the recognition of supreme values with an enhancement of spiritual experience, albeit no religious development occurs without conscious effort and positive and individual determinations. Religion is never a passive experience, a negative attitude. What is termed the "birth of religion" is not directly associated with so-called conversion experiences which usually characterize religious episodes occurring later in life as a result of mental conflict, emotional repression, and temperamental upheavals.

Page 1232:5   *Paper 112:5.4*

Human beings possess identity only in the material sense. Such qualities of the self are expressed by the material mind as it functions in the energy system of the intellect. When it is said that man has identity, it is recognized that he is in possession of a mind circuit which has been placed in subordination to the acts and choosing of the will of the human personality. **But this is a material and purely temporary manifestation, just as the human embryo is a transient parasitic stage of human life.** Human beings, from a cosmic perspective, are born, live, and die in a relative instant of time; they are not enduring. But mortal personality, through its own choosing, possesses the power of transferring its seat of identity from the passing material-intellect system to the higher morontia-soul system which, in association with the Thought Adjuster, is created as a new vehicle for personality manifestation.

---

5. The Assigned Sentinels

Four and Twenty Counselors

Machiventa, the four and twenty counselors, and Assigned Sentinels with regard to the Urantia advisory council—Parts I, II, and III.

Page 269:2   *Paper 24:5.3*

Within a local creation the Assigned Sentinels serve in rotation, being transferred from system to system. They are usually changed every millennium of local universe time. They are among the highest ranking personalities stationed on a system capital, but they never participate in deliberations concerned with system affairs. **In the local systems they serve as the ex officio heads of the four and twenty administrators hailing from the evolutionary worlds, but otherwise, ascending mortals have little contact with them.** The sentinels are almost exclusively concerned in keeping the Associate Inspector of their universe fully informed on all matters relating to the welfare and state of the systems of their assignment.

Page 513:4   *Paper 45:4.1*

At the center of the seven angelic residential circles on Jerusem is located the headquarters of the Urantia advisory council, the four and twenty counselors. John the Revelator called them the four and twenty elders: "And round about the throne were four and twenty seats, and upon the seats I saw four and twenty elders sitting, clothed in white raiment." The throne in the center of this group is the judgment seat of the presiding archangel, the throne of the resurrection roll call of mercy and justice for all Satania. This judgment seat has always been on Jerusem, but the twenty-four surrounding seats were placed in position no more than nineteen hundred years ago, soon after Christ Michael was elevated to the full sovereignty of Nebadon. These four and twenty counselors are his personal agents on Jerusem, and they have authority to represent the Master Son in all matters concerning the roll calls of Satania and in many other phases of the scheme of mortal ascension on the isolated worlds of the system. They are the designated agents for executing the special requests of Gabriel and the unusual mandates of Michael.

Page 1025:1   *Paper 93:10.5*

Machiventa continued as a planetary receiver up to the times of the triumph of Michael on Urantia. **Subsequently, he was attached to the Urantia service on Jerusem as one of the four and twenty directors,** only just recently having been elevated to the position of personal ambassador on Jerusem of the Creator Son, bearing the title Vicegerent Planetary Prince of Urantia. It is our belief that, as long as Urantia remains an inhabited planet, Machiventa Melchizedek will not be fully returned to the duties of his order of sonship but will remain, speaking in the terms of time, forever a planetary minister representing Christ Michael.

---

6. *THE GRADUATE GUIDES*

Bestowals of the Second Source and Center and the Original Michael

Page 270:2,6   *Paper 24:6.4,8*

See "Bestowals of the Second Source and Center and the Original Michael" on page 154 for full text references. (7:5.5)

PAPER 25: THE MESSENGER HOSTS OF SPACE

---

2. *THE UNIVERSAL CONCILIATORS*

Universal Conciliators
The traveling courts of the Conciliators—Parts I and II.

Page 276:3   *Paper 25:2.12*

These referee trios do not pass upon matters of eternal import; the soul, the eternal prospects of a creature of time, is never placed in jeopardy by their acts. Conciliators do not deal with questions extending beyond the temporal existence and the cosmic welfare of the creatures of time. But when a commission has once accepted jurisdiction of a problem, its rulings are final and always unanimous; there is no appeal from the decision of the judge-arbiter.

Page 414:1   *Paper 37:8.6*

The *Universal Conciliators* **are the traveling courts of the universes of time and space,** functioning from the evolutionary worlds up through every section of the local universe and on beyond. These referees are registered on

Uversa; the exact number operating in Nebadon is not of record, but I estimate that there are in the neighborhood of one hundred million conciliating commissions in our local universe.

---

3. THE FAR-REACHING SERVICE OF CONSILIATORS

Conciliators vs. Union of Souls

Conciliators for the supervising personalities, and *Union of Souls* for us.

Page 276:6   *Paper 25:3.3*

1. *Conciliators to the Worlds.* Whenever the supervising personalities of the individual worlds become greatly perplexed or actually deadlocked concerning the proper procedure under existing circumstances, **and if the matter is not of sufficient importance to be brought before the regularly constituted tribunals of the realm, then, upon the receipt of a petition of two personalities, one from each contention, a conciliating commission will begin to function forthwith.**

Page 311:5   *Paper 28:5.13*

3. *The Union of Souls.* Completing the triune staff of attachment to the Perfectors of Wisdom, are these reflectors of the ideals and status of ethical relationships. Of all the problems in the universe requiring an exercise of the consummate wisdom of experience and adaptability, none are more important than those arising out of the relationships and associations of intelligent beings. **Whether in human associations of commerce and trade, friendship and marriage, or in the liaisons of the angelic hosts, there continue to arise petty frictions, minor misunderstandings too trivial even to engage the attention of conciliators but sufficiently irritating and disturbing to mar the smooth working of the universe if they were allowed to multiply and continue.** Therefore do the Perfectors of Wisdom make available the wise experience of their order as the "oil of reconciliation" for an entire superuniverse. In all this work these wise men of the superuniverses are ably seconded by their reflective associates, the **Unions of Souls,** who make available current information regarding the status of the universe and concurrently portray the Paradise ideal of the best adjustment of these perplexing problems. When not specifically directionized elsewhere, these seconaphim remain in reflective liaison with the interpreters of ethics on Paradise.

PAPER 27: MINISTRY OF THE PRIMARY SUPERNAPHIM

6. MASTERS OF PHILOSOPHY

Philosophy of Perfection—Masters of Philosophy and the Soul of Philosophy
Philosophy of perfection only on Paradise, but...

Page 303:4   *Paper 27:6.6*

These intellectual pursuits of Paradise are not broadcast; **the philosophy of perfection is available only to those who are personally present.** The encircling creations know of these teachings only from those who have passed through this experience, and who have subsequently carried this wisdom out to the universes of space.

Page 311:3   *Paper 28:5.11*

2. The Soul of Philosophy. **These wonderful teachers are also attached to the Perfectors of Wisdom and, when not otherwise directionized, remain in focal synchrony with the masters of philosophy on Paradise.** Think of stepping up to a huge living mirror, as it were, but instead of beholding the likeness of your finite and material self, of perceiving a reflection of the wisdom of divinity and the philosophy of Paradise. And if it becomes desirable to "incarnate" this **philosophy of perfection,** so to dilute it as to make it practical of application to, and assimilation by, the lowly peoples of the lower worlds, these living mirrors have only to turn their faces downward to reflect the standards and needs of another world or universe.

PAPER 28: MINISTERING SPIRITS OF THE SUPERUNIVERSES

5. THE SECONDARY SECONAPHIM

Philosophy of Perfection—Masters of Philosophy and the Soul of Philosophy

Page 311:3   *Paper 28:5.11*

See "Philosophy of Perfection—Masters of Philosophy and the Soul of Philosophy" on page 200 for full text references. (27:6.6)

Conciliators vs. Union of Souls

Page 311:5   *Paper 28:5.13*

> See "Conciliators vs. Union of Souls" on page 199 for full text references. (25:3.3)

Personality—When Is it Bestowed?

Page 313:3   *Paper 28:5.21*

> See "Personality—When Is it Bestowed?" on page 195 for full text references. (24:2.8)

6. THE TERTIARY SECONAPHIM

God—Foreknowledge

Page 315:4   *Paper 28:6.11*

> See "God—Foreknowledge" on page 133 for full text references. (3:2.6)

7. MINISTRY OF THE SECONAPHIM

Lucifer Rebellion—Interplanetary Communication
Messages dispatched by seraphic agents and Solitary Messengers in lieu of the reflective circuits.

Page 318:2   *Paper 28:7.4*

We are denied the full privilege of using these angels of the reflective order on Urantia. They are frequent visitors on your world, accompanying assigned personalities, but here they cannot freely function. **This sphere is still under partial spiritual quarantine, and some of the circuits essential to their services are not here at present.** When your world is once more restored to the reflective circuits concerned, much of the work of interplanetary and interuniverse communication will be greatly simplified and expedited. Celestial workers on Urantia encounter many difficulties because of this functional curtailment of their reflective associates. But we go on joyfully conducting our affairs with the instrumentalities at hand, notwithstanding our local deprivation of many of the services of these marvelous beings, the living mirrors of space and the presence projectors of time.

Page 607:4   *Paper 53:7.3*

All secession propaganda had to be carried on by personal effort because the broadcast service and all other avenues of interplanetary communication were suspended by the action of the system circuit supervisors. **Upon the actual outbreak of the insurrection the entire system of Satania was isolated in both the constellation and the universe circuits. During this time all incoming and outgoing messages were dispatched by seraphic agents and Solitary Messengers.** The circuits to the fallen worlds were also cut off, so that Lucifer could not utilize this avenue for the furtherance of his nefarious scheme. And these circuits will not be restored so long as the archrebel lives within the confines of Satania.

PAPER 29: THE UNIVERSE POWER DIRECTORS

---

3. THE DOMAIN OF POWER CENTERS

Power Centers

Page 323:6   *Paper 29:3.8*

See "Power Centers" on page 170 for full text references. (15:8.2)

Supergravity Presence

Page 324:2   *Paper 29:3.12*

See "Supergravity Presence" on page 108 for full text references. (0:11.7)

---

4. THE MASTER PHYSICAL CONTROLLERS

Power Centers

Page 325:5   *Paper 29:4.5*

See "Power Centers" on page 170 for full text references. (15:8.2)

PAPER 30: PERSONALITIES OF THE GRAND UNIVERSE

Revelation—Unrevealed Concepts

Save some discoveries and revelation for those to follow—Parts I and III.

Page 330:2   *Paper 30:0.2*

It is not possible to formulate comprehensive and entirely consistent classifications of the personalities of the grand universe because *all* of the groups are not revealed. It would require numerous additional papers to cover the further revelation required to systematically classify all groups. **Such conceptual expansion would hardly be desirable as it would deprive the thinking mortals of the next thousand years of that stimulus to creative speculation which these partially revealed concepts supply. It is best that man not have an overrevelation; it stifles imagination.**

Page 1086:4   *Paper 99:1.1*

Mechanical inventions and the dissemination of knowledge are modifying civilization; certain economic adjustments and social changes are imperative if cultural disaster is to be avoided. This new and oncoming social order will not settle down complacently for a millennium. The human race must become reconciled to a procession of changes, adjustments, and readjustments. **Mankind is on the march toward a new and unrevealed planetary destiny.**

Page 1109:3   *Paper 101:4.2*

Mankind should understand that we who participate in the revelation of truth are very rigorously limited by the instructions of our superiors. **We are not at liberty to anticipate the scientific discoveries of a thousand years.** Revelators must act in accordance with the instructions which form a part of the revelation mandate. We see no way of overcoming this difficulty, either now or at any future time. We full well know that, while the historic facts and religious truths of this series of revelatory presentations will stand on the records of the ages to come, within a few short years many of our statements regarding the physical sciences will stand in need of revision in consequence of additional scientific developments and new discoveries. These new developments we even now foresee, but we are forbidden to include such humanly undiscovered facts in the revelatory records. Let it be made clear that revelations are not necessarily inspired. The cosmology of these revelations is *not inspired*. It is limited by our permission for the co-ordination and sorting of present-day knowledge. While divine or spiritual insight is a gift, *human wisdom must evolve.*

1. The Paradise Classification of Living Beings

Consummator of Universe Destiny

Page 333:12   *Paper 30:1.21*

See "Consummator of Universe Destiny" on page 109 for full text references. (0:12.7)

Qualified Vicegerents of the Ultimate

Page 333:12   *Paper 30:1.21*

See "Qualified Vicegerents of the Ultimate" on page 171 for full text references. (15:10.7)

4. The Ascending Mortals

Third Day—Third Period
Period and Days interchanged—Parts I, II, and III.

Page 341:1–2   *Paper 30:4.4–5*

2. *Sleeping Survivors.* All mortals of survival status, in the custody of personal guardians of destiny, **pass through the portals of natural death and, on the third period,** personalize on the mansion worlds. Those accredited beings who have, for any reason, been unable to attain that level of intelligence mastery and endowment of spirituality which would entitle them to personal guardians, cannot thus immediately and directly go to the mansion worlds. Such surviving souls must rest in unconscious sleep until the judgment day of a new epoch, a new dispensation, the coming of a Son of God to call the rolls of the age and adjudicate the realm, and this is the general practice throughout all Nebadon. It was said of Christ Michael that, when he ascended on high at the conclusion of his work on earth, "He led a great multitude of captives." And these captives were the sleeping survivors from the days of Adam to the day of the Master's resurrection on Urantia.

The passing of time is of no moment to sleeping mortals; they are wholly unconscious and oblivious to the length of their rest. On reassembly of personality at the end of an age, those who have slept five thousand years **will react no differently than those who have rested five days.** Aside from this time delay these survivors pass on through the ascension regime identically with those who avoid the longer or shorter sleep of death.

Page 569:4   *Paper 49:6.9*

Throughout the earlier ages of an evolutionary world, few mortals go to **judgment on the third day.** But as the ages pass, more and more the personal guardians of destiny are assigned to the advancing mortals, and thus increasing numbers of these evolving creatures are **repersonalized on the first mansion world on the third day after natural death.** On such occasions the return of the Adjuster signalizes the awakening of the human soul, and this is the repersonalization of the dead just as literally as when the en masse roll is called at the end of a dispensation on the evolutionary worlds.

Page 1232:1   *Paper 112:4.6*

If the human individual survives without delay, the Adjuster, so I am instructed, registers at Divinington, proceeds to the Paradise presence of the Universal Father, returns immediately and is embraced by the Personalized Adjusters of the superuniverse and local universe of assignment, receives the recognition of the chief Personalized Monitor of Divinington, and then, at once, passes into the "realization of identity transition," **being summoned therefrom on the third period** and on the mansion world in the actual personality form made ready for the reception of the surviving soul of the earth mortal as that form has been projected by the guardian of destiny.

Page 1247:2   *Paper 113:6.4*

But angels minister to evolutionary creatures in many ways aside from the services of personal and group guardianship. **Personal guardians whose subjects do not go immediately to the mansion worlds** do not tarry there in idleness awaiting the dispensational roll calls of judgment; they are reassigned to numerous ministering missions throughout the universe.

Cooperation or Resistance to Divine Leading— Survival of Immortal Soul

Can our spiritual birthday occur on the mansion worlds? Does it have to happen on Urantia? Only conscious resistance to the Adjuster's leading can prevent the survival of the evolving immortal soul. All will creatures are to experience one true opportunity to make one undoubted, self-conscious, and final choice. All four Parts touch on this.

Page 341:5   *Paper 30:4.8*

The physical body of mortal flesh is not a part of the reassembly of the sleeping survivor; the physical body has returned to dust. The seraphim of assignment sponsors the new body, the morontia form, as the new life vehicle for the immortal soul and for the indwelling of the returned Adjuster. The Adjuster is the custodian of the spirit transcript of the mind of the sleeping survivor. The assigned seraphim is the keeper of the surviving identity—the immortal soul—as far as it has evolved. And when these two, the Adjuster and the seraphim, reunite their personality trusts, the new individual constitutes the resurrection of the old personality, the survival of the evolving morontia identity of the soul. Such a reassociation of soul and Adjuster is quite properly called a resurrection, a reassembly of personality factors; but even this does not entirely explain the reappearance of the surviving *personality*. Though you will probably never understand the fact of such an inexplicable transaction, **you will sometime experientially know the truth of it if you do not reject the plan of mortal survival.**

Page 565:1   *Paper 49:4.9*

But mortal mind without immortal spirit cannot survive. The mind of man is mortal; only the bestowed spirit is immortal. Survival is dependent on spiritualization by the ministry of the Adjuster—on the birth and evolution of the immortal soul; **at least, there must not have developed an antagonism towards the Adjuster's mission of effecting the spiritual transformation of the material mind.**

Page 1130:6   *Paper 103:2.1*

Religion is functional in the human mind and has been realized in experience prior to its appearance in human consciousness. A child has been in existence about nine months before it experiences *birth*. But the "birth" of religion is not sudden; it is rather a gradual emergence. **Nevertheless, sooner or later there is a "birth day." You do not enter the kingdom of heaven unless you have been "born again"—born of the Spirit.** Many spiritual births are accompanied by much anguish of spirit and marked psychological perturbations, as many physical births are characterized by a "stormy labor" and other abnormalities of "delivery." Other spiritual births are a natural and normal growth of the recognition of supreme values with an enhancement of spiritual experience, albeit no religious development occurs without conscious effort and positive and individual determinations. Religion is never a passive experience, a negative attitude. What is termed the "birth of religion" is not directly associated with so-called conversion experiences which usually characterize religious episodes occurring later in life as a result of mental conflict, emotional repression, and temperamental upheavals.

Page 1206:3   *Paper 110:3.5*

Confusion, being puzzled, even sometimes discouraged and distracted, does not necessarily signify resistance to the leadings of the indwelling Adjuster. Such attitudes may sometimes connote lack of active co-operation with the divine Monitor and may, therefore, somewhat delay spiritual progress, but such intellectual emotional difficulties do not in the least interfere with the certain survival of the God-knowing soul. Ignorance alone can never prevent survival; neither can confusional doubts nor fearful uncertainty. **Only conscious resistance to the Adjuster's leading can prevent the survival of the evolving immortal soul.**

Page 1233:5   *Paper 112:5.9*

This does not mean that human beings are to enjoy a second opportunity in the face of the rejection of a first, not at all. **But it does signify that all will creatures are to experience one true opportunity to make one undoubted, self-conscious, and final choice.** The sovereign Judges of the universes will not deprive any being of personality status who has not finally and fully made the eternal choice; the soul of man must and will be given full and ample opportunity to reveal its true intent and real purpose.

Page 1774:6   *Paper 160:1.14*

In a continually changing world, in the midst of an evolving social order, it is impossible to maintain settled and established goals of destiny. Stability of personality can be experienced only by those who have discovered and embraced the living God as the eternal goal of infinite attainment. **And thus to transfer one's goal from time to eternity, from earth to Paradise, from the human to the divine, requires that man shall become regenerated, converted, be born again; that he shall become the re-created child of the divine spirit; that he shall gain entrance into the brotherhood of the kingdom of heaven.** All philosophies and religions which fall short of these ideals are immature. The philosophy which I teach, linked with the gospel which you preach, represents the new religion of maturity, the ideal of all future generations. And this is true because our ideal is final, infallible, eternal, universal, absolute, and infinite.

Page 1795:2   *Paper 162:5.3*

"I have already told you that I am going away, and that you will seek me and not find me, for where I am going you cannot come. You who would reject this light are from beneath; I am from above. You who prefer to sit in darkness are of this world; I am not of this world, and I live in the eternal light of the Father of lights. You all have had abundant opportunity to learn who I am, but you shall have still other evidence confirming the identity of the Son of Man. **I am the light of life, and every one who deliberately and with understanding rejects this saving light shall die in his sins.** Much I have to tell you, but you are unable to receive my words. However, he who sent me is true and faithful; my Father loves even his erring children. And all that my Father has spoken I also proclaim to the world.

Goodness and Evil

Page 342:4   *Paper 30:4.14*

See "Goodness and Evil" on page 140 for full text references. (4:3.6)

Corps of the Finality—Allegiance to the Paradise Trinity

Oath of allegiance expounded—Parts I and III, almost 1,000 pages apart.

Page 345:12   *Paper 31:0.6*

We have no idea as to the nature of the future organization of this extraordinary group, but the finaliters are now wholly a self-governing body. They choose their own permanent, periodic, and assignment leaders and directors. **No outside influence can ever be brought to bear upon their policies, and their oath of allegiance is only to the Paradise Trinity.**

Page 1292:1   *Paper 117:7.7*

When mortal ascenders are admitted to the finaliter corps of Paradise, they take an oath to the Paradise Trinity, and in taking this oath of allegiance, they are thereby pledging eternal fidelity to God the Supreme, who is the Trinity as comprehended by all finite creature personalities. Subsequently, as the finaliter companies function throughout the evolving universes, they are solely amenable to the mandates of Paradise origin until the eventful times of the settling of local universes in light and life. As the new governmental organizations of these perfected creations begin to be reflective of the emerging sovereignty of the Supreme, we observe that the outlying finaliter companies then acknowledge the jurisdictional authority of such new governments. It appears that God the Supreme is evolving as the unifier of the evolutionary Corps of the Finality, but it is highly probable that the eternal destiny of these seven corps will be directed by the Supreme as a member of the Ultimate Trinity.

---

1. THE HAVONA NATIVES

Havona Natives

Page 346:5   *Paper 31:1.4*

See "Havona Natives" on page 182 for full text references. (19:6.1)

4. Adopted Seraphim

Father Indwelt—Adjusters and Other Types of Spirit

Seraphic guardians can be Father indwelt (not Adjuster fused) at time of being mustered into the Mortal Corps of Finality—at the end of Part I and near the end of Part II.

Page 348:6   *Paper 31:4.1*

Many of the faithful seraphic guardians of mortals are permitted to go through the ascendant career with their human wards, and many of these guardian angels, after **becoming Father fused,** join their subjects in taking the finaliter oath of eternity and forever accept the destiny of their mortal associates. Angels who pass through the ascending experience of mortal beings may share the destiny of human nature; they may equally and eternally be mustered into this Corps of the Finality. Large numbers of the adopted and glorified seraphim are attached to the various nonmortal finaliter corps.

Page 629:9   *Paper 55:4.28*

On the evolutionary worlds a being must humanize to receive a Thought Adjuster. **All ascendant members of the Mortal Corps of Finaliters have been Adjuster indwelt and fused except seraphim, and they are Father indwelt by another type of spirit** at the time of being mustered into this corps.

10. The Ultimate Adventure

Corps of the Finality—Future Destiny

Parts I and III getting together on this topic.

Page 354:5   *Paper 31:10.12*

As we view this triune development, embracing creatures, universes, and Deity, can we be criticized for anticipating that something new and unrevealed is approaching culmination in the master universe? Is it not natural that we should associate this agelong mobilization and organization of physical universes on such a hitherto unknown scale and the personality emergence of the Supreme Being with this stupendous scheme of upstepping the mortals of time to divine perfection and with their subsequent mobilization on Paradise in the Corps of the Finality—a designation and destiny enshrouded in universe mystery? **It is increasingly the belief of all Uversa that the assembling Corps of the Finality are destined to some future** service in the universes of outer space, where we already are able to identify the clustering of at least seventy thousand aggregations of matter, each of which is greater than any one of the present superuniverses.

Page 1239:5   *Paper 112:7.16*

You have been instructed to a certain extent about the organization and personnel of the central universe, the superuniverses, and the local universes; you have been told something about the character and origin of some of the various personalities who now rule these far-flung creations. **You have also been informed that there are in process of organization vast galaxies of universes far out beyond the periphery of the grand universe, in the first outer space level. It has also been intimated in the course of these narratives that the Supreme Being is to disclose his unrevealed tertiary function in these now uncharted regions of outer space; and you have also been told that the finaliters of the Paradise corps are the experiential children of the Supreme.**

PAPER 32: THE EVOLUTION OF LOCAL UNIVERSES

---

2. UNIVERSE ORGANIZATION

Neighboring Universes
Avalon and Henselon share some things with us as noted in Parts II and III.

Page 360:1   *Paper 32:2.12*

The universe of Nebadon now swings far to the south and east in the superuniverse circuit of Orvonton. **The nearest neighboring universes are: Avalon, Henselon,** Sanselon, Portalon, Wolvering, Fanoving, and Alvoring.

Page 393:8   *Paper 35:9.8*

Of all the local universes in Orvonton, our universe has, with the exception of Henselon, lost the largest number of this order of Sons. On Uversa it is the consensus that we have had so much administrative trouble in Nebadon because our Sons of the Lanonandek order have been created with such a large degree of personal liberty in choosing and planning. I do not make this observation by way of criticism. The Creator of our universe has full authority and power to do this. It is the contention of our high rulers that, while such free-choosing Sons make excessive trouble in the earlier ages of the universe, when things are fully sifted and finally settled, the gains of higher loyalty and fuller volitional service on the part of these thoroughly tested Sons will far more than compensate for the confusion and tribulations of earlier times.

Page 420:6   *Paper 38:5.1*

Seraphim spend their first millennium as noncommissioned observers on Salvington and its associated world schools. The second millennium is spent on the seraphic worlds of the Salvington circuit. Their central training school is now presided over by the first one hundred thousand Nebadon seraphim, and at their head is the original or first-born angel of this local universe. **The first created group of Nebadon seraphim were trained by a corps of one thousand seraphim from Avalon;** subsequently our angels have been taught by their own seniors. The Melchizedeks also have a large part in the education and training of all local universe angels—seraphim, cherubim, and sanobim.

Page 742:7   *Paper 66:2.7*

Accordingly, fifty males and fifty females of the Andon and Fonta posterity, representing the survival of the best strains of that unique race, were chosen by the Life Carriers. With one or two exceptions these Andonite contributors to the advancement of the race were strangers to one another. They were assembled from widely separated places by co-ordinated Thought Adjuster direction and seraphic guidance at the threshold of the planetary headquarters of the Prince. **Here the one hundred human subjects were given into the hands of the highly skilled volunteer commission from Avalon, who directed the material extraction of a portion of the life plasm of these Andon descendants.** This living material was then transferred to the material bodies constructed for the use of the one hundred Jerusemite members of the Prince's staff. Meantime, these newly arrived citizens of the system capital were held in the sleep of seraphic transport.

Page 759:8   *Paper 67:6.5*

The affairs of Urantia were for a long time administered by a council of planetary receivers, twelve Melchizedeks, confirmed by the mandate of the senior constellation ruler, the Most High Father of Norlatiadek. **Associated with the Melchizedek receivers was an advisory council consisting of: one of the loyal aids of the fallen Prince, the two resident Life Carriers, a Trinitized Son in apprenticeship training, a volunteer Teacher Son, a Brilliant Evening Star of Avalon (periodically),** the chiefs of seraphim and cherubim, advisers from two neighboring planets, the director general of subordinate angelic life, and Van, the commander in chief of the midway creatures. And thus was Urantia governed and administered until the arrival of Adam. It is not strange that the courageous and loyal Van was assigned a place on the council of planetary receivers which for so long administered the affairs of Urantia.

Page 857:3   *Paper 77:2.6*

Thus did the Nodite peoples arise out of certain peculiar and unexpected modifications occurring in the life plasm which had been transferred from the bodies of the Andonite **contributors to those of the corporeal staff members by the Avalon surgeons.**

---

3. THE EVOLUTIONARY IDEA

Perfection—Innate and Evolved
Two types of finite will creatures—innate perfection and evolved perfection.

Page 361:5   *Paper 32:3.10*

The fact of **animal evolutionary origin** does not attach stigma to any personality in the sight of the universe as **that is the exclusive method of producing one of the two basic types of finite intelligent will creatures.** When the heights of perfection and eternity are attained, all the more honor to those who began at the bottom and joyfully climbed the ladder of life, round by round, and who, when they do reach the heights of glory, will have gained a personal experience which embodies an actual knowledge of every phase of life from the bottom to the top.

Page 362:1   *Paper 32:3.12*

In the universes encircling Havona there are provided only a sufficient number of perfect creatures to meet the need for pattern teacher guides for those who are ascending the evolutionary scale of life. The experiential nature of the evolutionary type of personality is the natural cosmic complement of the ever-perfect natures of the Paradise-Havona creatures. **In reality, both perfect and perfected creatures are incomplete as regards finite totality.** But in the complemental association of the existentially perfect creatures of the Paradise-Havona system with the experientially perfected finaliters ascending from the evolutionary universes, both types find release from inherent limitations and thus may conjointly attempt to reach the sublime heights of the ultimate of creature status.

Page 362:4   *Paper 32:3.15*

The two prime manifestations of finite reality, innate perfection and evolved perfection, be they personalities or universes, are co-ordinate, dependent, and integrated. Each requires the other to achieve completion of function, service, and destiny.

212

4. GOD'S RELATION TO A LOCAL UNIVERSE

Most Highs

Page 363:2   *Paper 32:4.4*

 See "Most Highs" on page 126 for full text references. (3:1.10)

Father Contact

Page 363:6   *Paper 32:4.8*

 See "Father Contact" on page 114 for full text references. (1:7.1)

Personality and Personality Circuit

Page 363:6   *Paper 32:4.8*

 See "Personality and Personality Circuit" on page 120 for full text references. (2:2.5–6)

5. THE ETERNAL AND DIVINE PURPOSE

Death

Page 364:6   *Paper 32:5.4*

 See "Death" on page 167 for full text references. (14:5.10)

PAPER 33: ADMINISTRATION OF THE LOCAL UNIVERSE

1. MICHAEL OF NEBADON

Father-Son and Spirit-Mother
Father-Son and Spirit-Mother as described in two papers in Part II.

Page 366:3   *Paper 33:1.2*

 Our Creator Son is not the Eternal Son, the existential Paradise associate of the Universal Father and the Infinite Spirit. Michael of Nebadon is not a member of the Paradise Trinity. Nevertheless our Master Son possesses in his realm all of the divine attributes and powers that the Eternal Son himself would manifest were he actually to be present on Salvington and functioning in Nebadon. Michael possesses even additional power and authority, for he not only personifies the Eternal Son but also fully represents and actually embodies the personality presence of the Universal Father to and in this local universe. **He even represents the Father-Son.** These relationships constitute a Creator Son the most powerful, versatile, and influential of all divine beings who are capable of direct administration of evolutionary universes and of personality contact with immature creature beings.

Page 406:1   *Paper 37:0.1*

At the head of all personality in Nebadon stands the Creator and Master Son, Michael, the universe father and sovereign. Co-ordinate in divinity and complemental in creative attributes is the local universe Mother Spirit, the Divine Minister of Salvington. **And these creators are in a very literal sense the Father-Son and the Spirit-Mother of all the native creatures of Nebadon.**

---

2. THE SOVEREIGN OF NEBADON

Love of the Creator Son

Page 367:3   *Paper 33:2.1*

See "Love of the Creator Son" on page 150 for full text references. (6:3.5)

---

3. THE UNIVERSE SON AND SPIRIT

Immanuel and Universe Mother Spirit—Share Same Number of Order

Immanuel and our Universe Mother Spirit have this in common.

Page 368:2   *Paper 33:3.2*

The Universe Mother Spirit of Salvington, the associate of Michael in the control and administration of Nebadon, **is of the sixth group of Supreme Spirits, being the 611,121st of that order.** She volunteered to accompany Michael on the occasion of his liberation from Paradise obligations and has ever since functioned with him in creating and governing his universe.

Page 370:7   *Paper 33:5.2*

Immanuel of Salvington, number 611,121 of the sixth order of Supreme Trinity Personalities, is a being of sublime dignity and of such superb condescension that he refuses the worship and adoration of all living creatures. He bears the distinction of being the only personality in all Nebadon who has never acknowledged subordination to his brother Michael. He functions as adviser to the Sovereign Son but gives counsel only on request. In the absence of the Creator Son he might preside over any high universe council but would not otherwise participate in the executive affairs of the universe except as requested.

---

4. GABRIEL—THE CHIEF EXECUTIVE

Gabriel—Commander in Chief
As commander in chief, Gabriel takes action.

Page 370:3   *Paper 33:4.6*

Gabriel is the chief officer of execution for superuniverse mandates relating to nonpersonal affairs in the local universe. Most matters pertaining to mass judgment and dispensational resurrections, adjudicated by the Ancients of Days, are also delegated to Gabriel and his staff for execution. Gabriel is thus the combined chief executive of both the super- and the local universe rulers. He has at his command an able corps of administrative assistants, created for their special work, who are unrevealed to evolutionary mortals. In addition to these assistants, Gabriel may employ any and all of the orders of celestial beings functioning in Nebadon, **and he is also the commander in chief of "the armies of heaven"**—the celestial hosts.

Page 605:8   *Paper 53:5.4*

Since Michael elected to remain aloof from the actual warfare of the Lucifer rebellion, **Gabriel called his personal staff together on Edentia and, in counsel with the Most Highs, elected to assume command of the loyal hosts of Satania.** Michael remained on Salvington while Gabriel proceeded to Jerusem, and establishing himself on the sphere dedicated to the Father—the same Universal Father whose personality Lucifer and Satan had questioned—in the presence of the forgathered hosts of loyal personalities, he displayed the banner of Michael, the material emblem of the Trinity government of all creation, the three azure blue concentric circles on a white background.

5. The Trinity Ambassadors

Immanuel and Universe Mother Spirit—Share Same Number of Order

Page 370:7   Paper 33:5.2

> See "Immanuel and Universe Mother Spirit—Share Same Number of Order" on page 214 for full text references. (33:3.2)

6. General Administration

Gabriel and Father Melchizedek

Immanuel was fully and efficiently responsible for the security and unbroken administration of our universe, Nebadon, during Michael's bestowal on Urantia.

Page 371:4   Paper 33:6.2

The Father Melchizedek is Gabriel's first assistant. **When the Bright and Morning Star is absent from Salvington, his responsibilities are assumed by this original Melchizedek Son.**

Page 385:1   Paper 35:1.2

In the universe of Nebadon the Father Melchizedek acts as the first executive associate of the Bright and Morning Star. Gabriel is occupied more with universe policies, Melchizedek with practical procedures. Gabriel presides over the regularly constituted tribunals and councils of Nebadon, Melchizedek over the special, extraordinary, and emergency commissions and advisory bodies. **Gabriel and the Father Melchizedek are never away from Salvington at the same time, for in Gabriel's absence the Father Melchizedek functions as the chief executive of Nebadon.**

Page 1326:1   Paper 120:1.4

"Throughout your Urantia bestowal you need be concerned with but one thing, the unbroken communion between you and your Paradise Father; and it will be by the perfection of such a relationship that the world of your bestowal, even all the universe of your creation, will behold a new and more understandable revelation of your Father and my Father, the Universal Father of all. Your concern, therefore, has only to do with your personal life on Urantia. **I will be fully and efficiently responsible for the security and unbroken administration of your universe from the moment of your**

216

voluntary relinquishment of authority until you return to us as **Universe Sovereign,** confirmed by Paradise, and receive back from my hands, not the vicegerent authority which you now surrender to me, but, instead, the supreme power over, and jurisdiction of, your universe.

Page 1753:4   *Paper 158:1.8*

When the three had been fast asleep for about half an hour, they were suddenly awakened by a near-by crackling sound, and much to their amazement and consternation, on looking about them, they beheld Jesus in intimate converse with two brilliant beings clothed in the habiliments of the light of the celestial world. And Jesus' face and form shone with the luminosity of a heavenly light. These three conversed in a strange language, but from certain things said, Peter erroneously conjectured that the beings with Jesus were Moses and Elijah; **in reality, they were Gabriel and the Father Melchizedek.** The physical controllers had arranged for the apostles to witness this scene because of Jesus' request.

Universe Broadcasts—Simultaneous Messages
Two different authors in Part II on simultaneous messages that are dispatched and received.

Page 371:7   *Paper 33:6.5*

From Salvington, broadcasts are simultaneously directed to the constellation headquarters, the system headquarters, and to individual planets. All higher orders of celestial beings are able to utilize this service for communication with their fellows scattered throughout the universe. **The universe broadcast is extended to all inhabited worlds regardless of their spiritual status. Planetary intercommunication is denied only those worlds under spiritual quarantine.**

Page 431:4   *Paper 39:2.15*

Broadcasters—receivers and dispatchers—are a specialized subdivision of the seraphic recorders, being concerned with the dispatch of records and with the dissemination of essential information. Their work is of a high order, being so multicircuited that **144,000 messages can simultaneously traverse the same lines of energy.** They adapt the higher ideographic techniques of the superaphic chief recorders and with these common symbols maintain reciprocal contact with both the intelligence co-ordinators of the tertiary supernaphim and the glorified intelligence co-ordinators of the Seraphic Corps of Completion.

PAPER 34: THE LOCAL UNIVERSE MOTHER SPIRIT

Creator Son and Creative Spirit

Page 374:1   *Paper 34:0.1*

> See "Creator Son and Creative Spirit" on page 157 for full text references. (8:3.4)

5. *THE MINISTRY OF THE SPIRIT*

Thought Adjusters—Preparation for Bestowal
The indwelling ministry of the adjutant mind-spirits and the mind being encircuited in the Holy Spirit are prerequisites for Thought Adjuster.

Page 379:3   *Paper 34:5.3*

Mortal man first experiences the ministry of the Spirit in conjunction with mind when the purely animal mind of evolutionary creatures develops reception capacity for the adjutants of worship and of wisdom. This ministry of the sixth and seventh adjutants indicates mind evolution crossing the threshold of spiritual ministry. And immediately are such minds of worship- and wisdom-function included in the spiritual circuits of the Divine Minister.

Page 1003:4   *Paper 92:0.4*

3. *The Holy Spirit*—this is the initial supermind bestowal, and it unfailingly appears in all bona fide human personalities. This ministry to a worship-craving and wisdom-desiring mind creates the capacity to self-realize the postulate of human survival, both in theologic concept and as an actual and factual personality experience.

Page 1187:1   *Paper 108:2.2*

The Adjusters cannot invade the mortal mind until it has been duly prepared by the indwelling ministry of the adjutant mind-spirits and encircuited in the Holy Spirit. And it requires the co-ordinate function of all seven adjutants to thus qualify the human mind for the reception of an Adjuster. Creature mind must exhibit the worship outreach and indicate wisdom function by exhibiting the ability to choose between the emerging values of good and evil—moral choice.

Page 1187:4   *Paper 108:2.5*

Before the times of the pouring out of the Spirit of Truth upon the inhabitants of an evolutionary world, the Adjusters' bestowal appears to be determined by many spirit influences and personality attitudes. We do not fully comprehend the laws governing such bestowals; we do not understand just what determines the release of the Adjusters who have volunteered to indwell such evolving minds. But we do observe numerous influences and conditions which appear to be associated with the arrival of the Adjusters in such minds prior to the bestowal of the Spirit of Truth, and they are:

Page 1198:2   *Paper 109:3.7*

In many of the early evolutionary races of Urantia, three groups of beings existed. There were those who were so animalistic that they were utterly lacking in Adjuster capacity. **There were those who exhibited undoubted capacity for Adjusters and promptly received them when the age of moral responsibility was attained.** There was a third class who occupied a borderline position; they had capacity for Adjuster reception, but the Monitors could only indwell the mind on the personal petition of the individual.

6. *The Spirit in Man*

Kingdom Within

The *new kingdom*—where is it leading us and how are we to get there? These **bold** passages below are insightful and uplifting. Nothing but good can come from a sincere study of these passages covering Parts II, III, and IV—spans 14 papers with over 1700 pages between the first and last reference.

Page 381:1   *Paper 34:6.7*

Those who have received and recognized the indwelling of God have been born of the Spirit. "You are the temple of God, and the spirit of God dwells in you." It is not enough that this spirit be poured out upon you; the divine **Spirit must dominate and control every phase of human experience.**

Page 804:15   *Paper 71:4.3*

And this progress in the arts of civilization leads directly to the realization of the highest human and divine goals of mortal endeavor—the social achievement of the brotherhood of man and the personal status of God-consciousness, **which becomes revealed in the supreme desire of every individual to do the will of the Father in heaven.**

Page 1124:3   *Paper 102:6.1*

The philosophic elimination of religious fear and the steady progress of science add greatly to the mortality of false gods; and even though these casualties of man-made deities may momentarily befog the spiritual vision, they eventually destroy that ignorance and superstition which so long obscured the living God of eternal love. The relation between the creature and the Creator is a living experience, a dynamic religious faith, which is not subject to precise definition. To isolate part of life and call it religion is to disintegrate life and to distort religion. **And this is just why the God of worship claims all allegiance or none.**

Page 1175:1   *Paper 106:9.12*

To material, evolutionary, finite creatures, **a life predicated on the living of the Father's will leads directly to the attainment of spirit supremacy in the personality arena and brings such creatures one step nearer the comprehension of the Father-Infinite.** Such a Father life is one predicated on truth, sensitive to beauty, and dominated by goodness. Such a God-knowing person is inwardly illuminated by worship and outwardly devoted to the wholehearted service of the universal brotherhood of all personalities, a service ministry which is filled with mercy and motivated by love, while all these life qualities are unified in the evolving personality on ever-ascending levels of cosmic wisdom, self-realization, God-finding, and Father worship.

Page 1434:2   *Paper 130:4.3*

The highest level to which a finite creature can progress is the recognition of the Universal Father and the knowing of the Supreme. And even then such beings of finality destiny go on experiencing change in the motions of the physical world and in its material phenomena. Likewise do they remain aware of selfhood progression in their continuing ascension of the spiritual universe and of growing consciousness in their deepening appreciation of, and response to, the intellectual cosmos. Only in the perfection, harmony, and unanimity of will can the creature become as one with the Creator; and such a state of divinity is attained and maintained only by the creature's continuing to live in time and eternity by consistently conforming his finite personal will to the divine will of the Creator. **Always must the desire to do the Father's will be supreme in the soul and dominant over the mind of an ascending son of God.**

Page 1521:1   *Paper 136:8.6*

Jesus chose to establish the kingdom of heaven in the hearts of mankind by natural, ordinary, difficult, and trying methods, just such procedures as his earth children must subsequently follow in their work of enlarging and extending that heavenly kingdom. For well did the Son of Man know that it would be "through much tribulation that many of the children of all ages would enter into the kingdom." Jesus was now passing through the great test of civilized man, to have power and steadfastly refuse to use it for purely selfish or personal purposes.

Page 1522:1   *Paper 136:9.2*

The Jews envisaged a deliverer who would come in miraculous power to cast down Israel's enemies and establish the Jews as world rulers, free from want and oppression. Jesus knew that this hope would never be realized. **He knew that the kingdom of heaven had to do with the overthrow of evil in the hearts of men, and that it was purely a matter of spiritual concern.** He thought out the advisability of inaugurating the spiritual kingdom with a brilliant and dazzling display of power—and such a course would have been permissible and wholly within the jurisdiction of Michael—but he fully decided against such a plan. He would not compromise with the revolutionary techniques of Caligastia. He had won the world in potential by submission to the Father's will, and he proposed to finish his work as he had begun it, and as the Son of Man.

Page 1522:5   *Paper 136:9.6*

Rome was mistress of the Western world. The Son of Man, now in isolation and achieving these momentous decisions, with the hosts of heaven at his command, represented the last chance of the Jews to attain world dominion; but this earthborn Jew, who possessed such tremendous wisdom and power, declined to use his universe endowments either for the aggrandizement of himself or for the enthronement of his people. He saw, as it were, "the kingdoms of this world," and he possessed the power to take them. The Most Highs of Edentia had resigned all these powers into his hands, but he did not want them. The kingdoms of earth were paltry things to interest the Creator and Ruler of a universe. **He had only one objective, the further revelation of God to man, the establishment of the kingdom, the rule of the heavenly Father in the hearts of mankind.**

Page 1535:5   *Paper 137:7.13*

While Jesus later directed that the apostles should go forth, as John had, preaching the gospel and instructing believers, he laid emphasis on the proclamation of the "good tidings of the kingdom of heaven." He unfailingly impressed upon his associates that they must "show forth love, compassion, and sympathy." **He early taught his followers that the kingdom of heaven was a spiritual experience having to do with the enthronement of God in the hearts of men.**

Page 1537:2   *Paper 137:8.15*

"Entrance into the Father's kingdom waits not upon marching armies, upon overturned kingdoms of this world, nor upon the breaking of captive yokes. **The kingdom of heaven is at hand, and all who enter therein shall find abundant liberty and joyous salvation.**

Page 1568:5   *Paper 140:1.2*

"**The new kingdom which my Father is about to set up in the hearts of his earth children is to be an everlasting dominion.** There shall be no end of this rule of my Father in the hearts of those who desire to do his divine will. I declare to you that my Father is not the God of Jew or gentile. Many shall come from the east and from the west to sit down with us in the Father's kingdom, while many of the children of Abraham will refuse to enter this new brotherhood of the rule of the Father's spirit in the hearts of the children of men.

Page 1588:4   *Paper 141:2.1*

The night before they left Pella, Jesus gave the apostles some further instruction with regard to the new kingdom. Said the Master: "You have been taught to look for the coming of the kingdom of God, and now I come announcing that this long-looked-for kingdom is near at hand, even that it is already here and in our midst. In every kingdom there must be a king seated upon his throne and decreeing the laws of the realm. And so have you developed a concept of the kingdom of heaven as a glorified rule of the Jewish people over all the peoples of the earth with Messiah sitting on David's throne and from this place of miraculous power promulgating the laws of all the world. But, my children, you see not with the eye of faith, and you hear not with the understanding of the spirit. **I declare that the kingdom of heaven is the realization and acknowledgment of God's rule within the hearts of men.** True, there is a King in this kingdom, and that King is my Father and your Father. We are indeed his loyal subjects, but far transcending that fact is the transforming truth that we are his *sons*. In my life this truth is to become manifest to all. Our Father also sits upon a throne, but not one made with hands. The throne of the Infinite is the eternal dwelling place of the Father in the heaven of heavens; he fills all things and proclaims his laws to universes upon universes. **And the Father also rules within the hearts of his children on earth by the spirit which he has sent to live within the souls of mortal men.**

Page 1609:5   *Paper 143:2.4*

"By the old way you seek to suppress, obey, and conform to the rules of living; **by the new way you are first *transformed* by the Spirit of Truth** and thereby strengthened in your inner soul by the constant spiritual renewing of your mind, and so are you endowed with the power of the certain and joyous performance of the gracious, acceptable, and perfect will of God. **Forget not—it is your personal faith in the exceedingly great and precious promises of God that ensures your becoming partakers of the divine nature.** Thus by your faith and the spirit's transformation, you become in reality the temples of God, and his spirit actually dwells within you. If, then, the spirit dwells within you, you are no longer bondslaves of the flesh but free and liberated sons of the spirit. The new law of the spirit endows you with the liberty of self-mastery in place of the old law of the fear of self-bondage and the slavery of self-denial.

Page 1865:1   *Paper 170:5.11*

The kingdom, to the Jews, was the Israelite *community*; to the gentiles it became the Christian *church*. **To Jesus the kingdom was the sum of those *individuals* who had confessed their faith in the fatherhood of God, thereby declaring their wholehearted dedication to the doing of the will of God, thus becoming members of the spiritual brotherhood of man.**

Page 1866:2   *Paper 170:5.19*

Sooner or later another and greater John the Baptist is due to arise proclaiming "the kingdom of God is at hand"—**meaning a return to the high spiritual concept of Jesus, who proclaimed that the kingdom is the will of his heavenly Father dominant and transcendent in the heart of the believer**—and doing all this without in any way referring either to the visible church on earth or to the anticipated second coming of Christ. There must come a revival of the *actual* teachings of Jesus, such a restatement as will undo the work of his early followers who went about to create a sociophilosophical system of belief regarding the fact of Michael's sojourn on earth. In a short time the teaching of this story *about* Jesus nearly supplanted the preaching of Jesus' gospel of the kingdom. In this way a historical religion displaced that teaching in which Jesus had blended man's highest moral ideas and spiritual ideals with man's most sublime hope for the future—eternal life. And that was the gospel of the kingdom.

Page 1951:2   *Paper 180:6.1*

After Peter, James, John, and Matthew had asked the Master numerous questions, he continued his farewell discourse by saying: "And I am telling you about all this before I leave you in order that you may be so prepared for what is coming upon you that you will not stumble into serious error. The authorities will not be content with merely putting you out of the synagogues; I warn you the hour draws near when they who kill you will think they are doing a service to God. And all of these things they will do to you and to those whom you lead into the kingdom of heaven because they do not know the Father. They have refused to know the Father by refusing to receive me; and they refuse to receive me when they reject you, provided you have kept my new commandment that you love one another even as I have loved you. I am telling you in advance about these things so that, when your hour comes, as mine now has, you may be strengthened in the knowledge that all was known to me, and that my spirit shall be with you in all your sufferings for my sake and the gospel's. It was for this purpose that I have been talking so plainly to you from the very beginning. I have even warned you that a man's foes may be those of his own household. Although this gospel of the kingdom never fails to bring great peace to the soul of the individual believer, **it will not bring peace on earth until man is willing to believe my teaching wholeheartedly and to establish the practice of doing the Father's will as the chief purpose in living the mortal life.**

Page 2061:6   *Paper 194:2.8*

Jesus lived a life which is a revelation of man submitted to the Father's will, not an example for any man literally to attempt to follow. This life in the flesh, together with his death on the cross and subsequent resurrection, presently became a new gospel of the ransom which had thus been paid in order to purchase man back from the clutch of the evil one—from the condemnation of an offended God. Nevertheless, even though the gospel did become greatly distorted, it remains a fact that this new message about Jesus carried along with it many of the fundamental truths and teachings of his earlier gospel of the kingdom. **And, sooner or later, these concealed truths of the fatherhood of God and the brotherhood of men will emerge to effectually transform the civilization of all mankind.**

Page 2088:3–4   *Paper 196:0.8–9*

The faith of Jesus visualized all spirit values as being found in the kingdom of God; therefore he said, "Seek first the kingdom of heaven." Jesus saw in the advanced and ideal fellowship of the kingdom the achievement and fulfillment of the "will of God." The very heart of the prayer which he taught his disciples was, "Your kingdom come; your will be done." Having thus conceived of the kingdom as comprising the will of God, he devoted himself to the cause of its realization with amazing self-forgetfulness and unbounded enthusiasm. But in all his intense mission and throughout his extraordinary life there never appeared the fury of the fanatic nor the superficial frothiness of the religious egotist.

The Master's entire life was consistently conditioned by this living faith, this sublime religious experience. **This spiritual attitude wholly dominated his thinking and feeling, his believing and praying, his teaching and preaching.** This personal faith of a son in the certainty and security of the guidance and protection of the heavenly Father imparted to his unique life a profound endowment of spiritual reality. And yet, despite this very deep consciousness of close relationship with divinity, this Galilean, God's Galilean, when addressed as Good Teacher, instantly replied, "Why do you call me good?" When we stand confronted by such splendid self-forgetfulness, we begin to understand how the Universal Father found it possible so fully to manifest himself to him and reveal himself through him to the mortals of the realms.

---

7. THE SPIRIT AND THE FLESH

Lucifer Rebellion—Caligastic

Caligastic rebellion—phrase used in Parts II and III, over 500 pages apart.

Page 382:6   *Paper 34:7.6*

Notwithstanding this double disaster to man's nature and his environment, present-day mortals would experience less of this apparent warfare between the flesh and the spirit if they would enter the spirit kingdom, wherein the faith sons of God enjoy comparative deliverance from the slave-bondage of the flesh in the enlightened and liberating service of wholehearted devotion to doing the will of the Father in heaven. Jesus showed mankind the new way of mortal living whereby human beings may very largely escape the dire consequences of the **Caligastic rebellion** and most effectively compensate for the deprivations resulting from the Adamic default. "The spirit of the life of Christ Jesus has made us free from the law of animal living and the temptations of evil and sin." "This is the victory that overcomes the flesh, even your faith."

Page 962:1   *Paper 87:4.7*

The tragedy of all this lies in the fact that, when these ideas were taking root in the primitive mind of man, there really were no bad or disharmonious spirits in all the world. Such an unfortunate situation did not develop until after the **Caligastic rebellion** and only persisted until Pentecost. The concept of good and evil as cosmic co-ordinates is, even in the twentieth century, very much alive in human philosophy; most of the world's religions still carry this cultural birthmark of the long-gone days of the emerging ghost cults.

PAPER 35: THE LOCAL UNIVERSE SONS OF GOD

---

1. *THE FATHER MELCHIZEDEK*

Gabriel and Father Melchizedek

Page 385:1   *Paper 35:1.2*

See "Gabriel and Father Melchizedek" on page 216 for full text references. (33:6.2)

---

2. *THE MELCHIZEDEK SONS*

Avonal Sons—Magisterial Visitations

Page 386:4   *Paper 35:2.6*

See "Avonal Sons—Magisterial Visitations" on page 182 for full text references. (20:4.1–2)

---

5. *THE VORONDADEK SONS*

Most Highs on Urantia—Times of Crisis
The Most High observer and his awesome powers!—Parts II and III, over 800 pages apart.

Page 390:2–3   *Paper 35:5.5–6*

The service of the Vorondadeks in the local universes is extensive and varied. They serve as ambassadors to other universes and as consuls representing constellations within their native universe. Of all orders of local universe sonship they are the most **often intrusted with the full delegation of sovereign powers to be exercised in critical universe situations.**

On those worlds segregated in spiritual darkness, those spheres which have, through rebellion and default, suffered planetary isolation, an observer Vorondadek is usually present pending the restoration of normal status. **In certain emergencies this Most High observer could exercise absolute and arbitrary authority over every celestial being assigned to that planet.** It is of record on Salvington that the Vorondadeks have sometimes exercised such authority as Most High regents of such planets. And this has also been true even of inhabited worlds that were untouched by rebellion.

Page 490:7–13   *Paper 43:5.2–8*

The present government of the constellation, however, has been expanded to include twelve Sons of the Vorondadek order. These twelve are as follows:

1. The Constellation Father. The present Most High ruler of Norlatiadek is number 617,318 of the Vorondadek series of Nebadon. He saw service in many constellations throughout our local universe before taking up his Edentia responsibilities.
2. The senior Most High associate.
3. The junior Most High associate.
4. The Most High adviser, the personal representative of Michael since his attainment of the status of a Master Son.
5. The Most High executive, the personal representative of Gabriel stationed on Edentia ever since the Lucifer rebellion.
6. **The Most High chief of planetary observers, the director of the Vorondadek observers stationed on the isolated worlds of Satania.**

Page 1201:8   *Paper 109:7.7*

When the planetary Vorondadek observer of Urantia—**the Most High custodian who not long since assumed an emergency regency of your world**—asserted his authority in the presence of the resident governor general, he began his emergency administration of Urantia with a full staff of his own choosing. He immediately assigned to all his associates and assistants their planetary duties. But he did not choose the three Personalized Adjusters who appeared in his presence the instant he assumed the regency. He did not even know they would thus appear, for they did not so manifest their divine presence at the time of a previous regency. And the Most High regent did not assign service or designate duties for these volunteer Personalized Adjusters. Nevertheless, these three omnipersonal beings were among the most active of the numerous orders of celestial beings then serving on Urantia.

Page 1253:7   *Paper 114:4.4*

A Most High observer is empowered, at his discretion, to seize the planetary government in times of grave planetary crises, and it is of record that this has happened thirty-three times in the history of Urantia. At such times the Most High observer functions as the Most High regent, exercising unquestioned authority over all ministers and administrators resident on the planet excepting only the divisional organization of the archangels.

Most Highs

Page 390:3   *Paper 35:5.6*

See "Most Highs" on page 126 for full text references. (3:1.10)

---

9. THE LANONANDEK RULERS

Neighboring Universes

Page 393:8   *Paper 35:9.8*

See "Neighboring Universes" on page 210 for full text references. (32:2.12)

PAPER 36: THE LIFE CARRIERS

Life—Appearance on Planets

Page 396:1   *Paper 36:0.1*

Life does not originate spontaneously. **Life is constructed according to plans formulated by the (unrevealed) Architects of Being and appears on the inhabited planets either by direct importation or as a result of the operations of the Life Carriers** of the local universes. These carriers of life are among the most interesting and versatile of the diverse family of universe Sons. They are intrusted with designing and carrying creature life to the planetary spheres. And after planting this life on such new worlds, they remain there for long periods to foster its development.

Page 664:1   *Paper 58:0.1*

In all Satania there are only sixty-one worlds similar to Urantia, life-modification planets. The majority of inhabited worlds are peopled in accordance with established techniques; on such spheres the Life Carriers are afforded little leeway in their plans for life implantation. But about one world in ten is designated as a *decimal planet* and assigned to the special registry of the Life Carriers; and on such planets we are permitted to undertake certain life experiments in an effort to modify or possibly improve the standard universe types of living beings.

Page 667:5   *Paper 58:4.1*

That we are called Life Carriers should not confuse you. We can and do carry life to the planets, **but we brought no life to Urantia.** Urantia life is unique, original with the planet. This sphere is a life-modification world; all life appearing hereon was formulated by us right here on the planet; and there is no other world in all Satania, even in all Nebadon, that has a life existence just like that of Urantia.

---

1. ORIGIN AND NATURE OF LIFE CARRIERS

Magisterial Sons—Judges of Survival

Page 396:2   *Paper 36:1.1*

See "Magisterial Sons—Judges of Survival" on page 155 for full text references. (7:6.5)

---

2. THE LIFE CARRIER WORLDS

Germ Plasm
A little about the germ plasm in two papers.

Page 398:1   *Paper 36:2.7*

There are over one million fundamental or cosmic chemical formulas which constitute the parent patterns and the numerous basic functional variations of life manifestations. Satellite number one of the life-planning sphere is the realm of the universe physicists and electrochemists who serve as technical assistants to the Life Carriers in the work of capturing, organizing, **and manipulating the essential units of energy which are employed in building up the material vehicles of life transmission, the so-called germ plasm.**

Page 857:2   *Paper 77:2.5*

These mutant traits appearing in the first Nodite generation resulted from certain changes which had been wrought in the configuration and in the chemical constituents of the inheritance factors of the **Andonic germ plasm.** These changes were caused by the presence in the bodies of the staff members of the powerful life-maintenance circuits of the Satania system. These life circuits caused the chromosomes of the specialized Urantia pattern to reorganize more after the patterns of the standardized Satania specialization of the ordained Nebadon life manifestation. **The technique of this germ** plasm metamorphosis by the action of the system life currents is not unlike those procedures whereby Urantia scientists modify the germ plasm of plants and animals by the use of X rays.

---

4. MELCHIZEDEK LIFE CARRIERS

Dematerialization—Salvable and Unsalvable Personalities

Dematerialization technique is made for the entire salvable population. To consider who the salvable population is, it is helpful to look at the description of unsalvable, morally degenerate, and nonsalvable personalities.

Page 400:7   *Paper 36:4.3*

The midsonite creatures live and function as reproducing beings on their magnificent worlds until they are one thousand standard years of age; whereupon they are translated by seraphic transport. Midsoniters are nonreproducing beings thereafter **because the technique of dematerialization** which they pass through in preparation for enseraphiming forever deprives them of reproductive prerogatives.

Page 526:1   *Paper 46:5.19*

4. *The circles of the Master Physical Controllers.* The various orders of the Master Physical Controllers are concentrically arranged around the vast temple of power, wherein presides the power chief of the system in association with the chief of the Morontia Power Supervisors. This temple of power is one of two sectors on Jerusem where ascending mortals and midway creatures are not permitted. **The other one is the dematerializing sector in the area of the Material Sons,** a series of laboratories wherein the transport seraphim transform material beings into a state quite like that of the morontia order of existence.

Page 569:2   *Paper 49:6.7*

Thus are the sleeping survivors of a planetary age repersonalized in the dispensational roll calls. **But with regard to the nonsalvable personalities of a realm,** no immortal spirit is present to function with the group guardians of destiny, and this constitutes cessation of creature existence. While some of your records have pictured these events as taking place on the planets of mortal death, they all really occur on the mansion worlds.

Page 582:3   *Paper 51:2.3*

While there is this dematerializing technique for preparing the Adams for transit from Jerusem to the evolutionary worlds, there is no equivalent method for taking them away from such worlds unless the entire planet is to be emptied, **in which event emergency installation of the dematerialization technique is made for the entire salvable population.** If some physical catastrophe should doom the planetary residence of an evolving race, the Melchizedeks and the Life Carriers would install the technique of dematerialization for all survivors, and by seraphic transport these beings would be carried away to the new world prepared for their continuing existence. The evolution of a human race, once initiated on a world of space, must proceed quite independently of the physical survival of that planet, but during the evolutionary ages it is not otherwise intended that a Planetary Adam or Eve shall leave their chosen world.

Page 592:5   *Paper 52:2.12*

It is neither tenderness nor altruism to bestow futile sympathy upon degenerated human beings, unsalvable abnormal and inferior mortals. There exist on even the most normal of the evolutionary worlds sufficient differences between individuals and between numerous social groups to provide for the full exercise of all those noble traits of altruistic sentiment and unselfish mortal ministry without perpetuating the **socially unfit and the morally degenerate strains of evolving humanity.** There is abundant opportunity for the exercise of tolerance and the function of altruism in behalf of **those unfortunate and needy individuals who have not irretrievably lost their moral heritage and forever destroyed their spiritual birthright.**

PAPER 37: PERSONALITIES OF THE LOCAL UNIVERSE

Father-Son and Spirit-Mother

Page 406:1   *Paper 37:0.1*

See "Father-Son and Spirit-Mother" on page 213 for full text references. (33:1.2)

2. THE BRILLIANT EVENING STARS

"Be About Your Father's Business"

Stages of progressive self-realization of the Master's divinity included "Be about your Father's business"—four related papers.

Page 407:8   *Paper 37:2.8*

One of the high duties of the Evening Stars is to accompany the Avonal bestowal Sons on their planetary missions, even as Gabriel accompanied Michael on his Urantia bestowal. The two attending superangels are the ranking personalities of such missions, serving as cocommanders of the archangels and all others assigned to these undertakings. **It is the senior of these superangel commanders who, at the significant time and age, bids the Avonal bestowal Son, "Be about your brother's business."**

Page 1376:1   *Paper 124:6.15*

On the day before the Passover Sabbath, flood tides of spiritual illumination swept through the mortal mind of Jesus and filled his human heart to overflowing with affectionate pity for the spiritually blind and morally ignorant multitudes assembled for the celebration of the ancient Passover commemoration. This was one of the most extraordinary days that the Son of God spent in the flesh; and during the night, for the first time in his earth career, **there appeared to him an assigned messenger from Salvington, commissioned by Immanuel, who said: "The hour has come. It is time that you began to be about your Father's business."**

Page 1389:8   *Paper 126:3.5*

This year Jesus was much troubled with confused thinking. Family responsibility had **quite effectively removed all thought of immediately carrying out any plan for responding to the Jerusalem visitation directing him to "be about his Father's business."** Jesus rightly reasoned that the watchcare of his earthly father's family must take precedence of all duties; that the support of his family must become his first obligation.

Page 2091:2   *Paper 196:1.6*

Just as men must progress from the consciousness of the human to the realization of the divine, so did Jesus ascend from the nature of man to the consciousness of the nature of God. And the Master made this great ascent from the human to the divine by the conjoint achievement of the faith of his mortal intellect and the acts of his indwelling Adjuster. The fact-realization of the attainment of totality of divinity (all the while fully conscious of the reality of humanity) was attended by seven stages of faith consciousness of progressive divinization. **These stages of progressive self-realization were marked off by the following extraordinary events in the Master's bestowal experience:**

1. The arrival of the Thought Adjuster.
2. **The messenger of Immanuel who appeared to him at Jerusalem when he was about twelve years old.**
3. The manifestations attendant upon his baptism.
4. The experiences on the Mount of Transfiguration.
5. The morontia resurrection.
6. The spirit ascension.
7. The final embrace of the Paradise Father, conferring unlimited sovereignty of his universe.

Evening Stars—Liaison Between Mortals and Teacher Sons

Page 408:1   *Paper 37:2.9*

> See "Evening Stars—Liaison Between Mortals and Teacher Sons" on page 189 for full text references. (20:9.3)

3. THE ARCHANGELS

Archangels' Circuit on Urantia

Four papers about the archangels circuit and their divisional headquarters' location.

Page 408:6   *Paper 37:3.3*

The archangel corps of Nebadon is directed by the first-born of this order, **and in more recent times a divisional headquarters of the archangels has been maintained on Urantia.** It is this unusual fact that soon arrests the attention of extra-Nebadon student visitors. Among their early observations of intrauniverse transactions is the discovery that many ascendant activities of the Brilliant Evening Stars are directed from the capital of a local system, Satania. On further examination they discover that certain archangel activities are directed from a small and apparently insignificant inhabited world called Urantia. And then ensues the revelation of Michael's bestowal on Urantia and their immediately quickened interest in you and your lowly sphere.

Page 490:7–16   *Paper 43:5.2–11*

The present government of the constellation, however, has been expanded to include twelve Sons of the Vorondadek order. These twelve are as follows:

1. The Constellation Father. The present Most High ruler of Norlatiadek is number 617,318 of the Vorondadek series of Nebadon. He saw service in many constellations throughout our local universe before taking up his Edentia responsibilities.
2. The senior Most High associate.
3. The junior Most High associate.
4. The Most High adviser, the personal representative of Michael since his attainment of the status of a Master Son.
5. The Most High executive, the personal representative of Gabriel stationed on Edentia ever since the Lucifer rebellion.
6. The Most High chief of planetary observers, the director of the Vorondadek observers stationed on the isolated worlds of Satania.
7. The Most High referee, the Vorondadek Son intrusted with the duty of adjusting all difficulties consequential to rebellion within the constellation.
8. The Most High emergency administrator, the Vorondadek Son charged with the task of adapting the emergency enactments of the Norlatiadek legislature to the rebellion-isolated worlds of Satania.
9. **The Most High mediator,** the Vorondadek Son assigned to harmonize the special bestowal adjustments on Urantia with the routine administration of the constellation. **The presence of certain archangel activities and numerous other irregular ministrations on Urantia, together with the special activities of the Brilliant Evening Stars on Jerusem, necessitates the functioning of this Son.**

Page 522:4   *Paper 46:3.4*

The Jerusem sending station is located at the opposite pole of the sphere. All broadcasts to the individual worlds are relayed from the system capitals except the **Michael messages, which sometimes go direct to their destinations over the archangels' circuit.**

Page 2024:4   *Paper 189:3.2*

The circuit of the archangels then operated for the first time from Urantia. Gabriel and the archangel hosts moved to the place of the spiritual polarity of the planet; and when Gabriel gave the signal, there flashed to the first of the system mansion worlds the voice of Gabriel, saying: "By the mandate of Michael, let the dead of a Urantia dispensation rise!" Then all the survivors of the human races of Urantia who had fallen asleep since the days of Adam, and who had not already gone on to judgment, appeared in the resurrection halls of mansonia in readiness for morontia investiture. And in an instant of time the seraphim and their associates made ready to depart for the mansion worlds. Ordinarily these seraphic guardians, onetime assigned to the group custody of these surviving mortals, would have been present at the moment of their awaking in the resurrection halls of mansonia, but they were on this world itself at this time because of the necessity of Gabriel's presence here in connection with the morontia resurrection of Jesus.

Personality—When Is it Bestowed?

Page 409:4   *Paper 37:3.7*

> See "Personality—When Is it Bestowed?" on page 195 for full text references. (24:2.8)

8. HIGHER SPIRIT ORDERS OF ASSIGNMENT

Personality—When Is it Bestowed?

Page 413:7   *Paper 37:8.4*

> See "Personality—When Is it Bestowed?" on page 195 for full text references. (24:2.8)

Universal Conciliators

Page 414:1   *Paper 37:8.6*

> See "Universal Conciliators" on page 198 for full text references. (25:2.12)

9. PERMANENT CITIZENS OF THE LOCAL UNIVERSE

Abandonters—Permanent Citizens of our Superuniverse
Skills of the abandonters........

Page 415:4   *Paper 37:9.8*

In like manner, all divisions of the administrative organization of the local universes and superuniverses have their more or less permanent populations, inhabitants of citizenship status. As Urantia has its midwayers, Jerusem, your system capital, has the Material Sons and Daughters; Edentia, your constellation headquarters, has the univitatia, while the citizens of Salvington are twofold, the created susatia and the evolved Spirit-fused mortals. The administrative worlds of the minor and major sectors of the superuniverses do not have permanent citizens. **But the Uversa headquarters spheres are continuously fostered by an amazing group of beings known as the** *abandonters,* **the creation of the unrevealed agents of the Ancients of Days and the seven Reflective Spirits resident on the capital of Orvonton.** These residential citizens on Uversa are at present administering the routine affairs of their world under the immediate supervision of the Uversa corps of the Son-fused mortals. Even Havona has its native beings, and the central Isle of Light and Life is the home of the various groups of Paradise Citizens.

Page 452:4   *Paper 40:10.4*

We have analyzed this problem and have reached the undoubted conclusion that the consignment of all mortals to an ultimate Paradise destiny would be unfair to the time-space universes inasmuch as the courts of the Creator Sons and of the Ancients of Days would then be wholly dependent on the services of those who were in transit to higher realms. **And it does seem to be no more than fitting that the local and the superuniverse governments should each be provided with a permanent group of ascendant citizenship; that the functions of these administrations should be enriched by the efforts of certain groups of glorified mortals who are of permanent status, evolutionary complements of the abandonters and of the susatia.** Now it is quite obvious that the present ascension scheme effectively provides the time-space administrations with just such groups of ascendant creatures; and we have many times wondered: Does all this represent an intended part of the all-wise plans of the Architects of the Master Universe designed to provide the Creator Sons and the Ancients of Days with a permanent ascendant population? with evolved orders of citizenship that will become increasingly competent to carry forward the affairs of these realms in the universe ages to come?

Page 493:5   *Paper 43:7.4*

In the various courtesy colonies, ascending morontia mortals predominate among the reversion directors, but the univitatia represent the largest group associated with the Nebadon corps of celestial artisans. **In all Orvonton no extra-Havona beings excepting the Uversa abandonters can equal the univitatia in artistic skill, social adaptability, and co-ordinating cleverness.**

Page 631:2   *Paper 55:6.6*

We observe that on these highly evolved and long seventh-stage worlds human beings fully learn the local universe language before they are translated; **and I have visited a few very old planets where abandonters were** teaching the older mortals the tongue of the superuniverse. And on these worlds I have observed the technique whereby the absonite personalities reveal the presence of the finaliters in the morontia temple.

PAPER 38: MINISTERING SPIRITS OF THE LOCAL UNIVERSE

---

5. SERAPHIC TRAINING

Neighboring Universes

Page 420:6   *Paper 38:5.1*

See "Neighboring Universes" on page 210 for full text references. (32:2.12)

---

6. SERAPHIC ORGANIZATION

Angels—Twelve Legions
A host of over 71 million seraphim—covered in Parts II and IV.

Page 421:5   *Paper 38:6.2*

Twelve legions of angels comprise a host numbering 2,985,984 pairs or 5,971,968 individuals, and twelve such hosts (35,831,808 pairs or 71,663,616 individuals) make up the largest operating organization of seraphim, an angelic army. A seraphic host is commanded by an archangel or by some other personality of co-ordinate status, while the angelic armies are directed by the Brilliant Evening Stars or by other immediate lieutenants of Gabriel. And Gabriel is the "supreme commander of the armies of heaven," the chief executive of the Sovereign of Nebadon, "the Lord God of hosts."

Page 1516:1   *Paper 136:5.1*

On the third day after beginning this conference with himself and his Personalized Adjuster, Jesus was presented with the vision of the assembled celestial hosts of Nebadon sent by their commanders to wait upon the will of their beloved Sovereign. **This mighty host embraced twelve legions of seraphim and proportionate numbers of every order of universe intelligence.** And the first great decision of Jesus' isolation had to do with whether or not he would make use of these mighty personalities in connection with the ensuing program of his public work on Urantia.

Page 1974:5   *Paper 183:3.7*

Jesus was ready to go back to Jerusalem with the guards, and the captain of the soldiers was altogether willing to allow the three apostles and their associates to go their way in peace. But before they were able to get started, as Jesus stood there awaiting the captain's orders, one Malchus, the Syrian bodyguard of the high priest, stepped up to Jesus and made ready to bind his hands behind his back, although the Roman captain had not directed that Jesus should be thus bound. When Peter and his associates saw their Master being subjected to this indignity, they were no longer able to restrain themselves. Peter drew his sword and with the others rushed forward to smite Malchus. But before the soldiers could come to the defense of the high priest's servant, Jesus raised a forbidding hand to Peter and, speaking sternly, said: "Peter, put up your sword. They who take the sword shall perish by the sword. Do you not understand that it is the Father's will that I drink this cup? **And do you not further know that I could even now command more than twelve legions of angels and their associates,** who would deliver me from the hands of these few men?"

---

9. THE MIDWAY CREATURES

Secondary Midwayers—Not an Accident
Secondary order of midwayers, no accident about these strangely behaving offspring.

Page 424:5   *Paper 38:9.5*

Neither of these groups is an evolutionary accident; both are essential features in the predetermined plans of the universe architects, and their appearance on the evolving worlds at the opportune juncture is in accordance with the original designs and developmental plans of the supervising Life Carriers.

Page 861:5–6   *Paper 77:5.5–6*

A company of twenty-seven followed Adamson northward in quest of these people of his childhood fantasies. In a little over three years Adamson's party actually found the object of their adventure, and among these people he discovered a wonderful and beautiful woman, twenty years old, who claimed to be the last pure-line descendant of the Prince's staff. This woman, Ratta, said that her ancestors were all descendants of two of the fallen staff of the Prince. She was the last of her race, having no living brothers or sisters. **She had about decided not to mate,** had about made up her mind to die without issue, but she lost her heart to the majestic Adamson. And when she heard the story of Eden, how the predictions of Van and Amadon had really come to pass, and as she listened to the recital of the Garden default, she was encompassed with but a single thought—to marry this son and heir of Adam. And quickly the idea grew upon Adamson. In a little more than three months they were married.

Adamson and Ratta had a family of sixty-seven children. They gave origin to a great line of the world's leadership, but they did something more. It should be remembered that both of these beings were really superhuman. Every fourth child born to them was of a unique order. It was often invisible. Never in the world's history had such a thing occurred. Ratta was greatly perturbed—even superstitious—**but Adamson well knew of the existence of the primary midwayers, and he concluded that something similar was transpiring before his eyes.** When the second strangely behaving offspring arrived, he decided to mate them, since one was male and the other female, and this is the origin of the secondary order of midwayers. Within one hundred years, before this phenomenon ceased, almost two thousand were brought into being.

Ministering Spirits from the Conjoint Actor— Relation with Humans

Page 425:1 *Paper 38:9.9*

> See "Ministering Spirits from the Conjoint Actor—Relation with Humans" on page 145 for full text references. (5:3.5)

Inspired Spirits and Our Divine Fragment

Page 425:1 *Paper 38:9.9*

> See "Inspired Spirits and Our Divine Fragment" on page 180 for full text references. (19:5.6–8)

PAPER 39: THE SERAPHIC HOSTS

1. SUPREME SERAPHIM

Creator Son—Fourth Creature Bestowal

Paper 119 elaborates on Part II's mention of Creator Son's fourth creature bestowal.

Page 429:1   *Paper 39:1.15*

The fourth creature bestowal of the Creator Son was in the likeness of a teaching counselor of the supreme seraphim of Nebadon.

Page 1313:5–6   *Paper 119:4.2–3*

On the third day after this bestowal disappearance we observed, in the universe broadcasts to Uversa, this significant news item from the seraphic headquarters of Nebadon: "Reporting the unannounced arrival of an unknown seraphim, accompanied by a solitary supernaphim and Gabriel of Salvington. This unregistered seraphim qualifies as of the Nebadon order and bears credentials from the Uversa Ancients of Days, certified by Immanuel of Salvington. **This seraphim tests out as belonging to the supreme order of the angels of a local universe and has already been assigned to the corps of the teaching counselors.**"

Michael was absent from Salvington during this, the seraphic bestowal, for a period of over forty standard universe years. **During this time he was attached as a seraphic teaching counselor, what you might denominate a private secretary, to twenty-six different master teachers, functioning on twenty-two different worlds.** His last or terminal assignment was as counselor and helper attached to a bestowal mission of a Trinity Teacher Son on world 462 in system 84 of constellation 3 in the universe of Nebadon.

2. SUPERIOR SERAPHIM

Communication Between Superuniverses—Paradise Clearinghouse

Page 429:7   *Paper 39:2.4*

See "Communication Between Superuniverses—Paradise Clearinghouse" on page 172 for full text references. (15:10.9)

Conscious, Unconscious, and Superconscious— Technique

Page 430:7   *Paper 39:2.11*

> See "Conscious, Unconscious, and Superconscious—Technique" on page 181 for full text references. (19:5.9)

Universe Broadcasts—Simultaneous Messages

Page 431:4   *Paper 39:2.15*

> See "Universe Broadcasts—Simultaneous Messages" on page 217 for full text references. (33:6.5)

4. ADMINISTRATOR SERAPHIM

Cosmic Citizenship

Cosmic citizenship, universe-conscious citizens, morontian awareness on Urantia, and related passages. Seven Papers to whet the appetite for this discussion.

Page 435:2   *Paper 39:4.9*

The seraphic interpreters of cosmic citizenship guide the new citizens of the system capitals and quicken their appreciation of the responsibilities of universe government. These seraphim are also closely associated with the Material Sons in the system administration, **while they portray the responsibility and morality of cosmic citizenship to the material mortals on the inhabited worlds.**

Page 1100:5   *Paper 100:6.3*

The marks of human response to the religious impulse embrace the qualities of nobility and grandeur. **The sincere religionist is conscious of universe citizenship and is aware of making contact with sources of superhuman power.** He is thrilled and energized with the assurance of belonging to a superior and ennobled fellowship of the sons of God. The consciousness of self-worth has become augmented by the stimulus of the quest for the highest universe objectives—supreme goals.

Page 1112:4   *Paper 101:6.8*

The teachings of Jesus constituted the first Urantian religion which so fully embraced a harmonious co-ordination of knowledge, wisdom, faith, truth, and love as completely and simultaneously to provide temporal tranquillity, intellectual certainty, moral enlightenment, philosophic stability, ethical sensitivity, God-consciousness, and the positive assurance of personal survival. The faith of Jesus pointed the way to finality of human salvation, to the ultimate of mortal universe attainment, since it provided for:

1. Salvation from material fetters in the personal realization of sonship with God, who is spirit.
2. Salvation from intellectual bondage: man shall know the truth, and the truth shall set him free.
3. Salvation from spiritual blindness, the human realization of the fraternity of mortal beings and **the morontian awareness of the brotherhood of all universe creatures;** the service-discovery of spiritual reality and the ministry-revelation of the goodness of spirit values.

Page 1206:4–8   *Paper 110:3.6–10*

You must not regard co-operation with your Adjuster as a particularly conscious process, for it is not; but your motives and your decisions, your faithful determinations and your supreme desires, do constitute real and effective co-operation. **You can consciously augment Adjuster harmony by:**

1. Choosing to respond to divine leading; sincerely basing the human life on the highest consciousness of truth, beauty, and goodness, and then co-ordinating these qualities of divinity through wisdom, worship, faith, and love.
2. Loving God and desiring to be like him—genuine recognition of the divine fatherhood and loving worship of the heavenly Parent.
3. Loving man and sincerely desiring to serve him—wholehearted recognition of the brotherhood of man coupled with an intelligent and wise affection for each of your fellow mortals.
4. **Joyful acceptance of cosmic citizenship**—honest recognition of your progressive obligations to the Supreme Being, awareness of the interdependence of evolutionary man and evolving Deity. This is the birth of cosmic morality and the dawning realization of universal duty.

Page 1211:1   *Paper 110:6.16*

Perhaps these psychic circles of mortal progression would be better denominated *cosmic levels*—actual meaning grasps and value realizations of progressive approach to the **morontia consciousness** of initial relationship of the evolutionary soul with the emerging Supreme Being. And it is this very relationship that makes it forever impossible fully to explain the significance of the cosmic circles to the material mind. These circle attainments are only relatively related to God-consciousness. **A seventh or sixth circler can be almost as truly God-knowing—sonship conscious—as a second or first circler, but such lower circle beings are far less conscious of experiential relation to the Supreme Being, universe citizenship.** The attainment of these cosmic circles will become a part of the ascenders' experience on the mansion worlds if they fail of such achievement before natural death.

Page 1258:5   *Paper 114:7.10*

(The **cosmic reserve corps of universe-conscious citizens on Urantia** now numbers over one thousand mortals whose insight of cosmic citizenship far transcends the sphere of their terrestrial abode, but I am forbidden to reveal the real nature of the function of this unique group of living human beings.)

Page 1284:5–6   *Paper 117:4.9–10*

The evolution of Adjuster progress in the spiritualizing and eternalizing of a human personality is directly productive of an enlargement of the sovereignty of the Supreme. Such achievements in human evolution are at the same time achievements in the evolutionary actualization of the Supreme. While it is true that creatures could not evolve without the Supreme, it is probably also true that the evolution of the Supreme can never be fully attained independent of the completed evolution of all creatures. **Herein lies the great cosmic responsibility of self-conscious personalities:** That Supreme Deity is in a certain sense dependent on the choosing of the mortal will. And the mutual progression of creature evolution and of Supreme evolution is faithfully and fully indicated to the Ancients of Days over the inscrutable mechanisms of universe reflectivity.

The great challenge that has been given to mortal man is this: Will you decide to personalize the experiencible value meanings of the cosmos into your own evolving selfhood? or by rejecting survival, will you allow these secrets of Supremacy to lie dormant, awaiting the action of another creature at some other time who will in his *way* attempt a creature contribution to the evolution of the finite God? But that will be his contribution to the Supreme, not yours.

Page 2082:9   *Paper 195:9.4*

Religion does need new leaders, spiritual men and women who will dare to depend solely on Jesus and his incomparable teachings. If Christianity persists in neglecting its spiritual mission while it continues to busy itself with social and material problems, **the spiritual renaissance must await the coming of these new teachers of Jesus' religion who will be exclusively devoted to the spiritual regeneration of men.** And then will these spirit-born souls quickly supply the leadership and inspiration requisite for the social, moral, economic, and political reorganization of the world.

3. TRANSLATED MIDWAYERS

Secondary Midwayers

A little more about the Adamic midwayers in Parts II and III.

Page 444:3   *Paper 40:3.1*

Although deprived of the immediate benefits of the planetary bestowals of the descending Sons of God, though the Paradise ascent is long deferred, nevertheless, soon after an evolutionary planet has attained the intermediate epochs of light and life (if not before), both groups of midway creatures are released from planetary duty. Sometimes the majority of them are translated, along with their human cousins, on the day of the descent of the temple of light and the elevation of the Planetary Prince to the dignity of Planetary Sovereign. Upon being relieved of planetary service, both orders are registered in the local universe as ascending Sons of God and immediately begin the long Paradise ascent by the very routes ordained for the progression of the mortal races of the material worlds. **The primary group are destined to various finaliter corps, but the secondary or Adamic midwayers are all routed for enrollment in the Mortal Corps of Finality.**

Page 583:5   *Paper 51:3.6*

The secondary midway creatures are indigenous to the Adamic missions. As with the corporeal staff of the Planetary Prince, the descendants of the Material Sons and Daughters are of two orders: their physical children and the secondary order of midway creatures. **These material but ordinarily invisible planetary ministers contribute much to the advancement of civilization and even to the subjection of insubordinate minorities who may seek to subvert social development and spiritual progress.**

Page 861:5–6   *Paper 77:5.5–6*

A company of twenty-seven followed Adamson northward in quest of these people of his childhood fantasies. In a little over three years Adamson's party actually found the object of their adventure, and among these people he discovered a wonderful and beautiful woman, twenty years old, who claimed to be the last pure-line descendant of the Prince's staff. This woman, Ratta, said that her ancestors were all descendants of two of the fallen staff of the Prince. She was the last of her race, having no living brothers or sisters. **She had about decided not to mate, had about made up her mind to die without issue, but she lost her heart to the majestic Adamson. And when she heard** the story of Eden, how the predictions of Van and Amadon had really come to pass, and as she listened to the recital of the Garden default, she was encompassed with but a single thought—to marry this son and heir of Adam. And quickly the idea grew upon Adamson. In a little more than three months they were married.

Adamson and Ratta had a family of sixty-seven children. They gave origin to a great line of the world's leadership, but they did something more. It should be remembered that both of these beings were really superhuman. **Every fourth child born to them was of a unique order. It was often invisible.** Never in the world's history had such a thing occurred. Ratta was greatly perturbed—even superstitious—but Adamson well knew of the existence of the primary midwayers, and he concluded that something similar was transpiring before his eyes. When the second strangely behaving offspring arrived, he decided to mate them, since one was male and the other female, and this is the origin of the secondary order of midwayers. Within one hundred years, before this phenomenon ceased, almost two thousand were brought into being.

---

5. MORTALS OF TIME AND SPACE

Thought Adjusters—God Adjusts with Mind of Imperfection

Page 447:2   *Paper 40:5.14*

See "Thought Adjusters—God Adjusts with Mind of Imperfection" on page 132 for full text references. (3:2.5)

Survival—Co-operation with Indwelling Adjuster
Only conscious resistance can prevent survival—related topics in five papers.

Page 447:4   *Paper 40:5.16*

As to the chances of mortal survival, let it be made forever clear: **All souls of every possible phase of mortal existence will survive provided they manifest willingness to co-operate with their indwelling Adjusters and exhibit a desire to find God and to attain divine perfection, even though these desires be but the first faint flickers** of the primitive comprehension of that "true light which lights every man who comes into the world."

Page 448:9   *Paper 40:7.2*

Your own races of surviving mortals belong to this group of the ascending Sons of God. You are now planetary sons, evolutionary creatures derived from the Life Carrier implantations and modified by the Adamic-life infusion, hardly yet ascending sons; but you are indeed sons of ascension potential—even to the highest heights of glory and divinity attainment—**and this spiritual status of ascending sonship you may attain by faith and by freewill co-operation with the spiritualizing activities of the indwelling Adjuster.** When you and your Adjusters are finally and forever fused, when you two are made one, even as in Christ Michael the Son of God and the Son of Man are one, then in fact have you become the ascending sons of God.

Page 565:1   *Paper 49:4.9*

But mortal mind without immortal spirit cannot survive. The mind of man is mortal; only the bestowed spirit is immortal. Survival is dependent on spiritualization by the ministry of the Adjuster—on the birth and evolution of the immortal soul; **at least, there must not have developed an antagonism towards the Adjuster's mission of effecting the spiritual transformation of the material mind.**

Page 1206:3   *Paper 110:3.5*

Confusion, being puzzled, even sometimes discouraged and distracted, does not necessarily signify resistance to the leadings of the indwelling Adjuster. Such attitudes may sometimes connote lack of active co-operation with the divine Monitor and may, therefore, somewhat delay spiritual progress, but such intellectual emotional difficulties do not in the least interfere with the certain survival of the God-knowing soul. Ignorance alone can never prevent survival; neither can confusional doubts nor fearful uncertainty. **Only conscious resistance to the Adjuster's leading can prevent the survival of the evolving immortal soul.**

Page 1233:5   *Paper 112:5.9*

This does not mean that human beings are to enjoy a second opportunity in the face of the rejection of a first, not at all. **But it does signify that all will creatures are to experience one true opportunity to make one undoubted, self-conscious, and final choice.** The sovereign Judges of the universes will not deprive any being of personality status who has not finally and fully made the eternal choice; the soul of man must and will be given full and ample opportunity to reveal its true intent and real purpose.

Page 1733:5   *Paper 155:6.17*

Now, mistake not, my Father will ever respond to the faintest flicker of faith. He takes note of the physical and superstitious emotions of the primitive man. And with those honest but fearful souls whose faith is so weak that it amounts to little more than an intellectual conformity to a passive attitude of assent to religions of authority, the Father is ever alert to honor and foster even all such feeble attempts to reach out for him. But you who have been called out of darkness into the light are expected to believe with a whole heart; your faith shall dominate the combined attitudes of body, mind, and spirit.

---

6. THE FAITH SONS OF GOD

"Behold, what manner of love..."

Page 448:1   *Paper 40:6.2*

See ""Behold, what manner of love..."" on page 122 for full text references. (2:5.4)

---

7. FATHER-FUSED MORTALS

Survival—Co-operation with Indwelling Adjuster

Page 448:9   *Paper 40:7.2*

See "Survival—Co-operation with Indwelling Adjuster" on page 245 for full text references. (40:5.16)

---

8. SON-FUSED MORTALS

Seventy Times and Seven
Peter, how about 77 times!—three Papers.

Page 449:5   *Paper 40:8.2*

Such mortals have been deemed worthy of survival by the adjudicational authorities, and even their Adjusters, by returning from Divinington, have concurred in their ascension to the mansion worlds. Such beings have ascended through a system, a constellation, and through the educational worlds of the Salvington circuit; they have enjoyed the "seventy times seven" opportunities for fusion and still have been unable to attain oneness with their Adjusters.

Page 1551:1 *Paper 139:2.5*

The one trait which Peter most admired in Jesus was his supernal tenderness. Peter never grew weary of contemplating Jesus' forbearance. **He never forgot the lesson about forgiving the wrongdoer, not only seven times but seventy times and seven.** He thought much about these impressions of the Master's forgiving character during those dark and dismal days immediately following his thoughtless and unintended denial of Jesus in the high priest's courtyard.

Page 1763:1 *Paper 159:1.4*

Simon Peter was the apostle in charge of the workers at Hippos, and when he heard Jesus thus speak, he asked: "Lord, how often shall my brother sin against me, and I forgive him? Until seven times?" **And Jesus answered Peter: "Not only seven times but even to seventy times and seven.** Therefore may the kingdom of heaven be likened to a certain king who ordered a financial reckoning with his stewards. And when they had begun to conduct this examination of accounts, one of his chief retainers was brought before him confessing that he owed his king ten thousand talents. Now this officer of the king's court pleaded that hard times had come upon him, and that he did not have wherewith to pay this obligation. And so the king commanded that his property be confiscated, and that his children be sold to pay his debt. When this chief steward heard this stern decree, he fell down on his face before the king and implored him to have mercy and grant him more time, saying, `Lord, have a little more patience with me, and I will pay you all.' And when the king looked upon this negligent servant and his family, he was moved with compassion. He ordered that he should be released, and that the loan should be wholly forgiven.

---

9. SPIRIT-FUSED MORTALS

Personality Relationships
Personality relationships in Parts I and II.

Page 451:3 *Paper 40:9.7*

Even with Adjuster-fusion candidates, only those human experiences which were of spiritual value are common possessions of the surviving mortal and the returning Adjuster and hence are immediately remembered subsequent to mortal survival. Concerning those happenings which were not of spiritual significance, even these Adjuster-fusers must depend upon the attribute of recognition-response in the surviving soul. **And since any one event may have a spiritual connotation to one mortal but not to another, it becomes possible** for a group of contemporary ascenders from the same planet to pool their store of Adjuster-remembered events and thus to reconstruct any experience which they had in common, and which was of spiritual value in the life of any one of them.

Page 1228:3   *Paper 112:2.4*

All mortal concepts of reality are based on the assumption of the actuality of human personality; all concepts of superhuman realities are based on the experience of the human personality with and in the cosmic realities of certain associated spiritual entities and divine personalities. Everything nonspiritual in human experience, excepting personality, is a means to an end. **Every true relationship of mortal man with other persons—human or divine—is an end in itself.** And such fellowship with the personality of Deity is the eternal goal of universe ascension.

Page 1235:4   *Paper 112:5.22*

The Thought Adjuster will recall and rehearse for you only those memories and experiences which are a part of, and essential to, your universe career. If the Adjuster has been a partner in the evolution of aught in the human mind, then will these worth-while experiences survive in the eternal consciousness of the Adjuster. But much of your past life and its memories, having neither spiritual meaning nor morontia value, will perish with the material brain; much of material experience will pass away as onetime scaffolding which, having bridged you over to the morontia level, no longer serves a purpose in the universe. **But personality and the relationships between personalities are never scaffolding; mortal memory of personality relationships has cosmic value and will persist.** On the mansion worlds you will know and be known, and more, you will remember, and be remembered by, your onetime associates in the short but intriguing life on Urantia.

---

10. ASCENDANT DESTINIES

Spirit- and Son-fused Mortals

Page 452:2   *Paper 40:10.2*

See "Spirit- and Son-fused Mortals" on page 194 for full text references. (22:6.1)

Abandonters—Permanent Citizens of our Superuniverse

Page 452:4   *Paper 40:10.4*

> See "Abandonters—Permanent Citizens of our Superuniverse" on page 235 for full text references. (37:9.8)

PAPER 41: PHYSICAL ASPECTS OF THE LOCAL UNIVERSE

2. THE SATANIA PHYSICAL CONTROLLERS

Solar System— in Contradistinction to System

Page 457:1   *Paper 41:2.2*

> See "Solar System— in Contradistinction to System" on page 168 for full text references. (15:2.2–3)

6. CALCIUM—THE WANDERER OF SPACE

Sun and Earth Core Temperature

Center of earth's temperature slightly above the surface temperature of the sun—references in Parts I and II.

Page 462:5   *Paper 41:6.7*

It should be remembered that spectral analyses show only sun-surface compositions. For example: Solar spectra exhibit many iron lines, but iron is not the chief element in the sun. **This phenomenon is almost wholly due to the present temperature of the sun's surface, a little less than 6,000 degrees, this temperature being very favorable to the registry of the iron spectrum.**

Page 463:2   *Paper 41:7.2*

The surface temperature of your sun is almost 6,000 degrees, but it rapidly increases as the interior is penetrated until it attains the unbelievable height of about 35,000,000 degrees in the central regions. (All of these temperatures refer to your Fahrenheit scale.)

Page 668:3   *Paper 58:5.1*

The continental land drift continued. The earth's core had become as dense and rigid as steel, being subjected to a pressure of almost 25,000 tons to the square inch, and owing to the enormous gravity pressure, it was and still is very hot in the deep interior. **The temperature increases from the surface downward until at the center it is slightly above the surface temperature of the sun.**

---

7. SOURCES OF SOLAR ENERGY

Sun and Earth Core Temperature

Page 463:2   *Paper 41:7.2*

> See "Sun and Earth Core Temperature" on page 250 for full text references. (41:6.7)

---

10. ORIGIN OF INHABITED WORLDS

Solar System— in Contradistinction to System

Page 466:1   *Paper 41:10.2*

> See "Solar System— in Contradistinction to System" on page 168 for full text references. (15:2.2–3)

PAPER 42: ENERGY—MIND AND MATTER

---

1. PARADISE FORCES AND ENERGIES

Energy—Undiscovered Nonspiritual

Page 467:5 Page 468:1   *Paper 42:1.3–4*

> See "Energy—Undiscovered Nonspiritual" on page 131 for full text references. (3:2.3)

2. UNIVERSAL NONSPIRITUAL ENERGY SYSTEMS (PHYSICAL ENERGIES)

Paucity of Language

Page 469:1   *Paper 42:2.1*

See "Paucity of Language" on page 105 for full text references. (0:6.2)

Prereality
Prereality—over 600 pages apart.

Page 469:5   *Paper 42:2.5*

Space potency is a prereality; it is the domain of the Unqualified Absolute and is responsive only to the personal grasp of the Universal Father, notwithstanding that it is seemingly modifiable by the presence of the Primary Master Force Organizers.

Page 1153:2   *Paper 105:1.5*

To the finite mind there simply must be a beginning, and though there never was a real beginning to reality, still there are certain source relationships which reality manifests to infinity. **The prereality,** primordial, eternity situation may be thought of something like this: At some infinitely distant, hypothetical, past-eternity moment, the I AM may be conceived as both thing and no thing, as both cause and effect, as both volition and response. At this hypothetical eternity moment there is no differentiation throughout all infinity. Infinity is filled by the Infinite; the Infinite encompasses infinity. This is the hypothetical static moment of eternity; actuals are still contained within their potentials, and potentials have not yet appeared within the infinity of the I AM. But even in this conjectured situation we must assume the existence of the possibility of self-will.

Page 1154:2   *Paper 105:2.3*

This self-metamorphosis of the I AM culminates in the multiple differentiation of deified reality and of undeified reality, of potential and actual reality, and of certain other realities that can hardly be so classified. These differentiations of the theoretical monistic I AM are eternally integrated by simultaneous relationships arising within the same I AM—the prepotential, preactual, prepersonal, **monothetic prereality** which, though infinite, is revealed as absolute in the presence of the First Source and Center and as personality in the limitless love of the Universal Father.

4. Energy and Matter Transmutations

Energy—Undiscovered Nonspiritual

Page 472:12   *Paper 42:4.1*

See "Energy—Undiscovered Nonspiritual" on page 131 for full text references. (3:2.3)

Relativity

Page 474:1   *Paper 42:4.11*

See "Relativity" on page 152 for full text references. (7:1.2)

Paper 43: The Constellations

1. The Constellation Headquarters

Jesus—Morontia
Morontia gas, morontia body, and morontia Jesus—Parts II and IV.

Page 486:2   *Paper 43:1.3*

Edentia and its associated worlds have a true atmosphere, the usual three-gas mixture which is characteristic of such architectural creations, **and which embodies the two elements of Urantian atmosphere plus that morontia gas suitable for the respiration of morontia creatures.** But while this atmosphere is both material and morontial, there are no storms or hurricanes; neither is there summer nor winter. This absence of atmospheric disturbances and of seasonal variation makes it possible to embellish all outdoors on these especially created worlds.

Page 2029:3   *Paper 190:0.3*

The mortals of the realms will arise in the morning of the resurrection with the same type of transition or morontia body that Jesus had when he arose from the tomb on this Sunday morning. These bodies do not have circulating blood, and **such beings do not partake of ordinary material food;** nevertheless, these morontia forms are *real*. When the various believers saw Jesus after his resurrection, they really saw him; they were not the self-deceived victims of visions or hallucinations.

Page 2035:2   *Paper 190:5.5*

By this time they had come near to the village where these brothers dwelt. Not a word had these two men spoken since Jesus began to teach them as they walked along the way. Soon they drew up in front of their humble dwelling place, and Jesus was about to take leave of them, going on down the road, but they constrained him to come in and abide with them. They insisted that it was near nightfall, and that he tarry with them. Finally Jesus consented, and very soon after they went into the house, they sat down to eat. **They gave him the bread to bless, and as he began to break and hand to them, their eyes were opened, and Cleopas recognized that their guest was the Master himself. And when he said, "It is the Master—," the morontia Jesus vanished from their sight.**

Page 2047:1   *Paper 192:1.8*

Jesus spoke to them, saying: "Come now, all of you, to breakfast. Even the twins should sit down while I visit with you; John Mark will dress the fish." John Mark brought seven good-sized fish, which the Master put on the fire, and when they were cooked, the lad served them to the ten. Then Jesus broke the bread and handed it to John, who in turn served it to the hungry apostles. When they had all been served, Jesus bade John Mark sit down while he himself served the fish and the bread to the lad. **And as they ate, Jesus visited with them and recounted their many experiences in Galilee and by this very lake.**

---

4. MOUNT ASSEMBLY—THE FAITHFUL OF DAYS

Lucifer Rebellion—Satan
The father of sin and the meaning of "beheld Satan fall as lightning from heaven"—four related Papers.

Page 490:4   *Paper 43:4.9*

Since the triumph of Christ, all Norlatiadek is being cleansed of sin and rebels. **Sometime before Michael's death in the flesh the fallen Lucifer's associate, Satan, sought to attend such an Edentia conclave,** but the solidification of sentiment against the archrebels had reached the point where the doors of sympathy were so well-nigh universally closed that there could be found no standing ground for the Satania adversaries. When there exists no open door for the reception of evil, there exists no opportunity for the entertainment of sin. The doors of the hearts of all Edentia closed against Satan; **he was unanimously rejected by the assembled System Sovereigns, and it was at this time that the Son of Man "beheld Satan fall as lightning from heaven."**

Page 490:6   *Paper 43:5.1*

The rotation of the Most Highs on Edentia was suspended at the time of the Lucifer rebellion. We now have the same rulers who were on duty at that time. **We infer that no change in these rulers will be made until Lucifer and his associates are finally disposed of.**

Page 609:6   *Paper 53:8.3*

The bestowal of Michael terminated the Lucifer rebellion in all Satania aside from the planets of the apostate Planetary Princes. **And this was the significance of Jesus' personal experience, just before his death in the flesh, when he one day exclaimed to his disciples, "And I beheld Satan fall as lightning from heaven."** He had come with Lucifer to Urantia for the last crucial struggle.

Page 615:5   *Paper 54:3.3*

But if this universe rebel against the reality of truth and goodness refuses to approve the verdict, **and if the guilty one knows in his heart the justice of his condemnation but refuses to make such confession, then must the execution of sentence be delayed in accordance with the discretion of the Ancients of Days.** And the Ancients of Days refuse to annihilate any being until all moral values and all spiritual realities are extinct, **both in the evildoer** and in all related supporters and possible sympathizers.

Page 1972:1   *Paper 183:1.2*

The Father in heaven desired the bestowal Son to finish his earth career *naturally*, just as all mortals must finish up their lives on earth and in the flesh. Ordinary men and women cannot expect to have their last hours on earth and the supervening episode of death made easy by a special dispensation. Accordingly, Jesus elected to lay down his life in the flesh in the manner which was in keeping with the outworking of natural events, and he steadfastly refused to extricate himself from the cruel clutches of a wicked conspiracy of inhuman events which swept on with horrible certainty toward his unbelievable humiliation and ignominious death. And every bit of all this astounding manifestation of hatred and this unprecedented demonstration of cruelty was the work of evil men and wicked mortals. God in heaven did not will it, neither did the archenemies of Jesus dictate it, though they did much to insure that unthinking and evil mortals would thus reject the bestowal Son. **Even the father of sin turned his face away from the excruciating horror of the scene of the crucifixion.**

5. The Edentia Fathers Since the Lucifer Rebellion

Lucifer Rebellion—Satan

Page 490:6   *Paper 43:5.1*

See "Lucifer Rebellion—Satan" on page 254 for full text references. (43:4.9)

Most Highs on Urantia—Times of Crisis

Page 490:7–13   *Paper 43:5.2–8*

See "Most Highs on Urantia—Times of Crisis" on page 226 for full text references. (35:5.5–6)

Archangels' Circuit on Urantia

Page 490:7–16   *Paper 43:5.2–11*

See "Archangels' Circuit on Urantia" on page 233 for full text references. (37:3.3)

Most Highs

Page 491:12–13   *Paper 43:5.16–17*

See "Most Highs" on page 126 for full text references. (3:1.10)

6. The Gardens of God

Eden—Named After Edentia
How the Garden of Eden was named.

Page 492:2   *Paper 43:6.2*

About one half of **Edentia** is devoted to the exquisite gardens of the Most Highs, and these gardens are among the most entrancing morontia creations of the local universe. **This explains why the extraordinarily beautiful places on the inhabited worlds of Norlatiadek are so often called "the garden of Eden."**

Page 823:7   *Paper 73:4.1*

When Material Sons, the biologic uplifters, begin their sojourn on an evolutionary world, their place of abode is often called the Garden of Eden because it is characterized by the floral beauty and the botanic grandeur of Edentia, the constellation capital. Van well knew of these customs and accordingly provided that the entire peninsula be given over to the Garden. Pasturage and animal husbandry were projected for the adjoining mainland. Of animal life, only the birds and the various domesticated species were to be found in the park. Van's instructions were that Eden was to be a garden, and only a garden. No animals were ever slaughtered within its precincts. All flesh eaten by the Garden workers throughout all the years of construction was brought in from the herds maintained under guard on the mainland.

Tree of Life—Color

Did the shrub of Edentia on Urantia have a characteristic green color?—Parts II and III.

Page 492:6   *Paper 43:6.6*

The vegetable life is also very different from that of Urantia, consisting of both material and morontia varieties. The material growths have a characteristic green coloration, **but the morontia equivalents of vegetative life have a violet or orchid tinge of varying hue and reflection.** Such morontia vegetation is purely an energy growth; when eaten there is no residual portion.

Page 745:3   *Paper 66:4.13*

These antidotal complements of the Satania life currents were derived from the fruit of the tree of life, a shrub of Edentia which was sent to Urantia by the Most Highs of Norlatiadek at the time of Caligastia's arrival. In the days of Dalamatia this tree grew in the central courtyard of the temple of the unseen Father, and it was **the fruit of the tree of life** that enabled the material and otherwise mortal beings of the Prince's staff to live on indefinitely as long as they had access to it.

---

7. THE UNIVITATIA

Abandonters—Permanent Citizens of our Superuniverse

Page 493:5   *Paper 43:7.4*

See "Abandonters—Permanent Citizens of our Superuniverse" on page 235 for full text references. (37:9.8)

8. THE EDENTIA TRAINING WORLDS

Morontia Bodies—Ascending Changes

Seventy-one ascending morontia changes during Edentia age—seventy on training worlds plus one.

Page 494:1   *Paper 43:8.2*

The time spent on the **seventy training worlds of transition morontia culture associated with the Edentia age** of mortal ascension, is the most settled period in an ascending mortal's career up to the status of a finaliter; this is really the typical morontia life. **While you are re-keyed each time you pass from one major cultural world to another, you retain the same morontia body,** and there are no periods of personality unconsciousness.

Page 542:2   *Paper 48:1.5*

The Morontia Power Supervisors are able to effect a union of material and of spiritual energies, thereby organizing a morontia form of materialization which is receptive to the superimposition of a controlling spirit. When you traverse the morontia life of Nebadon, these same patient and skillful Morontia Power Supervisors will successively provide you with 570 morontia bodies, each one a phase of your progressive transformation. From the time of leaving the material worlds until you are constituted a first-stage spirit on Salvington, **you will undergo just 570 separate and ascending morontia changes.** Eight of these occur in the system, **seventy-one in the constellation,** and 491 during the sojourn on the spheres of Salvington.

Page 544:4   *Paper 48:2.14*

6. *Selective Assorters.* **As you progress from one class or phase of a morontia world to another, you must be re-keyed or advance-tuned,** and it is the task of the selective assorters to keep you in progressive synchrony with the morontia life.

9. CITIZENSHIP ON EDENTIA

Education—Purpose

Page 495:6   *Paper 43:9.4*

See "Education—Purpose" on page 126 for full text references. (2:7.12)

Most Highs

Page 495:6   *Paper 43:9.4*

See "Most Highs" on page 126 for full text references. (3:1.10)

PAPER 44: THE CELESTIAL ARTISANS

---

8. MORTAL ASPIRATIONS AND MORONTIA ACHIEVEMENTS

Ability
Ability—over 1200 pages apart in Parts II and IV.

Page 508:1   *Paper 44:8.3*

There is no caste in the ranks of spirit artisans. No matter how lowly your origin, **if you have ability and the gift of expression,** you will gain adequate recognition and receive due appreciation as you ascend upward in the scale of morontia experience and spiritual attainment. There can be no handicap of human heredity or deprivation of mortal environment which the morontia career will not fully compensate and wholly remove. And all such satisfactions of artistic achievement and expressionful self-realization will be effected by your own personal efforts in progressive advancement. At last the aspirations of evolutionary mediocrity may be realized. While the Gods do not arbitrarily bestow talents and ability upon the children of time, they do provide for the attainment of the satisfaction of all their noble longings and for the gratification of all human hunger for supernal self-expression.

Page 1779:3   *Paper 160:4.5*

Ability is that which you inherit, while skill is what you acquire. Life is not real to one who cannot do some one thing well, expertly. Skill is one of the real sources of the satisfaction of living. **Ability implies the gift of foresight, farseeing vision.** Be not deceived by the tempting rewards of dishonest achievement; be willing to toil for the later returns inherent in honest endeavor. The wise man is able to distinguish between means and ends; otherwise, sometimes overplanning for the future defeats its own high purpose. As a pleasure seeker you should aim always to be a producer as well as a consumer.

PAPER 45: THE LOCAL SYSTEM ADMINISTRATION

1. TRANSITIONAL CULTURE WORLDS

Lucifer Rebellion—Status of Personalities
Status of all personalities concerned in the Lucifer rebellion—located in two papers.

Page 510:9   *Paper 45:1.11*

As a sojourner on the seventh mansion world, you have access to the seventh transition world, the sphere of the Universal Father, and are also permitted to visit the Satania prison worlds surrounding this planet, **whereon are now confined Lucifer and the majority of those personalities who followed him in rebellion against Michael.** And this sad spectacle has been observable during these recent ages and will continue to serve as a solemn warning to all Nebadon until the Ancients of Days shall adjudicate the sin of Lucifer and his fallen associates who rejected the salvation proffered by Michael, their universe Father.

Page 607:2   *Paper 53:7.1*

The Lucifer rebellion was system wide. Thirty-seven seceding Planetary Princes swung their world administrations largely to the side of the archrebel. Only on Panoptia did the Planetary Prince fail to carry his people with him. On this world, under the guidance of the Melchizedeks, the people rallied to the support of Michael. Ellanora, a young woman of that mortal realm, grasped the leadership of the human races, and not a single soul on that strife-torn world enlisted under the Lucifer banner. **And ever since have these loyal Panoptians served on the seventh Jerusem transition world as the caretakers and builders on the Father's sphere and its surrounding seven detention worlds.** The Panoptians not only act as the literal custodians of these worlds, but they also execute the personal orders of Michael for the embellishment of these spheres for some future and unknown use. They do this work as they tarry en route to Edentia.

Page 610:6, 611:2   *Paper 53:9.1,3*

Early in the days of the Lucifer rebellion, salvation was offered all rebels by Michael. To all who would show proof of sincere repentance, he offered, upon his attainment of complete universe sovereignty, forgiveness and reinstatement in some form of universe service. None of the leaders accepted this merciful proffer. But thousands of the angels and the lower orders of celestial beings, including hundreds of the Material Sons and Daughters, accepted the mercy proclaimed by the Panoptians and were given rehabilitation at the time of Jesus' resurrection nineteen hundred years ago. **These beings have since been transferred to the Father's world of Jerusem, where they must be held, technically, until the Uversa courts hand down a decision in the matter of Gabriel vs. Lucifer.** But no one doubts that, when the annihilation verdict is issued, these repentant and salvaged personalities will be exempted from the decree of extinction. These probationary souls now labor with the Panoptians in the work of caring for the Father's world....

Michael, upon assuming the supreme sovereignty of Nebadon, **petitioned the Ancients of Days for authority to intern all personalities concerned in the Lucifer rebellion pending the rulings of the superuniverse tribunals in the case of Gabriel vs. Lucifer,** placed on the records of the Uversa supreme court almost two hundred thousand years ago, as you reckon time. Concerning the system capital group, the Ancients of Days granted the Michael petition with but a single exception: Satan was allowed to make periodic visits to the apostate princes on the fallen worlds until another Son of God should be accepted by such apostate worlds, or until such time as the courts of Uversa should begin the adjudication of the case of Gabriel *vs.* Lucifer.

---

2. THE SYSTEM SOVEREIGN

Lanaforge

Lanaforge's authority after Michael's bestowal—three papers over 500 pages apart.

Page 511:4   *Paper 45:2.4*

While all the affairs of the isolated worlds of Satania have not been returned to his jurisdiction, Lanaforge discloses great interest in their welfare, and he is a frequent visitor on Urantia. As in other and normal systems, the Sovereign presides over the system council of world rulers, the Planetary Princes and the resident governors general of the isolated worlds. This planetary council assembles from time to time on the headquarters of the system—"When the Sons of God come together."

Page 808:1   *Paper 72:0.1*

By permission of Lanaforge and with the approval of the Most Highs of Edentia, I am authorized to narrate something of the social, moral, and political life of the most advanced human race living on a not far-distant planet belonging to the Satania system.

Page 1016:4   *Paper 93:3.2*

Melchizedek taught the concept of one God, a universal Deity, but he allowed the people to associate this teaching with the Constellation Father of Norlatiadek, whom he termed El Elyon—the Most High. Melchizedek remained all but silent as to the status of Lucifer and the state of affairs on Jerusem. **Lanaforge, the System Sovereign, had little to do with Urantia until after the completion of Michael's bestowal.** To a majority of the Salem students Edentia was heaven and the Most High was God.

4. THE FOUR AND TWENTY COUNSELORS

Four and Twenty Counselors

Page 513:4   *Paper 45:4.1*

See "Four and Twenty Counselors" on page 197 for full text references. (24:5.3)

Mansant—a Nodite?

Mansant, a Nodite descendant of the rebel members of the Prince's staff?—three papers on the Four and Twenty Counsel.

Page 513:5–7   *Paper 45:4.2–4*

These twenty-four counselors have been recruited from the eight Urantia races, and the last of this group were assembled at the time of the resurrection roll call of Michael, nineteen hundred years ago. This Urantia advisory council is made up of the following members:

1. *Onagar,* **the master mind of the pre-Planetary Prince age,** who directed his fellows in the worship of "The Breath Giver."
2. *Mansant,* **the great teacher of the post-Planetary Prince age** on Urantia, who pointed his fellows to the veneration of "The Great Light."

Page 715:8   *Paper 63:6.1*

As the Andonic dispersion extended, the cultural and spiritual status of the clans retrogressed for nearly ten thousand years until the days of Onagar, who assumed the leadership of these tribes, brought peace among them, and for the first time, led all of them in the worship of the "Breath Giver to men and animals."

Page 821:6   *Paper 73:1.3*

The *Nodites* were the descendants of the rebel members of the Prince's staff, their name deriving from their first leader, Nod, onetime chairman of the Dalamatia commission on industry and trade. The *Amadonites* were the descendants of those Andonites who chose to remain loyal with Van and Amadon. "Amadonite" is more of a cultural and religious designation than a racial term; racially considered the Amadonites were essentially Andonites. **"Nodite" is both a cultural and racial term, for the Nodites themselves constituted the eighth race of Urantia.**

Onamonalonton—a Red Man

Hesunanin Onamonalonton, the spiritual deliverer of the red men—in Parts II and III.

Page 513:5,8   *Paper 45:4.2,5*

These twenty-four counselors have been recruited from the eight Urantia races, and the last of this group were assembled at the time of the resurrection roll call of Michael, nineteen hundred years ago. This Urantia advisory council is made up of the following members:

3. *Onamonalonton,* a far-distant leader of the red man and the one who directed this race from the worship of many gods to the veneration of "The Great Spirit."

Page 723:6   *Paper 64:6.7*

Because of this great retrogression the red men seemed doomed when, about **sixty-five thousand years ago, Onamonalonton appeared as their leader and spiritual deliverer.** He brought temporary peace among the American red men and revived their worship of the "Great Spirit." Onamonalonton lived to be ninety-six years of age and maintained his headquarters among the great redwood trees of California. Many of his later descendants have come down to modern times among the Blackfoot Indians.

Page 884:2   *Paper 79:5.8*

The red and the yellow races are the only human stocks that ever achieved a high degree of civilization apart from the influences of the Andites. **The oldest Amerindian culture was the Onamonalonton center in California, but** this had long since vanished by 35,000 B.C. In Mexico, Central America, and in the mountains of South America the later and more enduring civilizations were founded by a race predominantly red but containing a considerable admixture of the yellow, orange, and blue.

Page 1008:8   *Paper 92:5.3*

Many races have conceived of their leaders as being born of virgins; their careers are liberally sprinkled with miraculous episodes, and their return is always expected by their respective groups. In central Asia the tribesmen still look for the return of Genghis Khan; in Tibet, China, and India it is Buddha; in Islam it is Mohammed; **among the Amerinds it was Hesunanin Onamonalonton;** with the Hebrews it was, in general, Adam's return as a material ruler. In Babylon the god Marduk was a perpetuation of the Adam legend, the son-of-God idea, the connecting link between man and God. Following the appearance of Adam on earth, so-called sons of God were common among the world races.

Enoch
Enoch and the land of Nod—in Parts II and III.

Page 514:3   *Paper 45:4.13*

11. *Enoch,* the first of the mortals of Urantia to fuse with the Thought Adjuster during the mortal life in the flesh.

Page 849:3   *Paper 76:2.9*

And so Cain departed for the land of Nod, east of the second Eden. He became a great leader among one group of his father's people and did, to a certain degree, fulfill the predictions of Serapatatia, for he did promote peace between this division of the Nodites and the Adamites throughout his lifetime. **Cain married Remona, his distant cousin, and their first son, Enoch, became the head of the Elamite Nodites.** And for hundreds of years the Elamites and the Adamites continued to be at peace.

Page 859:8   *Paper 77:4.5*

After the establishment of the second garden it was customary to allude to this near-by Nodite settlement as **"the land of Nod"**; and during the long period of relative peace between this Nodite group and the Adamites, the two races were greatly blended, for it became more and more the custom for the Sons of God (the Adamites) to intermarry with the daughters of men (the Nodites).

Elijah
Elijah and monotheism—Parts II and III.

Page 514:5   *Paper 45:4.15*

13. *Elijah,* **a translated soul** of brilliant spiritual achievement during the post-Material Son age.

Page 1064:3   *Paper 97:2.2*

When Elijah was called away, Elisha, his faithful associate, took up his work and, with the invaluable assistance of the little-known Micaiah, kept the light of truth alive in Palestine.

Page 1065:2–3   *Paper 97:3.5–6*

Out of this basic difference in the regard for land, there evolved the bitter antagonisms of social, economic, moral, and religious attitudes exhibited by the Canaanites and the Hebrews. **This socioeconomic controversy did not become a definite religious issue until the times of Elijah.** From the days of this aggressive prophet the issue was fought out on more strictly religious lines—Yahweh *vs.* Baal—and it ended in the triumph of Yahweh and the subsequent drive toward monotheism.

Elijah shifted the Yahweh-Baal controversy from the land issue to the religious aspect of Hebrew and Canaanite ideologies. When Ahab murdered the Naboths in the intrigue to get possession of their land, Elijah made a moral issue out of the olden land mores and launched his vigorous campaign against the Baalites. This was also a fight of the country folk against domination by the cities. It was chiefly under Elijah that Yahweh became Elohim. The prophet began as an agrarian reformer and ended up by exalting Deity. Baals were many, Yahweh was *one*—monotheism won over polytheism.

John the Baptist—Distant Cousins
John and Elizabeth's distant cousins—Parts II and IV.

Page 514:7   *Paper 45:4.17*

15. *John the Baptist,* the forerunner of Michael's mission on Urantia and, in the flesh, **distant cousin** of the Son of Man.

Page 1346:1   *Paper 122:2.6*

Gabriel appeared to Mary about the middle of November, 8 B.C., while she was at work in her Nazareth home. Later on, after Mary knew without doubt that she was to become a mother, she persuaded Joseph to let her journey to the City of Judah, four miles west of Jerusalem, in the hills, to visit Elizabeth. Gabriel had informed each of these mothers-to-be of his appearance to the other. Naturally they were anxious to get together, compare experiences, and talk over the probable futures of their sons. Mary remained with **her distant cousin** for three weeks. Elizabeth did much to strengthen Mary's faith in the vision of Gabriel, so that she returned home more fully dedicated to the call to mother the child of destiny whom she was so soon to present to the world as a helpless babe, an average and normal infant of the realm.

---

6. ADAMIC TRAINING OF ASCENDERS

Infant-receiving Schools
Criteria for advancing to the infant receiving schools—in three papers.

Page 516:5   *Paper 45:6.7*

This probation nursery of Satania is maintained by certain morontia personalities on the finaliters' world, one half of the planet being devoted to this work of child rearing. Here are received and reassembled **certain children of surviving mortals,** such as those offspring who perished on the evolutionary worlds before acquiring spiritual status as individuals. The ascension of either of its natural parents insures that such a mortal child of the realms will be accorded repersonalization on the system finaliter planet and there be permitted to demonstrate by subsequent freewill choice whether or not it elects to follow the parental path of mortal ascension. Children here appear as on the nativity world except for the absence of sex differentiation. There is no reproduction of mortal kind after the life experience on the inhabited worlds.

Page 531:5–6   *Paper 47:2.1–2*

The infant-receiving schools of Satania are situated on the finaliter world, the first of the Jerusem transition-culture spheres. These infant-receiving schools are enterprises devoted to the nurture and training of the children of time, **including those who have died on the evolutionary worlds of space before the acquirement of individual status on the universe records. In the event of the survival of either or both of such a child's parents,** the guardian of destiny deputizes her associated cherubim as the custodian of the child's potential identity, charging the cherubim with the responsibility of delivering this undeveloped soul into the hands of the Mansion World Teachers in the probationary nurseries of the morontia worlds.

It is these same deserted cherubim who, as Mansion World Teachers, under the supervision of the Melchizedeks, maintain such extensive educational facilities for the training of the probationary wards of the finaliters. These wards of the finaliters, these **infants of ascending mortals, are always personalized as of their exact physical status at the time of death except for reproductive potential.** This awakening occurs at the exact time of the parental arrival on the first mansion world. And then are these children given every opportunity, as they are, to choose the heavenly way just as they would have made such a choice on the worlds where death so untimely terminated their careers.

Page 569:6, 570:1   *Paper 49:6.11–12*

3. *Mortals of the probationary-dependent orders of ascension.* The arrival of an Adjuster constitutes identity in the eyes of the universe, and all indwelt beings are on the roll calls of justice. But temporal life on the evolutionary worlds is uncertain, and many die in youth before choosing the Paradise career. **Such Adjuster-indwelt children and youths follow the parent of most advanced spiritual status,** thus going to the system finaliter world (the probationary nursery) on the third day, at a special resurrection, or at the regular millennial and dispensational roll calls.

Children who die when too young to have Thought Adjusters **are repersonalized on the finaliter world** of the local systems concomitant with the arrival of either parent on the mansion worlds. A child acquires physical entity at mortal birth, but in the matter of survival all Adjusterless children are reckoned as still attached to their parents.

PAPER 46: THE LOCAL SYSTEM HEADQUARTERS

---

3. THE JERUSEM BROADCASTS

Archangels' Circuit on Urantia

Page 522:4   *Paper 46:3.4*

See "Archangels' Circuit on Urantia" on page 233 for full text references. (37:3.3)

Dematerialization—Salvable and Unsalvable Personalities

Page 526:1   *Paper 46:5.19*

> See "Dematerialization—Salvable and Unsalvable Personalities" on page 230 for full text references. (36:4.3)

Seraphic Transport
Examples of seraphic transport in four papers.

Page 526:1   *Paper 46:5.19*

4. *The circles of the Master Physical Controllers.* The various orders of the Master Physical Controllers are concentrically arranged around the vast temple of power, wherein presides the power chief of the system in association with the chief of the Morontia Power Supervisors. This temple of power is one of two sectors on Jerusem where ascending mortals and midway creatures are not permitted. The other one is the dematerializing sector in the area of the Material Sons, **a series of laboratories wherein the transport seraphim transform material beings into a state quite like that of the morontia order of existence.**

Page 575:1   *Paper 50:3.6*

At the end of the prince's dispensation, when the time comes for this "reversion staff" to be returned to the system headquarters for the resumption of the Paradise career, these ascenders present themselves to the Life Carriers for the purpose of yielding up their material bodies. They enter the transition slumber and awaken delivered from their mortal investment and clothed with morontia forms, **ready for seraphic transportation back to the system capital, where their detached Adjusters await them.** They are a whole dispensation behind their Jerusem class, but they have gained a unique and extraordinary experience, a rare chapter in the career of an ascending mortal.

Page 582:3   *Paper 51:2.3*

While there is this dematerializing technique for preparing the Adams for transit from Jerusem to the evolutionary worlds, there is no equivalent method for taking them away from such worlds unless the entire planet is to be emptied, in which event emergency installation of the dematerialization technique is made for the entire salvable population. If some physical catastrophe should doom the planetary residence of an evolving race, **the Melchizedeks** and the Life Carriers would install the technique of dematerialization for all survivors, **and by seraphic transport these beings would be carried away to the new world prepared for their continuing existence.** The evolution of a human race, once initiated on a world of space, must proceed quite independently of the physical survival of that planet, but during the evolutionary ages it is not otherwise intended that a Planetary Adam or Eve shall leave their chosen world.

Page 825:8   *Paper 73:6.3*

The "tree of the knowledge of good and evil" may be a figure of speech, a symbolic designation covering a multitude of human experiences, but the "tree of life" was not a myth; it was real and for a long time was present on Urantia. When the Most Highs of Edentia approved the commission of Caligastia as Planetary Prince of Urantia and those of the one hundred Jerusem citizens as his administrative staff, **they sent to the planet, by the Melchizedeks, a shrub of Edentia, and this plant grew to be the tree of life on Urantia.** This form of nonintelligent life is native to the constellation headquarters spheres, being also found on the headquarters worlds of the local and superuniverses as well as on the Havona spheres, but not on the system capitals.

PAPER 47: THE SEVEN MANSION WORLDS

---

1. THE FINALITERS' WORLD

Transition Mortals
Transitional beings—in Parts II and III.

Page 530:5   *Paper 47:1.1*

Although only finaliters and certain groups of salvaged children and their caretakers are resident on transitional world number one, provision is made for the entertainment of all classes of spirit beings, **transition mortals,** and student visitors. The spornagia, who function on all of these worlds, are hospitable hosts to all beings whom they can recognize. They have a vague feeling concerning the finaliters but cannot visualize them. They must regard them much as you do the angels in your present physical state.

Page 533:2   *Paper 47:3.4*

If a **transitory personality** of mortal origin should never be thus reassembled, the spirit elements of the nonsurviving mortal creature would forever continue as an integral part of the individual experiential endowment of the onetime indwelling Adjuster.

Page 1233:3   *Paper 112:5.7*

If ever there is doubt as to the advisability of advancing a human identity to the mansion worlds, the universe governments invariably rule in the personal interests of that individual; **they unhesitatingly advance such a soul to the status of a transitional being,** while they continue their observations of the emerging morontia intent and spiritual purpose. Thus divine justice is certain of achievement, and divine mercy is accorded further opportunity for extending its ministry.

---

2. THE PROBATIONARY NURSERY

Infant-receiving Schools

Page 531:5–6   *Paper 47:2.1–2*

See "Infant-receiving Schools" on page 266 for full text references. (45:6.7)

---

3. THE FIRST MANSION WORLD

Transition Mortals

Page 533:2   *Paper 47:3.4*

See "Transition Mortals" on page 269 for full text references. (47:1.1)

8. The Sixth Mansion World

Forty Days

Forty days on Urantia and on mansion world—Parts II and IV, almost 1500 pages apart.

Page 538:3   *Paper 47:8.5*

Immediately upon the confirmation of Adjuster fusion the new morontia being is introduced to his fellows for the first time by his new name and is **granted the forty days of spiritual retirement** from all routine activities wherein to commune with himself and to choose some one of the optional routes to Havona and to select from the differential techniques of Paradise attainment.

Page 1512:2   *Paper 136:2.6*

When Jesus was baptized, he repented of no misdeeds; he made no confession of sin. His was the baptism of consecration to the performance of the will of the heavenly Father. At his baptism he heard the unmistakable call of his Father, the final summons to be about his Father's business, and he went away into private seclusion for forty days to think over these manifold problems. In thus retiring for a season from active personality contact with his earthly associates, Jesus, as he was and on Urantia, **was following the very procedure that obtains on the morontia worlds whenever an ascending mortal fuses with the inner presence of the Universal Father.**

Page 1512:6   *Paper 136:3.2*

After his baptism he **entered upon the forty days** of adjusting himself to the changed relationships of the world and the universe occasioned by the personalization of his Adjuster. During this isolation in the Perean hills he determined upon the policy to be pursued and the methods to be employed in the new and changed phase of earth life which he was about to inaugurate.

Visions of "Heaven"
John and Paul and their view of the second heaven of heavens.

Page 539:4–5   *Paper 47:10.2–3*

John the Revelator **saw a vision of the arrival of a class of advancing mortals from the seventh mansion world to their first heaven, the glories of Jerusem.** He recorded: "And I saw as it were a sea of glass mingled with fire; and those who had gained the victory over the beast that was originally in them and over the image that persisted through the mansion worlds and finally over the last mark and trace, standing on the sea of glass, having the harps of God, and singing the song of deliverance from mortal fear and death." (Perfected space communication is to be had on all these worlds; and your anywhere reception of such communications is made possible by carrying the "harp of God," a morontia contrivance compensating for the inability to directly adjust the immature morontia sensory mechanism to the reception of space communications.)

Paul **also had a view of the ascendant-citizen corps of perfecting mortals on Jerusem,** for he wrote: "But you have come to Mount Zion and to the city of the living God, the heavenly Jerusalem, and to an innumerable company of angels, to the grand assembly of Michael, and to the spirits of just men being made perfect."

Page 553:4   *Paper 48:6.12*

You should consider the statement about "heaven" and the "heaven of heavens." **The heaven conceived by most of your prophets was the first of the mansion worlds of the local system.** When the apostle spoke of being "caught up to the third heaven," he referred to that experience in which his Adjuster was detached during sleep and in this unusual state made a projection to the third of the seven mansion worlds. Some of your wise men saw the vision of the greater heaven, "the heaven of heavens," of which the sevenfold mansion world experience was but the first; the second being Jerusem; the third, Edentia and its satellites; the fourth, Salvington and the surrounding educational spheres; the fifth, Uversa; the sixth, Havona; and the seventh, Paradise.

1. MORONTIA MATERIALS

Morontia Bodies—Ascending Changes

Page 542:2   *Paper 48:1.5*

See "Morontia Bodies—Ascending Changes" on page 258 for full text references. (43:8.2)

2. MORONTIA POWER SUPERVISORS

Morontia Bodies—Ascending Changes

Page 544:4   *Paper 48:2.14*

See "Morontia Bodies—Ascending Changes" on page 258 for full text references. (43:8.2)

6. MORONTIA WORLD SERAPHIM—TRANSITION MINISTERS

Soul

Page 551:7   *Paper 48:6.2*

See "Soul" on page 104 for full text references. (0:5.10)

Race Commissioners
The active race commissioners—in Parts II and III.

Page 553:3   *Paper 48:6.11*

2. *Racial Interpreters.* All races of mortal beings are not alike. True, there is a planetary pattern running through the physical, mental, and spiritual natures and tendencies of the various races of a given world; but there are also distinct racial types, and very definite social tendencies characterize the offspring of these different basic types of human beings. On the worlds of time the seraphic **racial interpreters further the efforts of the race commissioners** to harmonize the varied viewpoints of the races, and they continue to function on the mansion worlds, where these same differences

tend to persist in a measure. On a confused planet, such as Urantia, these brilliant beings have hardly had a fair opportunity to function, but they are the skillful sociologists and the wise ethnic advisers of the first heaven.

Page 1253:5   *Paper 114:4.2*

There are certain groups of planetary problems which are still under the control of the Most Highs of Edentia, jurisdiction over them having been seized at the time of the Lucifer rebellion. Authority in these matters is exercised by a Vorondadek Son, the Norlatiadek observer, who maintains very close advisory relations with the planetary supervisors. **The race commissioners are very active on Urantia,** and their various group chiefs are informally attached to the resident Vorondadek observer, who acts as their advisory director.

Page 1255:8   *Paper 114:6.9*

5. The angels of the races. Those who work for the conservation of the evolutionary races of time, regardless of their political entanglements and religious groupings. On Urantia there are remnants of nine human races which have commingled and combined into the people of modern times. **These seraphim are closely associated with the ministry of the race commissioners,** and the group now on Urantia is the original corps assigned to the planet soon after the day of Pentecost.

Visions of "Heaven"

Page 553:4   *Paper 48:6.12*

See "Visions of "Heaven"" on page 272 for full text references. (47:10.2–3)

Mind—Exchange for the Mind of Jesus
How to go about exchanging your mind for the Master's mind—Parts II, III, and IV.

Page 553:7   *Paper 48:6.15*

Even on Urantia, these seraphim teach the everlasting truth: **If your own mind does not serve you well, you can exchange it for the mind of Jesus of Nazareth, who always serves you well.**

Page 1002:6–8   *Paper 91:9.1–3*

If you would engage in effective praying, you should bear in mind the laws of prevailing petitions:

1. You must qualify as a potent prayer by sincerely and courageously facing the problems of universe reality. You must possess cosmic stamina.

2. **You must have honestly exhausted the human capacity for human adjustment. You must have been industrious.**

Page 1123:1   *Paper 102:4.1*

Because of the presence in your minds of the Thought Adjuster, it is no more of a mystery for you to know the mind of God than for you to be sure of the consciousness of knowing any other mind, human or superhuman. Religion and social consciousness have this in common: They are predicated on the consciousness of other-mindness. **The technique whereby you can accept another's idea as yours is the same whereby you may "let the mind which was in Christ be also in you."**

Page 1639:5   *Paper 146:2.10*

9. "I have come forth from the Father; **if, therefore, you are ever in doubt as to what you would ask of the Father, ask in my name, and I will present your petition in accordance with your real needs** and desires and in accordance with my Father's will." Guard against the great danger of becoming self-centered in your prayers. Avoid praying much for yourself; pray more for the spiritual progress of your brethren. Avoid materialistic praying; pray in the spirit and for the abundance of the gifts of the spirit.

Truth Sensitivity

The substitute for the truth sensitivity mota is revelation—in four related papers.

Page 554:2   *Paper 48:6.17*

Mota is more than a superior philosophy; it is to philosophy as two eyes are to one; it has a stereoscopic effect on meanings and values. Material man sees the universe, as it were, with but one eye—flat. Mansion world students achieve cosmic perspective—depth—by superimposing the perceptions of the morontia life upon the perceptions of the physical life. And they are enabled to bring these material and morontial viewpoints into true focus largely through the untiring ministry of their seraphic counselors, who so patiently teach the mansion world students and the morontia progressors. Many of the teaching counselors of the supreme order of seraphim began their careers as advisers of the newly liberated souls of the mortals of time.

Page 740:2   *Paper 65:8.6*

When physical conditions are ripe, *sudden* mental evolutions may take place; when mind status is propitious, **sudden spiritual transformations may occur;** when spiritual values receive proper recognition, then cosmic meanings become discernible, and increasingly the personality is released from the handicaps of time and delivered from the limitations of space.

Page 1106:9   *Paper 101:2.8*

Faith reveals God in the soul. **Revelation, the substitute for morontia insight on an evolutionary world,** enables man to see the same God in nature that faith exhibits in his soul. Thus does revelation successfully bridge the gulf between the material and the spiritual, even between the creature and the Creator, between man and God.

Page 1136:3   *Paper 103:6.8*

Metaphysics has proved a failure; mota, man cannot perceive. **Revelation is the only technique which can compensate for the absence of the truth sensitivity of mota in a material world.** Revelation authoritatively clarifies the muddle of reason-developed metaphysics on an evolutionary sphere.

---

7. MORONTIA MOTA

Jesus—Perfecting Mortal Existence

As Jesus becomes a man, match some of the twenty-eight statements of human philosophy in Part II with Part IV as he depends on the guidance of his heavenly Father.

Page 556:2,7-Page 557:2,8,14   *Paper 48:7.2,7,18,24,30*

Not long since, while executing an assignment on the first mansion world of Satania, I had occasion to observe this method of teaching; and though I may not undertake to present the mota content of the lesson, **I am permitted to record the twenty-eight statements of human philosophy which this morontia instructor was utilizing as illustrative material designed to assist these new mansion world sojourners in their early efforts to grasp the significance and meaning of mota. These illustrations of human philosophy were:**

    5. Difficulties may challenge mediocrity and defeat the fearful, but they only stimulate the true children of the Most Highs.

   16. You cannot perceive spiritual truth until you feelingly experience it, and many truths are not really felt except in adversity.

22. The evolving soul is not made divine by what it does, but by what it strives to do.

28. The argumentative defense of any proposition is inversely proportional to the truth contained.

Page 1405:4   *Paper 127:6.12*

Jesus is rapidly becoming a man, not just a young man but an adult. **He has learned well to bear responsibility. He knows how to carry on in the face of disappointment. He bears up bravely when his plans are thwarted and his purposes temporarily defeated. He has learned how to be fair and just even in the face of injustice.** He is learning how to adjust his ideals of spiritual living to the practical demands of earthly existence. He is learning how to plan for the achievement of a higher and distant goal of idealism while he toils earnestly for the attainment of a nearer and immediate goal of necessity. He is steadily acquiring the art of adjusting his aspirations to the commonplace demands of the human occasion. He has very nearly mastered the technique of utilizing the energy of the spiritual drive to turn the mechanism of material achievement. **He is slowly learning how to live the heavenly life while he continues on with the earthly existence.** More and more he depends upon the ultimate guidance of his heavenly Father while he assumes the fatherly role of guiding and directing the children of his earth family. He is becoming experienced in the skillful wresting of victory from the very jaws of defeat; he is learning how to transform the difficulties of time into the triumphs of eternity.

Anxiety
Anxiety—covered in Parts II and IV.

Page 557:5   *Paper 48:7.21*

19. **Anxiety must be abandoned.** The disappointments hardest to bear are those which never come.

Page 1823:2   *Paper 165:5.2*

"Yes, Andrew, I will speak to you about these matters of wealth and self-support, but my words to you, the apostles, must be somewhat different from those spoken to the disciples and the multitude since you have forsaken everything, not only to follow me, but to be ordained as ambassadors of the kingdom. Already have you had several years' experience, and you know that the Father whose kingdom you proclaim will not forsake you. You have dedicated your lives to the ministry of the kingdom; therefore be not anxious or worried about the things of the temporal life, what you shall eat, nor yet for your body, what you shall wear. The welfare of the soul is more than food and drink; the progress in the spirit is far above the need of raiment. When you are tempted to doubt the sureness of your bread, consider the ravens; they sow not neither reap, they have no storehouses or barns, and yet the Father provides food for every one of them that seeks it. And of how much more value are you than many birds! **Besides, all of your anxiety or fretting doubts can do nothing to supply your material needs.** Which of you by anxiety can add a handbreadth to your stature or a day to your life? Since such matters are not in your hands, why do you give anxious thought to any of these problems?

PAPER 49: THE INHABITED WORLDS

---

2. PLANETARY PHYSICAL TYPES

Planets Suited to Harbor Life

Page 561:12   *Paper 49:2.6*

See "Planets Suited to Harbor Life" on page 170 for full text references. (15:6.10)

---

4. EVOLUTIONARY WILL CREATURES

Cooperation or Resistance to Divine Leading—Survival of Immortal Soul

Page 565:1   *Paper 49:4.9*

See "Cooperation or Resistance to Divine Leading—Survival of Immortal Soul" on page 205 for full text references. (30:4.8)

Survival—Co-operation with Indwelling Adjuster

Page 565:1   *Paper 49:4.9*

See "Survival—Co-operation with Indwelling Adjuster" on page 245 for full text references. (40:5.16)

5. THE PLANETARY SERIES OF MORTALS

Tabamantia—Agondonter of Finaliter Status
Tabamantia's role in Nebadon—in four papers.

Page 565:13   *Paper 49:5.5*

In the universe of Nebadon, all the life-modification worlds are serially linked together and constitute a special domain of universe affairs which is given attention by designated administrators; and all of these experimental worlds are periodically inspected by a corps of universe directors **whose chief is the veteran finaliter known in Satania as Tabamantia.**

Page 579:1   *Paper 50:7.2*

On Jerusem the ascenders from these isolated worlds occupy a residential sector by themselves and are known as the *agondonters*, meaning evolutionary will creatures who can believe without seeing, persevere when isolated, and triumph over insuperable difficulties even when alone. This functional grouping of the agondonters persists throughout the ascension of the local universe and the traversal of the superuniverse; it disappears during the sojourn in Havona but promptly reappears upon the attainment of Paradise and definitely persists in the Corps of the Mortal Finality. **Tabamantia is an** *agondonter* **of finaliter status,** having survived from one of the quarantined spheres involved in the first rebellion ever to take place in the universes of time and space.

Page 821:3   *Paper 73:0.3*

Tabamantia, sovereign supervisor of the series of decimal or experimental worlds, came to inspect the planet and, after his survey of racial progress, duly recommended that Urantia be granted Material Sons. In a little less than one hundred years from the time of this inspection, Adam and Eve, a Material Son and Daughter of the local system, arrived and began the difficult task of attempting to untangle the confused affairs of a planet retarded by rebellion and resting under the ban of spiritual isolation.

Page 1189:1   *Paper 108:3.5*

It is interesting to note that local universe inspectors always address themselves, when carrying out a planetary examination, to the planetary chief of Thought Adjusters, just as they deliver charges to the chiefs of seraphim and to the leaders of other orders of beings attached to the administration of an evolving world. **Not long since, Urantia underwent such a periodic inspection by Tabamantia, the sovereign supervisor of all life-experiment** planets in the universe of Nebadon. And the records reveal that, in addition to his admonitions and indictments delivered to the various chiefs of superhuman personalities, he also delivered the following acknowledgment to the chief of Adjusters, whether located on the planet, on Salvington, Uversa, or Divinington, we do not definitely know, but he said:

Light and Life
The succeeding paper referenced in Paper 49 is Paper 55, six papers later.

Page 567:7–8   *Paper 49:5.20–21*

As a result of the ministry of all the successive orders of divine sonship, the inhabited worlds and their advancing races begin to approach the apex of planetary evolution. Such worlds now become ripe for the culminating mission, the arrival of the Trinity Teacher Sons. **This epoch of the Teacher Sons is the vestibule to the final planetary age—evolutionary utopia—the age of light and life.**

This classification of human beings will receive particular attention in a succeeding paper.

Page 621:1   *Paper 55:0.1*

The age of light and life is the final evolutionary attainment of a world of time and space.

Dematerialization—Salvable and Unsalvable Personalities

Page 569:2   *Paper 49:6.7*

See "Dematerialization—Salvable and Unsalvable Personalities" on page 230 for full text references. (36:4.3)

Adam and Eve—Ascension
Eve waited nineteen years for Adam. Parts I and II discusses this type of ascension.

Page 569:3   *Paper 49:6.8*

2. *Mortals of the individual orders of ascension.* The individual progress of human beings is measured by their successive attainment and traversal (mastery) of the seven cosmic circles. These circles of mortal progression are

levels of associated intellectual, social, spiritual, and cosmic-insight values. Starting out in the seventh circle, mortals strive for the first, and all who have attained the third immediately have personal guardians of destiny assigned to them. **These mortals may be repersonalized in the morontia life independent of dispensational or other adjudications.**

Page 852:4   *Paper 76:5.5*

Adam lived for 530 years; he died of what might be termed old age. His physical mechanism simply wore out; the process of disintegration gradually gained on the process of repair, and the inevitable end came. **Eve had died nineteen years previously of a weakened heart.** They were both buried in the center of the temple of divine service which had been built in accordance with their plans soon after the wall of the colony had been completed. And this was the origin of the practice of burying noted and pious men and women under the floors of the places of worship.

Page 853:3   *Paper 76:6.2*

They did not long rest in the oblivion of the unconscious sleep of the mortals of the realm. On the third day after Adam's death, the second following his reverent burial, the orders of Lanaforge, sustained by the acting Most High of Edentia and concurred in by the Union of Days on Salvington, acting for Michael, were placed in Gabriel's hands, directing the special roll call of the distinguished survivors of the Adamic default on Urantia. And in accordance with this mandate of special resurrection, number twenty-six of the Urantia series, **Adam and Eve were repersonalized and reassembled in the resurrection halls of the mansion worlds of Satania together with 1,316 of their associates in the experience of the first garden.** Many other loyal souls had already been translated at the time of Adam's arrival, which was attended by a dispensational adjudication of both the sleeping survivors and of the living qualified ascenders.

Third Day—Third Period

Page 569:4   *Paper 49:6.9*

See "Third Day—Third Period" on page 204 for full text references. (30:4.4–5)

Infant-receiving Schools

Page 569:6, 570:1   *Paper 49:6.11–12*

See "Infant-receiving Schools" on page 266 for full text references. (45:6.7)

Personality—When Is it Bestowed?

Page 570:1   *Paper 49:6.12*

See "Personality—When Is it Bestowed?" on page 195 for full text references. (24:2.8)

PAPER 50: THE PLANETARY PRINCES

3. THE PRINCE'S CORPOREAL STAFF

Seraphic Transport

Page 575:1   *Paper 50:3.6*

See "Seraphic Transport" on page 268 for full text references. (46:5.19)

6. PLANETARY CULTURE

Ideal Government—Struggle For

Page 578:1   *Paper 50:6.1*

See "Ideal Government—Struggle For" on page 129 for full text references. (3:1.10)

7. THE REWARDS OF ISOLATION

Tabamantia—Agondonter of Finaliter Status

Page 579:1   *Paper 50:7.2*

See "Tabamantia—Agondonter of Finaliter Status" on page 279 for full text references. (49:5.5)

2. TRANSIT OF THE PLANETARY ADAMS

Dematerialization—Salvable and Unsalvable Personalities

Page 582:3   *Paper 51:2.3*

See "Dematerialization—Salvable and Unsalvable Personalities" on page 230 for full text references. (36:4.3)

Seraphic Transport

Page 582:3   *Paper 51:2.3*

See "Seraphic Transport" on page 268 for full text references. (46:5.19)

3. THE ADAMIC MISSIONS

Secondary Midwayers

Page 583:5   *Paper 51:3.6*

See "Secondary Midwayers" on page 244 for full text references. (40:3.1)

4. THE SIX EVOLUTIONARY RACES

Overpopulation and Human Degeneracy

The evolution of the multiplication of the average or stabilized human being and a serious look at degeneracy and the results of selective reproduction.

Page 585:4   *Paper 51:4.8*

These six evolutionary races are destined to be blended and exalted by amalgamation with the progeny of the Adamic uplifters. But before these peoples are blended, the inferior and unfit are largely eliminated. The Planetary Prince and the Material Son, with other suitable planetary authorities, pass upon the fitness of the reproducing strains. **The difficulty of executing such a radical program on Urantia consists in the absence of** competent judges to pass upon the biologic fitness or unfitness of the individuals of your world races. Notwithstanding this obstacle, it seems that you ought to be able to agree upon the biologic disfellowshiping of your more markedly unfit, defective, degenerate, and antisocial stocks.

Page 596:7   *Paper 52:5.9*

During this era the problems of disease and delinquency are virtually solved. **Degeneracy has already been largely eliminated by selective reproduction.** Disease has been practically mastered through the high resistant qualities of the Adamic strains and by the intelligent and world-wide application of the discoveries of the physical sciences of preceding ages. The average length of life, during this period, climbs well above the equivalent of three hundred years of Urantia time.

Page 599:1   *Paper 52:7.5*

Life during this era is pleasant and profitable. **Degeneracy and the antisocial end products of the long evolutionary struggle have been virtually obliterated.** The length of life approaches five hundred Urantia years, and the reproductive rate of racial increase is intelligently controlled. An entirely new order of society has arrived. There are still great differences among mortals, but the state of society more nearly approaches the ideals of social brotherhood and spiritual equality. Representative government is vanishing, and the world is passing under the rule of individual self-control. The function of government is chiefly directed to collective tasks of social administration and economic co-ordination. The golden age is coming on apace; the temporal goal of the long and intense planetary evolutionary struggle is in sight. The reward of the ages is soon to be realized; the wisdom of the Gods is about to be manifested.

Page 629:11   *Paper 55:5.2*

The advanced stages of a world settled in light and life represent the acme of evolutionary material development. On these cultured worlds, gone are the idleness and friction of the earlier primitive ages. **Poverty and social inequality have all but vanished, degeneracy has disappeared, and delinquency is rarely observed.** Insanity has practically ceased to exist, and feeble-mindedness is a rarity.

Page 726:4   *Paper 64:6.32*

2. Stronger and better races are to be had from the interbreeding of diverse peoples when these different races are carriers of superior inheritance factors. And the Urantia races would have benefited by such an early amalgamation provided such a conjoint people could have been subsequently

effectively upstepped by a thoroughgoing admixture with the superior Adamic stock. **The attempt to execute such an experiment on Urantia under present racial conditions would be highly disastrous.**

Page 770:8   *Paper 68:6.11*

From a world standpoint, overpopulation has never been a serious problem in the past, but if war is lessened and science increasingly controls human diseases, it may become a serious problem in the near future. At such a time the great test of the wisdom of world leadership will present itself. **Will Urantia rulers have the insight and courage to foster the multiplication of the average or stabilized human being instead of the extremes of the supernormal and the enormously increasing groups of the subnormal?** The normal man should be fostered; he is the backbone of civilization and the source of the mutant geniuses of the race. The subnormal man should be kept under society's control; no more should be produced than are required to administer the lower levels of industry, those tasks requiring intelligence above the animal level but making such low-grade demands as to prove veritable slavery and bondage for the higher types of mankind.

Page 786:2–8   *Paper 70:2.6–12*

The nations of Urantia have already entered upon the gigantic struggle between nationalistic militarism and industrialism, and in many ways this conflict is analogous to the agelong struggle between the herder-hunter and the farmer. But if industrialism is to triumph over militarism, it must avoid the dangers which beset it. **The perils of budding industry on Urantia are:**
1. The strong drift toward materialism, spiritual blindness.
2. The worship of wealth-power, value distortion.
3. The vices of luxury, cultural immaturity.
4. The increasing dangers of indolence, service insensitivity.
5. **The growth of undesirable racial softness, biologic deterioration.**
6. The threat of standardized industrial slavery, personality stagnation. Labor is ennobling but drudgery is benumbing.

Page 793:6–9   *Paper 70:8.14–17*

Classes in society, having naturally formed, will persist until man gradually achieves their evolutionary obliteration through intelligent manipulation of the biologic, intellectual, and spiritual resources of a progressing civilization, such as:
1. **Biologic renovation of the racial stocks—the selective elimination of inferior human strains. This will tend to eradicate many mortal inequalities.**
2. Educational training of the increased brain power which will arise out of such biologic improvement.
3. Religious quickening of the feelings of mortal kinship and brotherhood.

PAPER 52: PLANETARY MORTAL EPOCHS

1. PRIMITIVE MAN

Spirit Fusion—Early Evolutionary Times
Spirit fusion in the early days—Parts II and III.

Page 590:2   Paper 52:1.6

Man's acquirement of ethical judgment, moral will, is usually coincident with the appearance of early language. Upon attaining the human level, after this emergence of mortal will, these beings become receptive to the temporary indwelling of the divine Adjusters, and upon death many are duly elected as survivors and sealed by the archangels **for subsequent resurrection and Spirit fusion.** The archangels always accompany the Planetary Princes, and a dispensational adjudication of the realm is simultaneous with the prince's arrival.

Page 711:3   Paper 63:0.3

Andon is the Nebadon name which signifies "the first Fatherlike creature to exhibit human perfection hunger." Fonta signifies "the first Sonlike creature to exhibit human perfection hunger." Andon and Fonta never knew these names until they were **bestowed upon them at the time of fusion with their Thought Adjusters.** Throughout their mortal sojourn on Urantia they called each other Sonta-an and Sonta-en, Sonta-an meaning "loved by mother," Sonta-en signifying "loved by father." They gave themselves these names, and the meanings are significant of their mutual regard and affection.

Page 717:3   Paper 63:7.2

On Jerusem both Andon and Fonta were fused with their Thought Adjusters, as also were several of their children, including Sontad, **but the majority of even their immediate descendants only achieved Spirit fusion.**

2. POST-PLANETARY PRINCE MAN

Dematerialization—Salvable and Unsalvable Personalities

Page 592:5   Paper 52:2.12

See "Dematerialization—Salvable and Unsalvable Personalities" on page 230 for full text references. (36:4.3)

286

5. POST-BESTOWAL SON MAN

Overpopulation and Human Degeneracy

Page 596:7   *Paper 52:5.9*

See "Overpopulation and Human Degeneracy" on page 283 for full
text references. (51:4.8)

7. POST-TEACHER SON MAN

Overpopulation and Human Degeneracy

Page 599:1   *Paper 52:7.5*

See "Overpopulation and Human Degeneracy" on page 283 for full
text references. (51:4.8)

PAPER 53: THE LUCIFER REBELLION

4. OUTBREAK OF THE REBELLION

Lucifer Rebellion—Commencement
The time that the Lucifer rebellion began—collaborated in Parts I and II.

Page 604:3   *Paper 53:4.1*

The Lucifer manifesto was issued at the annual conclave of Satania on the
sea of glass, in the presence of the assembled hosts of Jerusem, on the last day
of the year, **about two hundred thousand years ago, Urantia time.** Satan
proclaimed that worship could be accorded the universal forces—physical,
intellectual, and spiritual—but that allegiance could be acknowledged only to
the actual and present ruler, Lucifer, the "friend of men and angels" and the
"God of liberty."

Page 701:7   *Paper 61:7.8*

200,000 years ago, during the advance of the last glacier, there occurred
an episode which had much to do with the march of events on Urantia—the
Lucifer rebellion.

Gabriel—Commander in Chief

Page 605:8   *Paper 53:5.4*

See "Gabriel—Commander in Chief" on page 215 for full text references. (33:4.6)

Three Concentric Circles
Symbol of paradise trinity expressed several ways—in Parts II and III.

Page 605:8   *Paper 53:5.4*

Since Michael elected to remain aloof from the actual warfare of the Lucifer rebellion, Gabriel called his personal staff together on Edentia and, in counsel with the Most Highs, elected to assume command of the loyal hosts of Satania. Michael remained on Salvington while Gabriel proceeded to Jerusem, and establishing himself on the sphere dedicated to the Father—the same Universal Father whose personality Lucifer and Satan had questioned—in the presence of the forgathered hosts of loyal personalities, he displayed the banner of Michael, **the material emblem of the Trinity government of all creation, the three azure blue concentric circles on a white background.**

Page 1015:5   *Paper 93:2.5*

In personal appearance, Melchizedek resembled the then blended Nodite and Sumerian peoples, being almost six feet in height and possessing a commanding presence. He spoke Chaldean and a half dozen other languages. He dressed much as did the Canaanite priests except that on his breast he **wore an emblem of three concentric circles, the Satania symbol of the Paradise Trinity.** In the course of his ministry this insignia of three concentric circles became regarded as so sacred by his followers that they never dared to use it, and it was soon forgotten with the passing of a few generations.

Page 1016:5   *Paper 93:3.3*

The symbol of the three concentric circles, which Melchizedek adopted as the **insignia of his bestowal,** a majority of the people interpreted as standing for the three kingdoms of men, angels, and God. And they were allowed to continue in that belief; very few of his followers ever knew that **these three circles were emblematic of the infinity, eternity, and universality of the Paradise Trinity of divine maintenance and direction;** even Abraham rather regarded this symbol as standing for the three Most Highs of Edentia, as he had been instructed that the three Most Highs functioned as one. To the extent that Melchizedek taught the Trinity concept symbolized in his insignia, he usually associated it with the three Vorondadek rulers of the constellation of Norlatiadek.

Page 1143:6   *Paper 104:1.3*

The third presentation of the Trinity was made by Machiventa Melchizedek, and this doctrine **was symbolized by the three concentric circles which the sage of Salem wore on his breast plate.** But Machiventa found it very difficult to teach the Palestinian Bedouins about the Universal Father, the Eternal Son, and the Infinite Spirit. Most of his disciples thought that the Trinity consisted of the three Most Highs of Norlatiadek; a few conceived of the Trinity as the System Sovereign, the Constellation Father, and the local universe Creator Deity; still fewer even remotely grasped the idea of the Paradise association of the Father, Son, and Spirit.

7. HISTORY OF THE REBELLION

Lucifer Rebellion—Status of Personalities

Page 607:2   *Paper 53:7.1*

See "Lucifer Rebellion—Status of Personalities" on page 260 for full text references. (45:1.11)

Lucifer Rebellion—Panoptians

Our planet, plus the other thirty-six, and the loyal Panoptians—over six hundred pages apart.

Page 607:2   *Paper 53:7.1*

The Lucifer rebellion was system wide. **Thirty-seven seceding Planetary Princes swung their world administrations largely to the side of the archrebel.** Only on Panoptia did the Planetary Prince fail to carry his people with him. On this world, under the guidance of the Melchizedeks, the people rallied to the support of Michael. Ellanora, a young woman of that mortal realm, grasped the leadership of the human races, and not a single soul on that strife-torn world enlisted under the Lucifer banner. **And ever since have these loyal Panoptians served** on the seventh Jerusem transition world as the caretakers and builders on the Father's sphere and its surrounding seven detention worlds. The Panoptians not only act as the literal custodians of these worlds, but they also execute the personal orders of Michael for the embellishment of these spheres for some future and unknown use. They do this work as they tarry en route to Edentia.

Page 610:6   *Paper 53:9.1*

Early in the days of the Lucifer rebellion, salvation was offered all rebels by Michael. To all who would show proof of sincere repentance, he offered, upon his attainment of complete universe sovereignty, forgiveness and reinstatement in some form of universe service. None of the leaders accepted this merciful proffer. But thousands of the angels and the lower orders of celestial beings, including hundreds of the Material Sons and Daughters, **accepted the mercy proclaimed by the Panoptians and were given rehabilitation at the time of Jesus' resurrection nineteen hundred years ago.** These beings have since been transferred to the Father's world of Jerusem, where they must be held, technically, until the Uversa courts hand down a decision in the matter of Gabriel *vs.* Lucifer. But no one doubts that, when the annihilation verdict is issued, these repentant and salvaged personalities will be exempted from the decree of extinction. These probationary souls now labor with the Panoptians in the work of caring for the Father's world.

Page 1252:2   *Paper 114:2.4*

The members of this same commission of former Urantians also act as advisory supervisors of the thirty-six other rebellion-isolated worlds of the system; they perform a very valuable service in keeping Lanaforge, the System Sovereign, in close and sympathetic touch with the affairs of these planets, which still remain more or less under the overcontrol of the Constellation Fathers of Norlatiadek. These twenty-four counselors make frequent trips as individuals to each of the quarantined planets, especially to Urantia.

Lucifer Rebellion—Interplanetary Communication

Page 607:4   *Paper 53:7.3*

See "Lucifer Rebellion—Interplanetary Communication" on page 201 for full text references. (28:7.4)

8. THE SON OF MAN ON URANTIA

Lucifer Rebellion—Satan

Page 609:6   *Paper 53:8.3*

See "Lucifer Rebellion—Satan" on page 254 for full text references. (43:4.9)

Lucifer Rebellion—Spiritual Isolation

Page 609:7   *Paper 53:8.4*

See "Lucifer Rebellion—Spiritual Isolation" on page 188 for full text references. (20:8.4)

Lucifer Rebellion—Demoniacal Possession
Demoniacal possession, before and after Pentecost.

Page 610:2   *Paper 53:8.6*

The last act of Michael before leaving Urantia was to offer mercy to Caligastia and Daligastia, but they spurned his tender proffer. Caligastia, your apostate Planetary Prince, is still free on Urantia to prosecute his nefarious designs, **but he has absolutely no power to enter the minds of men, neither can he draw near to their souls to tempt or corrupt them unless they really desire to be cursed with his wicked presence.**

Page 863:6, 864:1   *Paper 77:7.5,8*

On no world can evil spirits possess any mortal mind subsequent to the life of a Paradise bestowal Son. **But before the days of Christ Michael on Urantia**—before the universal coming of the Thought Adjusters and the pouring out of the Master's spirit upon all flesh—**these rebel midwayers were actually able to influence the minds of certain inferior mortals and somewhat to control their actions.** This was accomplished in much the same way as the loyal midway creatures function when they serve as efficient contact guardians of the human minds of the Urantia reserve corps of destiny at those times when the Adjuster is, in effect, detached from the personality during a season of contact with superhuman intelligences....

The entire group of rebel midwayers is at present held prisoner by order of the Most Highs of Edentia. No more do they roam this world on mischief bent. Regardless of the presence of the Thought Adjusters, the pouring out of the Spirit of Truth upon all flesh forever made it impossible for disloyal spirits of any sort or description ever again to invade even the most feeble of human minds. **Since the day of Pentecost there never again can be such a thing as demoniacal possession.**

Lucifer Rebellion—Status of Caligastia and Daligastia
Caligastia and Daligastia's status before and after the cross of Christ.

Page 610:5   *Paper 53:8.9*

In general, when weak and dissolute mortals are supposed to be under the influence of devils and demons, they are merely being dominated by their own inherent and debased tendencies, being led away by their own natural propensities. The devil has been given a great deal of credit for evil which does not belong to him. **Caligastia has been comparatively impotent since the cross of Christ.**

Page 822:9   *Paper 73:2.5*

Although Caligastia and Daligastia had been deprived of much of their power for evil, they did everything possible to frustrate and hamper the work of preparing the Garden. But their evil machinations were largely offset by the faithful activities of the almost ten thousand loyal midway creatures who so tirelessly labored to advance the enterprise.

---

9. PRESENT STATUS OF THE REBELLION

Lucifer Rebellion—Status of Personalities

Page 610:6, 611:2   *Paper 53:9.1*

See "Lucifer Rebellion—Status of Personalities" on page 260 for full text references. (45:1.11)

Lucifer Rebellion—Panoptians

Page 610:6   *Paper 53:9.1*

See "Lucifer Rebellion—Panoptians" on page 289 for full text references. (53:7.1)

Lucifer Rebellion—Spiritual Isolation

Page 611:3   *Paper 53:9.4*

See "Lucifer Rebellion—Spiritual Isolation" on page 188 for full text references. (20:8.4)

PAPER 54: PROBLEMS OF THE LUCIFER REBELLION

Sin

Sin—in Parts II, III, and IV, covering four papers.

Page 613:1   *Paper 54:0.1*

Evolutionary man finds it difficult fully to comprehend the significance and to grasp the meanings of evil, error, sin, and iniquity. Man is slow to perceive that contrastive perfection and imperfection produce potential evil; that conflicting truth and falsehood create confusing error; **that the divine endowment of freewill choice eventuates in the divergent realms of sin and righteousness;** that the persistent pursuit of divinity leads to the kingdom of God as contrasted with its continuous rejection, which leads to the domains of iniquity.

Page 754:5   *Paper 67:1.4*

There are many ways of looking at sin, but from the universe philosophic viewpoint sin is the attitude of a personality who is knowingly resisting cosmic reality. Error might be regarded as a misconception or distortion of reality. Evil is a partial realization of, or maladjustment to, universe realities. **But sin is a purposeful resistance to divine reality**—a conscious choosing to oppose spiritual progress—while iniquity consists in an open and persistent defiance of recognized reality and signifies such a degree of personality disintegration as to border on cosmic insanity.

Page 761:3–4   *Paper 67:7.4–5*

Sin is never purely local in its effects. The administrative sectors of the universes are organismal; the plight of one personality must to a certain extent be shared by all. Sin, being an attitude of the person toward reality, is destined to exhibit its inherent negativistic harvest upon any and all related levels of universe values. But the full consequences of erroneous thinking, evil-doing, or sinful planning are experienced only on the level of actual performance. The transgression of universe law may be fatal in the physical realm without seriously involving the mind or impairing the spiritual experience. Sin is

293

fraught with fatal consequences to personality survival only when it is the attitude of the whole being, when it stands for the choosing of the mind and the willing of the soul.

Evil and sin visit their consequences in material and social realms and may sometimes even retard spiritual progress on certain levels of universe reality, but never does the sin of any being rob another of the realization of the divine right of personality survival. Eternal survival can be jeopardized only by the **decisions of the mind and the choice of the soul** of the individual himself.

Page 1301:1   *Paper 118:7.4*

Sin in time-conditioned space clearly proves the temporal liberty—even license—of the finite will. **Sin depicts immaturity dazzled by the freedom of the relatively sovereign will of personality while failing to perceive the supreme obligations and duties of cosmic citizenship.**

Page 1660:3   *Paper 148:4.4*

"Sin is the conscious, knowing, and deliberate transgression of the divine law, the Father's will. Sin is the measure of unwillingness to be divinely led and spiritually directed.

---

1. TRUE AND FALSE LIBERTY

Self-assertion

On self-assertion—in Parts II and IV, over fourteen hundred pages away.

Page 614:1   *Paper 54:1.6*

True liberty is the associate of genuine self-respect; false liberty is the consort of self-admiration. True liberty is the fruit of self-control; **false liberty, the assumption of self-assertion.** Self-control leads to altruistic service; self-admiration tends towards the exploitation of others for the selfish aggrandizement of such a mistaken individual as is willing to sacrifice righteous attainment for the sake of possessing unjust power over his fellow beings.

Page 2065:6   *Paper 194:3.18*

Pentecost was designed to lessen the self-assertiveness of individuals, groups, nations, and races. **It is this spirit of self-assertiveness** which so increases in tension that it periodically breaks loose in destructive wars. Mankind can be unified only by the spiritual approach, and the Spirit of Truth is a world influence which is universal.

3. THE TIME LAG OF JUSTICE

Lucifer Rebellion—Satan

Page 615:5   *Paper 54:3.3*

See "Lucifer Rebellion—Satan" on page 254 for full text references. (43:4.9)

4. THE MERCY TIME LAG

Lucifer Rebellion—Spiritual Isolation

Page 616:7   *Paper 54:4.8*

See "Lucifer Rebellion—Spiritual Isolation" on page 188 for full text references. (20:8.4)

5. THE WISDOM OF DELAY

Patience and Forbearance
Difference between patience and forbearance—in Parts II, III, and IV.

Page 617:4   *Paper 54:5.4*

3. No affectionate father is ever precipitate in visiting punishment upon an erring member of his family. **Patience cannot function** independently of time.

Page 941:9   *Paper 84:7.28*

Marriage, with children and consequent family life, is stimulative of the highest potentials in human nature and simultaneously provides the ideal avenue for the expression of these quickened attributes of mortal personality. The family provides for the biologic perpetuation of the human species. The home is the natural social arena wherein the ethics of blood brotherhood may be grasped by the growing children. The family is the fundamental unit of fraternity in which parents and children learn those **lessons of patience**, altruism, tolerance, and **forbearance which are so essential to the realization of brotherhood among all men.**

Page 1295:6   *Paper 118:1.6*

Patience is exercised by those mortals whose time units are short; true maturity transcends patience by a **forbearance born of real understanding.**

Page 1574:4   *Paper 140:5.11*

3. *"Happy are the meek, for they shall inherit the earth."* Genuine meekness has no relation to fear. It is rather an attitude of man co-operating with God—"Your will be done." **It embraces patience and forbearance** and is motivated by an unshakable faith in a lawful and friendly universe. It masters all temptations to rebel against the divine leading. Jesus was the ideal meek man of Urantia, and he inherited a vast universe.

PAPER 55: THE SPHERES OF LIGHT AND LIFE

Light and Life

Page 621:1   *Paper 55:0.1*

See "Light and Life" on page 280 for full text references. (49:5.20–21)

---

3. THE GOLDEN AGES

Global Viewpoints
Describing heaven on earth, and what it takes to get there—Parts II and IV.

Page 624:7   *Paper 55:3.1*

During this age of light and life the world increasingly prospers under the fatherly rule of the Planetary Sovereign. By this time the worlds are progressing under the momentum of one language, **one religion,** and, on normal spheres, one race. But this age is not perfect. These worlds still have well-appointed hospitals, homes for the care of the sick. There still remain the problems of caring for accidental injuries and the inescapable infirmities attendant upon the decrepitude of old age and the disorders of senility. Disease has not been entirely vanquished, neither have the earth animals been subdued in perfection; but such worlds are like Paradise in comparison with the early times of primitive man during the pre-Planetary Prince age. You would instinctively describe such a realm—could you be suddenly transported to a planet in this stage of development—as heaven on earth.

Page 626:11   *Paper 55:3.15*

No evolutionary world can hope to progress beyond the first stage of settledness in light until it has achieved **one language, one religion, and one philosophy.** Being of one race greatly facilitates such achievement, but the many peoples of Urantia do not preclude the attainment of higher stages.

Page 1491:3,5   *Paper 134:6.9,11*

World peace cannot be maintained by treaties, diplomacy, foreign policies, alliances, balances of power, or any other type of makeshift juggling with the sovereignties of nationalism. World law must come into being and must be enforced by **world government**—the sovereignty of all mankind....

Under global government the national groups will be afforded a real opportunity to realize and enjoy the personal liberties of genuine democracy. The fallacy of self-determination will be ended. With global regulation of money and trade will come the new era of world-wide peace. Soon may a global language evolve, and there will be at least some hope of sometime having a **global religion—or religions with a global viewpoint.**

---

4. ADMINISTRATIVE READJUSTMENTS

Father Indwelt—Adjusters and Other Types of Spirit

Page 629:9   *Paper 55:4.28*

See "Father Indwelt—Adjusters and Other Types of Spirit" on page 209 for full text references. (31:4.1)

---

5. THE ACME OF MATERIAL DEVELOPMENT

Overpopulation and Human Degeneracy

Page 629:11   *Paper 55:5.2*

See "Overpopulation and Human Degeneracy" on page 283 for full text references. (51:4.8)

6. THE INDIVIDUAL MORTAL

Abandonters—Permanent Citizens of our Superuniverse

Page 631:2   *Paper 55:6.6*

See "Abandonters—Permanent Citizens of our Superuniverse" on page 235 for full text references. (37:9.8)

10. THE FOURTH OR LOCAL UNIVERSE STAGE

Light and Life—Jubilee Event

Page 634:1   *Paper 55:10.1*

See "Light and Life—Jubilee Event" on page 177 for full text references. (17:3.11)

11. THE MINOR AND MAJOR SECTOR STAGES

Minor Sector

Page 635:4   *Paper 55:11.2*

See "Minor Sector" on page 169 for full text references. (15:2.6)

PAPER 56: UNIVERSAL UNITY

2. INTELLECTUAL UNITY

Supreme Mind
The Supreme Mind—over six hundred pages apart.

Page 638:7   *Paper 56:2.3*

This infinite and universal mind is ministered in the universes of time and space as the cosmic mind; and though extending from the primitive ministry of the adjutant spirits up to the magnificent mind of the chief executive of a universe, even this cosmic mind is adequately unified in the supervision of the

Seven Master Spirits, who are **in turn co-ordinated with the Supreme Mind** of time and space and perfectly correlated with the all-embracing mind of the Infinite Spirit.

Page 641:3   *Paper 56:6.2*

The personality realities of the Supreme Being come forth from the Paradise Deities and on the pilot world of the outer Havona circuit unify with the power prerogatives of the Almighty Supreme coming up from the Creator divinities of the grand universe. God the Supreme as a person existed in Havona before the creation of the seven superuniverses, but he functioned only on spiritual levels. The evolution of the Almighty power of Supremacy by diverse divinity synthesis in the evolving universes eventuated in a new power presence of Deity which co-ordinated with the spiritual person of the Supreme in Havona **by means of the Supreme Mind,** which concomitantly translated from the potential resident in the infinite mind of the Infinite Spirit to the active functional mind of the Supreme Being.

Page 1125:5   *Paper 102:6.10*

Organic evolution is a fact; purposive or progressive evolution is a truth which makes consistent the otherwise contradictory phenomena of the ever-ascending achievements of evolution. The higher any scientist progresses in his chosen science, the more will he abandon the theories of materialistic fact in favor of the cosmic truth of the **dominance of the Supreme Mind.** Materialism cheapens human life; the gospel of Jesus tremendously enhances and supernally exalts every mortal. Mortal existence must be visualized as consisting in the intriguing and fascinating experience of the realization of the reality of the meeting of the human upreach and the divine and saving downreach.

Page 1269:1–2,4   *Paper 116:1.2–3,5*

The union of the power and personality attributes of Supremacy **is the function of Supreme Mind;** and the completed evolution of the Almighty Supreme will result in one unified and personal Deity—not in any loosely co-ordinated association of divine attributes. From the broader perspective, there will be no Almighty apart from the Supreme, no Supreme apart from the Almighty.

Throughout the evolutionary ages the physical power potential of the Supreme is vested in the Seven Supreme Power Directors, and the mind potential reposes in the Seven Master Spirits. The Infinite Mind is the function of the Infinite Spirit; the cosmic mind, the ministry of the Seven Master Spirits; **the Supreme Mind is in process of actualizing** in the co-ordination of the grand universe and in functional association with the revelation and attainment of God the Sevenfold....

We really know less about the mind of Supremacy than about any other aspect of this evolving Deity. It is unquestionably active throughout the grand universe and is believed to have a potential destiny of master universe function which is of vast extent. But this we do know: Whereas physique may attain completed growth, and whereas spirit may achieve perfection of development, mind never ceases to progress—it is the experiential technique of endless progress. The Supreme is an experiential Deity and therefore never achieves completion of mind attainment.

Page 1272:2　　　　　　　　　　　　　　　　　　　*Paper 116:4.3*

Early in the projection of the superuniverse scheme of creation, the Master Spirits joined with the ancestral Trinity in the cocreation of the forty-nine Reflective Spirits, and concomitantly the Supreme Being functioned creatively as the culminator of the conjoined acts of the Paradise Trinity and the creative children of Paradise Deity. Majeston appeared and **ever since has focalized the cosmic presence of the Supreme Mind,** while the Master Spirits continue as source-centers for the far-flung ministry of the cosmic mind.

Page 1303:5　　　　　　　　　　　　　　　　　　　*Paper 118:9.4*

The grand universe is mechanism as well as organism, mechanical and living—**a living mechanism activated by a Supreme Mind,** co-ordinating with a Supreme Spirit, and finding expression on maximum levels of power and personality unification as the Supreme Being. But to deny the mechanism of the finite creation is to deny fact and to disregard reality.

---

4. PERSONALITY UNIFICATION

Personality Circuit as Divine Love

Page 640:2　　　　　　　　　　　　　　　　　　　*Paper 56:4.3*

See "Personality Circuit as Divine Love" on page 149 for full text references. (5:6.12)

---

6. UNIFICATION OF EVOLUTIONARY DEITY

Supreme Mind

Page 641:3　　　　　　　　　　　　　　　　　　　*Paper 56:6.2*

See "Supreme Mind" on page 298 for full text references. (56:2.3)

10. TRUTH, BEAUTY, AND GOODNESS

Personality Circuit as Divine Love

Page 647:8   *Paper 56:10.17*

See "Personality Circuit as Divine Love" on page 149 for full text references. (5:6.12)

PAPER 58: LIFE ESTABLISHMENT ON URANTIA

Life—Appearance on Planets

Page 664:1   *Paper 58:0.1*

See "Life—Appearance on Planets" on page 228 for full text references. (36:0.1)

4. THE LIFE-DAWN ERA

Life—Appearance on Planets

Page 667:5   *Paper 58:4.1*

See "Life—Appearance on Planets" on page 228 for full text references. (36:0.1)

5. THE CONTINENTAL DRIFT

Sun and Earth Core Temperature

Page 668:3   *Paper 58:5.1*

See "Sun and Earth Core Temperature" on page 250 for full text references. (41:6.7)

PAPER 61: THE MAMMALIAN ERA ON URANTIA

3. THE MODERN MOUNTAIN STAGE—AGE OF THE ELEPHANT AND THE HORSE

Horse
Significance of the horse—in three papers.

Page 697:6   *Paper 61:3.11*

As Urantia is entering the so-called " horseless age," you should pause and ponder what this animal meant to your ancestors. Men first used horses for food, then for travel, and later in agriculture and war. The horse has long served mankind and has played an important part in the development of human civilization.

Page 871:6   *Paper 78:3.10*

It took so long for the earlier waves of Adamites to pass over Eurasia that their culture was largely lost in transit. **Only the later Andites moved with sufficient speed** to retain the Edenic culture at any great distance from Mesopotamia.

Page 892:7   *Paper 80:4.4*

But the horse was the evolutionary factor which determined the dominance of the Andites in the Occident. The horse gave the dispersing Andites the hitherto nonexistent advantage of mobility, enabling the last groups of Andite cavalrymen to progress quickly around the Caspian Sea to overrun all of Europe. All previous waves of Andites had moved so slowly that they tended to disintegrate at any great distance from Mesopotamia. But these later waves moved so rapidly that they reached Europe as coherent groups, still retaining some measure of higher culture.

7. The Continuing Ice Age

Sangik Races—Fifth Glacier

5th glacier—six colored races, Planetary Prince arrival, and subhuman types pushed south.

Page 701:3   *Paper 61:7.4*

500,000 years ago, during the fifth advance of the ice, a new development accelerated the course of human evolution. *Suddenly* and in one generation the six colored races mutated from the aboriginal human stock. This is a doubly important date since it also marks the arrival of the Planetary Prince.

Page 883:3   *Paper 79:5.2*

While the early Neanderthalers were spread out over the entire breadth of Eurasia, the eastern wing was the more contaminated with debased animal strains. **These subhuman types were pushed south by the fifth glacier,** the same ice sheet which so long blocked Sangik migration into eastern Asia. And when the red man moved northeast around the highlands of India, he found northeastern Asia free from these subhuman types. The tribal organization of the red races was formed earlier than that of any other peoples, and they were the first to migrate from the central Asian focus of the Sangiks. The inferior Neanderthal strains were destroyed or driven off the mainland by the later migrating yellow tribes. But the red man had reigned supreme in eastern Asia for almost one hundred thousand years before the yellow tribes arrived.

Lucifer Rebellion—Commencement

Page 701:7   *Paper 61:7.8*

See "Lucifer Rebellion—Commencement" on page 287 for full text references. (53:4.1)

PAPER 63: THE FIRST HUMAN FAMILY

Andon and Fonta—Receiving Universe Names
Didn't receive real universe names until fusion—related topic between 400 pages.

Page 711:2–3   *Paper 63:0.2–3*

" Man-mind has appeared on 606 of Satania, **and these parents of the new race shall be called** *Andon* **and** *Fonta.* And all archangels pray that these creatures may speedily be endowed with the personal indwelling of the gift of the spirit of the Universal Father."

Andon is the Nebadon name which signifies "the first Fatherlike creature to exhibit human perfection hunger." Fonta signifies "the first Sonlike creature to exhibit human perfection hunger." **Andon and Fonta never knew these names until they were bestowed upon them at the time of fusion with their Thought Adjusters.** Throughout their mortal sojourn on Urantia they called each other Sonta-an and Sonta-en, Sonta-an meaning "loved by mother," Sonta-en signifying "loved by father." They gave themselves these names, and the meanings are significant of their mutual regard and affection.

Page 1188:5   *Paper 108:3.3*

Human subjects are often known by the numbers of their Adjusters; **mortals do not receive real universe names until after Adjuster fusion,** which union is signalized by the bestowal of the new name upon the new creature by the destiny guardian.

Spirit Fusion—Early Evolutionary Times

Page 711:3   *Paper 63:0.3*

See "Spirit Fusion—Early Evolutionary Times" on page 286 for full text references. (52:1.6)

2. THE FLIGHT OF THE TWINS

Andon and Fonta—Discovery of Fire

Andon discovered fire, but the secret was unraveled by Fonta.

Page 712:4–5   *Paper 63:2.4–5*

On their northward journey they discovered an exposed flint deposit and, finding many stones suitably shaped for various uses, gathered up a supply for the future. In attempting to chip these flints so that they would be better adapted for certain purposes, **Andon discovered their sparking quality and conceived the idea of building fire.** But the notion did not take firm hold of him at the time as the climate was still salubrious and there was little need of fire.

But the autumn sun was getting lower in the sky, and as they journeyed northward, the nights grew cooler and cooler. Already they had been forced to make use of animal skins for warmth. Before they had been away from home one moon, Andon signified to his mate that he thought he could make fire with the flint. They tried for two months to utilize the flint spark for kindling a fire but only met with failure. Each day this couple would strike the flints and endeavor to ignite the wood. **Finally, one evening about the time of the setting of the sun, the secret of the technique was unraveled when it occurred to Fonta to climb a near-by tree to secure an abandoned bird's nest.** The nest was dry and highly inflammable and consequently flared right up into a full blaze the moment the spark fell upon it. They were so surprised and startled at their success that they almost lost the fire, but they saved it by the addition of suitable fuel, and then began the first search for firewood by the parents of all mankind.

Page 777:7   *Paper 69:6.4*

Though Andon, the discoverer of fire, avoided treating it as an object of worship, many of his descendants regarded the flame as a fetish or as a spirit. They failed to reap the sanitary benefits of fire because they would not burn refuse. Primitive man feared fire and always sought to keep it in good humor, hence the sprinkling of incense. Under no circumstances would the ancients spit in a fire, nor would they ever pass between anyone and a burning fire. Even the iron pyrites and flints used in striking fire were held sacred by early mankind.

---

4. THE ANDONIC CLANS

Early Language
Tongue of Urantia—Insights from five papers.

Page 714:4   *Paper 63:4.6*

Before the extensive dispersion of the Andonic clans a well-developed language had evolved from their early efforts to intercommunicate. This language continued to grow, and almost daily additions were made to it because of the new inventions and adaptations to environment which were developed by these active, restless, and curious people. **And this language became the word of Urantia, the tongue of the early human family, until the later appearance of the colored races.**

Page 829:4   *Paper 74:2.2*

The tongue of Eden was an Andonic dialect as spoken by Amadon. Van and Amadon had markedly improved this language by creating a new alphabet of twenty-four letters, and they had hoped to see it become the tongue of Urantia as the Edenic culture would spread throughout the world. Adam and Eve had fully mastered this human dialect before they departed from Jerusem so that this son of Andon heard the exalted ruler of his world address him in his own tongue.

Page 860:2   *Paper 77:4.7*

And all this explains how the Sumerians appeared so suddenly and mysteriously on the stage of action in Mesopotamia. Investigators will never be able to trace out and follow these tribes back to the beginning of the Sumerians, who had their origin two hundred thousand years ago after the submergence of Dalamatia. Without a trace of origin elsewhere in the world, these ancient tribes suddenly loom upon the horizon of civilization with a full-grown and superior culture, embracing temples, metalwork, agriculture, animals, pottery, weaving, commercial law, civil codes, religious ceremonial, and an old system of writing. At the beginning of the historical era they had long since lost the alphabet of Dalamatia, having adopted the peculiar writing system originating in Dilmun. **The Sumerian language, though virtually lost to the world, was not Semitic; it had much in common with the so-called Aryan tongues.**

Page 862:8   *Paper 77:6.4*

Each of the eight couples eventually produced 248 midwayers, and thus did the original secondary corps—1,984 in number—come into existence. There are eight subgroups of secondary midwayers. They are designated as A-B-C the first, second, third, and so on. **And then there are D-E-F the first, second, and so on. [What do you think the 24 letters would be—V-W-X the first, second...?]**

Page 872:7   *Paper 78:5.3*

The civilization of Turkestan was constantly being revived and refreshed by the newcomers from Mesopotamia, especially by the later Andite cavalrymen. The so-called Aryan mother tongue was in process of formation in the highlands of Turkestan; it was a blend of the Andonic dialect of that region with the language of the Adamsonites and later Andites. Many modern languages are derived from this early speech of these central Asian tribes who conquered Europe, India, and the upper stretches of the Mesopotamian plains. **This ancient language gave the Occidental tongues all of that similarity which is called Aryan.**

Page 896:7   *Paper 80:8.2*

These descendants of Andon were dispersed through most of the mountainous regions of central and southeastern Europe. They were often reinforced by arrivals from Asia Minor, which region they occupied in considerable strength. **The ancient Hittites stemmed directly from the Andonite stock;** their pale skins and broad heads were typical of that race. This strain was carried in Abraham's ancestry and contributed much to the characteristic facial appearance of his later Jewish descendants who, while having a culture and religion derived from the Andites, spoke a very different language. **Their tongue was distinctly Andonite.**

---

6. *Onagar—The First Truth Teacher*

Mansant—a Nodite?

Page 715:8   *Paper 63:6.1*

See "Mansant—a Nodite?" on page 262 for full text references. (45:4.2–4)

7. THE SURVIVAL OF ANDON AND FONTA

Spirit Fusion—Early Evolutionary Times

Page 717:3   *Paper 63:7.2*

See "Spirit Fusion—Early Evolutionary Times" on page 286 for full text references. (52:1.6)

PAPER 64: THE EVOLUTIONARY RACES OF COLOR

5. ORIGIN OF THE COLORED RACES

India—Highlands
Highlands northwest of India.

Page 722:2–3   *Paper 64:5.1–2*

500,000 years ago the Badonan tribes of the **northwestern highlands of India** became involved in another great racial struggle. For more than one hundred years this relentless warfare raged, and when the long fight was finished, only about one hundred families were left. But these survivors were the most intelligent and desirable of all the then living descendants of Andon and Fonta.

And now, among these highland Badonites there was a new and strange occurrence. A man and woman living in the **northeastern part of the then inhabited highland region** began *suddenly* to produce a family of unusually intelligent children. This was the *Sangik family*, the ancestors of all of the six colored races of Urantia.

Page 879:7   *Paper 79:2.1*

India is the only locality where all the Urantia races were blended, the Andite invasion adding the last stock. **In the highlands northwest of India the Sangik races came into existence,** and without exception members of each penetrated the subcontinent of India in their early days, leaving behind them the most heterogeneous race mixture ever to exist on Urantia. Ancient India acted as a catch basin for the migrating races. The base of the peninsula was formerly somewhat narrower than now, much of the deltas of the Ganges and Indus being the work of the last fifty thousand years.

6. THE SIX SANGIK RACES OF URANTIA

Overpopulation and Human Degeneracy

Page 726:4   *Paper 64:6.32*

See "Overpopulation and Human Degeneracy" on page 283 for full text references. (51:4.8)

Onamonalonton—a Red Man

Page 723:6   *Paper 64:6.7*

See "Onamonalonton—a Red Man" on page 263 for full text references. (45:4.2,5)

7. DISPERSION OF THE COLORED RACES

Polynesian Andites
Polynesian Andites—two authors

Page 727:2   *Paper 64:7.5*

When the relatively pure-line remnants of the red race forsook Asia, there were eleven tribes, and they numbered a little over seven thousand men, women, and children. These tribes were accompanied by three small groups of mixed ancestry, the largest of these being a combination of the orange and blue races. These three groups never fully fraternized with the red man and early journeyed southward to Mexico and Central America, where they were later joined by a small group of mixed yellows and reds. These peoples all intermarried and founded a new and amalgamated race, one which was much less warlike than the pure-line red men. Within five thousand years this amalgamated race broke up into three groups, establishing the civilizations respectively of Mexico, Central America, and South America. **The South American offshoot did receive a faint touch of the blood of Adam.**

Page 884:3   *Paper 79:5.9*

These civilizations were evolutionary products of the Sangiks, notwithstanding that traces of Andite blood reached Peru. **Excepting the Eskimos in North America and a few Polynesian Andites in South America, the peoples of the Western Hemisphere had no contact with the rest of the world until the end of the first millennium after Christ.** In the original

309

Melchizedek plan for the improvement of the Urantia races it had been stipulated that one million of the pure-line descendants of Adam should go to upstep the red men of the Americas.

PAPER 65: THE OVERCONTROL OF EVOLUTION

---

8. EVOLUTION IN TIME AND SPACE

Truth Sensitivity

Page 740:2   *Paper 65:8.6*

See "Truth Sensitivity" on page 275 for full text references. (48: 6.17)

PAPER 66: THE PLANETARY PRINCE OF URANTIA

---

2. THE PRINCE'S STAFF

Neighboring Universes

Page 742:7   *Paper 66:2.7*

See "Neighboring Universes" on page 210 for full text references. (32:2.12)

---

4. EARLY DAYS OF THE ONE HUNDRED

Primary Midwayers—Discovery of Production
Accidental discovery of a phenomenon—covered over a hundred pages apart.

Page 744:9   *Paper 66:4.10*

In conformity to their instructions the staff did not engage in sexual reproduction, but they did painstakingly study their personal constitutions, and they carefully explored every imaginable phase of intellectual (mind) and morontia (soul) liaison. **And it was during the thirty-third year of their sojourn in Dalamatia, long before the wall was completed, that number two and number seven of the Danite group accidentally discovered a phenomenon attendant upon the liaison of their morontia selves (supposedly nonsexual and nonmaterial); and the result of this adventure proved to be the** first of the primary midway creatures. This new being was wholly visible to the planetary staff and to their celestial associates but was not visible to the men and women of the various human tribes. **Upon authority of the Planetary Prince the entire corporeal staff undertook the production of similar beings,** and all were successful, following the instructions of the pioneer Danite pair. Thus did the Prince's staff eventually bring into being the original corps of 50,000 primary midwayers.

Page 855:5   *Paper 77:1.3*

It was immediately discovered that a creature of this order, midway between the mortal and angelic levels, would be of great service in carrying on the affairs of the Prince's headquarters, **and each couple of the corporeal staff was accordingly granted permission to produce a similar being.** This effort resulted in the first group of fifty midway creatures.

Tree of Life—Color

Page 745:3   *Paper 66:4.13*

See "Tree of Life—Color" on page 257 for full text references. (43:6.6)

---

7. LIFE IN DALAMATIA

Commandments

Many of the commandments have a common thread stemming from Dalamatia—Parts III and IV.

Page 751:3–11   *Paper 66:7.8–16*

Hap presented the early races with a **moral law.** This code was known as "The Father's Way" and consisted of the **following seven commands:**

1. You shall not fear nor serve any God but the Father of all.
2. You shall not disobey the Father's Son, the world's ruler, nor show disrespect to his superhuman associates.
3. You shall not speak a lie when called before the judges of the people.
4. You shall not kill men, women, or children.
5. You shall not steal your neighbor's goods or cattle.
6. You shall not touch your friend's wife.
7. You shall not show disrespect to your parents or to the elders of the tribe.

This was the law of Dalamatia for almost three hundred thousand years. And many of the stones on which this law was inscribed now lie beneath the waters off the shores of Mesopotamia and Persia. It became the custom to hold one of these commands in mind for each day of the week, using it for salutations and mealtime thanksgiving.

Page 835:14, 836:1–8   *Paper 74:7.4–12*

The schools, in fact every activity of the Garden, were always open to visitors. Unarmed observers were freely admitted to Eden for short visits. To sojourn in the Garden a Urantian had to be "adopted." He received instructions in the plan and purpose of the Adamic bestowal, signified his intention to adhere to this mission, **and then made declaration of loyalty to the social rule of Adam and the spiritual sovereignty of the Universal Father.**

The laws of the Garden were based on the older codes of Dalamatia and were promulgated under seven heads:

1. The laws of health and sanitation.
2. The social regulations of the Garden.
3. The code of trade and commerce.
4. The laws of fair play and competition.
5. The laws of home life.
6. The civil codes of the golden rule.
7. The seven commands of supreme moral rule.

Page 975:1   *Paper 89:1.4*

The seven commandments of Dalamatia and Eden, as well as the ten injunctions of the Hebrews, were definite taboos, all expressed in the same negative form as were the most ancient prohibitions. But these newer codes were truly emancipating in that they took the place of thousands of pre-existent taboos. And more than this, these later commandments definitely promised something in return for obedience.

Page 1017:8–15   *Paper 93:4.6–13*

The seven commandments promulgated by Melchizedek were patterned along the lines of the ancient Dalamatian supreme law and very much resembled the seven commands taught in the first and second Edens. These commands of the Salem religion were:

1. You shall not serve any God but the Most High Creator of heaven and earth.
2. You shall not doubt that faith is the only requirement for eternal salvation.
3. You shall not bear false witness.
4. You shall not kill.
5. You shall not steal.
6. You shall not commit adultery.
7. You shall not show disrespect for your parents and elders.

Page 1056:6　　　　　　　　　　　　　　　　　*Paper 96:4.4*

The fact that Yahweh was the god of the fleeing Hebrews explains why they tarried so long before the holy mountain of Sinai, and **why they there received the Ten Commandments** which Moses promulgated in the name of Yahweh, the god of Horeb. During this lengthy sojourn before Sinai the religious ceremonials of the newly evolving Hebrew worship were further perfected.

Page 1392:7　　　　　　　　　　　　　　　　*Paper 126:4.9*

This Sabbath afternoon Jesus climbed the Nazareth hill with James and, when they returned home, **wrote out the Ten Commandments in Greek on two smooth boards in charcoal.** Subsequently Martha colored and decorated these boards, and for long they hung on the wall over James's small workbench.

Page 1446:1　　　　　　　　　　　　　　　　*Paper 131:2.12*

"Says God, the creator of heaven and earth: `Great peace have they who love my law. My commandments are: You shall love me with all your heart; you shall have no gods before me; you shall not take my name in vain; remember the Sabbath day to keep it holy; honor your father and mother; you shall not kill; you shall not commit adultery; you shall not steal; you shall not bear false witness; you shall not covet.'

Page 1599:2–13　　　　　　　　　　　　　　*Paper 142:3.10–21*

"Again should you have discerned the growth of the understanding of divine law in perfect keeping with these enlarging concepts of divinity. When the children of Israel came out of Egypt in the days before the enlarged revelation of Yahweh, **they had ten commandments which served as their law right up to the times when they were encamped before Sinai. And these ten commandments were:**
"1. You shall worship no other god, for the Lord is a jealous God.
"2. You shall not make molten gods.
"3. You shall not neglect to keep the feast of unleavened bread.
"4. Of all the males of men or cattle, the first-born are mine, says the Lord.
"5. Six days you may work, but on the seventh day you shall rest.
"6. You shall not fail to observe the feast of the first fruits and the feast of the ingathering at the end of the year.
"7. You shall not offer the blood of any sacrifice with leavened bread.
"8. The sacrifice of the feast of the Passover shall not be left until morning.

"9. The first of the first fruits of the ground you shall bring to the house of the Lord your God.

"10 You shall not seethe a kid in its mother's milk.

"**And then, amidst the thunders and lightnings of Sinai, Moses gave them the new Ten Commandments, which you will all allow are more worthy utterances** to accompany the enlarging Yahweh concepts of Deity. And did you never take notice of these commandments as twice recorded in the Scriptures, that in the first case deliverance from Egypt is assigned as the reason for Sabbath keeping, while in a later record the advancing religious beliefs of our forefathers demanded that this be changed to the recognition of the fact of creation as the reason for Sabbath observance?

Page 1901:2   *Paper 174:4.2*

Then came forward one of the groups of the Pharisees to ask harassing questions, and the spokesman, signaling to Jesus, said: "Master, I am a lawyer, and I would like to ask you which, in your opinion, is the greatest commandment?" Jesus answered: "**There is but one commandment, and that one is the greatest of all, and that commandment is:** `Hear O Israel, the Lord our God, the Lord is one; and you shall love the Lord your God with all your heart and with all your soul, with all your mind and with all your strength.' **This is the first and great commandment. And the second commandment is like this first; indeed, it springs directly therefrom, and it is:** `You shall love your neighbor as yourself.' There is no other commandment greater than these; on these two commandments hang all the law and the prophets."

Page 1939:4-5   *Paper 179:3.6-7*

As the Master made ready to begin washing Peter's feet, he said: "He who is already clean needs only to have his feet washed. You who sit with me tonight are clean—but not all. But the dust of your feet should have been washed away before you sat down at meat with me. And besides, **I would perform this service for you as a parable to illustrate the meaning of a new commandment which I will presently give you.**"

In like manner the Master went around the table, in silence, washing the feet of his twelve apostles, not even passing by Judas...

Page 1944:4   *Paper 180:1.1*

After a few moments of informal conversation, Jesus stood up and said: "When I enacted for you a parable indicating how you should be willing to serve one another, **I said that I desired to give you a new commandment; and I would do this now as I am about to leave you.** You well know the commandment which directs that you love one another; that you love your neighbor even as yourself. But I am not wholly satisfied with even that sincere devotion on the part of my children. I would have you perform still greater acts of love in the kingdom of the believing brotherhood. **And so I give you this new commandment: That you love one another even as I have loved you. And by this will all men know that you are my disciples if you thus love one another.**

PAPER 67: THE PLANETARY REBELLION

---

1. THE CALIGASTIA BETRAYAL

Sin

Page 754:5   *Paper 67:1.4*

See "Sin" on page 293 for full text references. (54:0.1)

---

3. THE SEVEN CRUCIAL YEARS

Cosmic Mind

Page 757:2   *Paper 67:3.9*

See "Cosmic Mind" on page 174 (16:9.1–2)

---

6. VAN—THE STEADFAST

Neighboring Universes

Page 759:8   *Paper 67:6.5*

See "Neighboring Universes" on page 210 for full text references. (32:2.12)

---

7. REMOTE REPERCUSSIONS OF SIN

Sin

Page 761:3–4   *Paper 67:7.4–5*

See "Sin" on page 293 for full text references. (54:0.1)

PAPER 68: THE DAWN OF CIVILIZATION

2. FACTORS IN SOCIAL PROGRESSION

Overindulgence—Perils
The perils of overindulgence—3 papers.

Page 765:2, 766:1   *Paper 68:2.5,11*

History is but the record of man's agelong food struggle. *Primitive man only thought when he was hungry;* food saving was his first self-denial, self-discipline. With the growth of society, food hunger ceased to be the only incentive for mutual association. Numerous other sorts of hunger, the realization of various needs, all led to the closer association of mankind. But today society is top-heavy with the overgrowth of supposed human needs. Occidental civilization of the **twentieth century groans wearily under the tremendous overload of luxury** and the inordinate multiplication of human desires and longings. **Modern society is enduring the strain of one of its most dangerous phases** of far-flung interassociation and highly complicated interdependence....

Vanity contributed mightily to the birth of society; but at the time of these revelations the devious strivings of a vainglorious generation threaten to swamp and submerge the whole complicated structure of a highly specialized civilization. **Pleasure-want has long since superseded hunger-want;** the legitimate social aims of self-maintenance are rapidly translating themselves into base and threatening forms of self-gratification. Self-maintenance builds society; unbridled self-gratification unfailingly destroys civilization.

Page 772:8   *Paper 69:1.5*

3. *The institutions of self-gratification.* These are the practices growing out of vanity proclivities and pride emotions; and they embrace customs in dress and personal adornment, social usages, war for glory, dancing, amusement, games, and other phases of sensual gratification. **But civilization has never evolved distinctive institutions of self-gratification.**

Page 942:2–3, 943:1   *Paper 84:8.1–2,6*

The great threat against family life is the menacing rising tide of self-gratification, the modern pleasure mania. The prime incentive to marriage used to be economic; sex attraction was secondary. Marriage, founded on self-maintenance, led to self-perpetuation and concomitantly provided one of the most desirable forms of self-gratification. It is the only institution of human society which embraces all three of the great incentives for living.

Originally, property was the basic institution of self-maintenance, while marriage functioned as the unique institution of self-perpetuation. Although food satisfaction, play, and humor, along with periodic sex indulgence, were means of self-gratification, it remains a fact that the evolving mores have failed to build any distinct institution of self-gratification. And it is due to this failure to evolve specialized techniques of pleasurable enjoyment that all human institutions are so completely shot through with this pleasure pursuit. Property accumulation is becoming an instrument for augmenting all forms of self-gratification, while marriage is often viewed only as a means of pleasure. **And this overindulgence, this widely spread pleasure mania, now constitutes the greatest threat that has ever been leveled at the social evolutionary institution of family life, the home....**

Let man enjoy himself; let the human race find pleasure in a thousand and one ways; let evolutionary mankind explore all forms of legitimate self-gratification, the fruits of the long upward biologic struggle. Man has well earned some of his present-day joys and pleasures. **But look you well to the goal of destiny!** Pleasures are indeed suicidal if they succeed in destroying property, which has become the institution of self-maintenance; and self-gratifications have indeed cost a fatal price if they bring about the collapse of marriage, the decadence of family life, and the destruction of the home—man's supreme evolutionary acquirement and civilization's only hope of survival.

6. Evolution of Culture

Overpopulation and Human Degeneracy

Page 770:8   *Paper 68:6.11*

See "Overpopulation and Human Degeneracy" on page 283 for full text references. (51:4.8)

PAPER 69: PRIMITIVE HUMAN INSTITUTIONS

1. BASIC HUMAN INSTITUTIONS

Overindulgence—Perils

Page 772:8   *Paper 69:1.5*

See "Overindulgence—Perils" on page 316 for full text references. (68:2.5)

3. THE SPECIALIZATION OF LABOR

Work—Roles of Men and Women
Working attitude between men and women.

Page 774:2   *Paper 69:3.3*

All down through the ages the taboos have operated to keep woman strictly in her own field. Man has most selfishly chosen the more agreeable work, leaving the routine drudgery to woman. **Man has always been ashamed to do woman's work, but woman has never shown any reluctance to doing man's work.** But strange to record, both men and women have always worked together in building and furnishing the home.

Page 934:5–6   *Paper 84:3.6–7*

The herdsman looked to his flocks for sustenance, but throughout these pastoral ages woman must still provide the vegetable food. Primitive man shunned the soil; it was altogether too peaceful, too unadventuresome. There was also an old superstition that women could raise better plants; they were mothers. In many backward tribes today, the men cook the meat, the women the vegetables, **and when the primitive tribes of Australia are on the march, the women never attack game, while a man would not stoop to dig a root.**

Woman has always had to work; at least right up to modern times the female has been a real producer. **Man has usually chosen the easier path,** and this inequality has existed throughout the entire history of the human race. Woman has always been the burden bearer, carrying the family property and tending the children, thus leaving the man's hands free for fighting or hunting.

6. FIRE IN RELATION TO CIVILIZATION

Andon and Fonta—Discovery of Fire

Page 777:7   *Paper 69:6.4*

See "Andon and Fonta—Discovery of Fire" on page 305 for full text references. (63:2.4–5)

Parsees

Page 778:1   *Paper 69:6.6*

The early myths about how fire came down from the gods grew out of the observations of fire caused by lightning. These ideas of supernatural origin led directly to fire worship, and fire worship led to the custom of "passing through fire," a practice carried on up to the times of Moses. And there still persists the idea of passing through fire after death. **The fire myth was a great bond in early times and still persists in the symbolism of the Parsees.**

Page 967:6   *Paper 88:1.4*

If an animal ate human flesh, it became a fetish. **In this way the dog came to be the sacred animal of the Parsees.** If the fetish is an animal and the ghost is permanently resident therein, then fetishism may impinge on reincarnation. In many ways the savages envied the animals; they did not feel superior to them and were often named after their favorite beasts.

Page 1050:4   *Paper 95:6.8*

But it is a far cry from the exalted teachings and noble psalms of Zoroaster to **the modern perversions of his gospel by the Parsees with their great fear of the dead,** coupled with the entertainment of beliefs in sophistries which Zoroaster never stooped to countenance.

PAPER 70: THE EVOLUTION OF HUMAN GOVERNMENT

---

2. THE SOCIAL VALUE OF WAR

Overpopulation and Human Degeneracy

Page 786:2–8   *Paper 70:2.6–12*

See "Overpopulation and Human Degeneracy" on page 283 for full text references. (51:4.8)

3. EARLY HUMAN ASSOCIATIONS

International Trade
International trade and the traveling trader.

Page 787:4   *Paper 70:3.4*

The peace of Urantia will be promoted far more by international trade organizations than by all the sentimental sophistry of visionary peace planning. Trade relations have been facilitated by development of language and by improved methods of communication as well as by better transportation.

Page 904:3   *Paper 81:3.7*

The traveling trader and the roving explorer did more to advance historic civilization than all other influences combined. Military conquests, colonization, and missionary enterprises fostered by the later religions were also factors in the spread of culture; but these were all secondary to the trading relations, which were ever accelerated by the rapidly developing arts and sciences of industry.

8. SOCIAL CLASSES

Overpopulation and Human Degeneracy

Page 793:6–9   *Paper 70:8.14–17*

See "Overpopulation and Human Degeneracy" on page 283 for full text references. (51:4.8)

11. LAWS AND COURTS

Courts and Judges—Integrity
Integrity of its judges and integrity of its courts—Parts III and IV

Page 797:12   *Paper 70:11.10*

The whole idea of primitive justice was not so much to be fair as to dispose of the contest and thus prevent public disorder and private violence. But primitive man did not so much resent what would now be regarded as an injustice; it was taken for granted that those who had power would use it selfishly. **Nevertheless, the status of any civilization may be very accurately determined by the thoroughness and equity of its courts and by the integrity of its judges.**

Page 1462:1   *Paper 132:4.8*

Meeting a poor man who had been falsely accused, Jesus went with him before the magistrate and, having been granted special permission to appear in his behalf, made that superb address in the course of which he said: "Justice makes a nation great, and the greater a nation the more solicitous will it be to see that injustice shall not befall even its most humble citizen. Woe upon any nation when only those who possess money and influence can secure ready justice before its courts! **It is the sacred duty of a magistrate to acquit the innocent as well as to punish the guilty. Upon the impartiality, fairness, and integrity of its courts the endurance of a nation depends.** Civil government is founded on justice, even as true religion is founded on mercy." The judge reopened the case, and when the evidence had been sifted, he discharged the prisoner. Of all Jesus' activities during these days of personal ministry, this came the nearest to being a public appearance.

---

12. ALLOCATION OF CIVIL AUTHORITY

Ideal Government—Struggle For

Page 799:1   *Paper 70:12.8*

See "Ideal Government—Struggle For" on page 129 for full text references. (3:1.10)

PAPER 71: DEVELOPMENT OF THE STATE

---

4. PROGRESSIVE CIVILIZATION

Kingdom Within

Page 804:15   *Paper 71:4.3*

See "Kingdom Within" on page 219 for full text references. (34:6.7)

8. The Character of Statehood

Light and Life—Early Stages
Entering the early stages of light and life.

Page 806:15–17–Page 807:1–10   *Paper 71:8.2–14*

The evolution of statehood entails progress from level to level, as follows:
1. The creation of a threefold government of executive, legislative, and judicial branches.
2. The freedom of social, political, and religious activities.
3. The abolition of all forms of slavery and human bondage.
4. The ability of the citizenry to control the levying of taxes.
5. The establishment of universal education—learning extended from the cradle to the grave.
6. The proper adjustment between local and national governments.
7. The fostering of science and the conquest of disease.
8. The due recognition of sex equality and the co-ordinated functioning of men and women in the home, school, and church, with specialized service of women in industry and government.
9. The elimination of toiling slavery by machine invention and the subsequent mastery of the machine age.
10. The conquest of dialects—the triumph of a universal language.
11. The ending of war—international adjudication of national and racial differences by continental courts of nations presided over by a supreme planetary tribunal automatically recruited from the periodically retiring heads of the continental courts. The continental courts are authoritative; the world court is advisory—moral.
12. **The world-wide vogue of the pursuit of wisdom—the exaltation of philosophy. The evolution of a world religion, which will presage the entrance of the planet upon the earlier phases of settlement in light and life.**

Page 820:3   *Paper 72:12.5*

The pouring out of the Spirit of Truth provides the spiritual foundation for the realization of great achievements in the interests of the human race of the bestowal world. Urantia is therefore far better prepared for the more immediate realization of a planetary government with its laws, mechanisms, symbols, conventions, and language—**all of which could contribute so mightily to the establishment of world-wide peace under law and could lead to the sometime dawning of a real age of spiritual striving; and such an age is the planetary threshold to the utopian ages of light and life.**

PAPER 72: GOVERNMENT ON A NEIGHBORING PLANET

Lanaforge

Page 808:1   *Paper 72:0.1*

See "Lanaforge" on page 261 for full text references. (45:2.4)

Spirt of Truth and Divine Fragment

Page 808:3   *Paper 72:0.3*

See "Spirt of Truth and Divine Fragment" on page 118 for full text references. (2:0.3)

3. THE HOME LIFE

Religion—Will of God and Serving Human Brotherhood
Religion on a neighboring planet and the religion that Jesus founded.

Page 811:5   *Paper 72:3.5*

All sex instruction is administered in the home by parents or by legal guardians. Moral instruction is offered by teachers during the rest periods in the school shops, but not so with religious training, which is deemed to be the exclusive privilege of parents, religion being looked upon as an integral part of home life. Purely religious instruction is given publicly only in the temples of philosophy, no such exclusively religious institutions as the Urantia churches having developed among this people. **In their philosophy, religion is the striving to know God and to manifest love for one's fellows through service for them, but this is not typical of the religious status of the other nations on this planet.** Religion is so entirely a family matter among these people that there are no public places devoted exclusively to religious assembly. Politically, church and state, as Urantians are wont to say, are entirely separate, but there is a strange overlapping of religion and philosophy.

Page 2092:4   *Paper 196:2.6*

Jesus founded the religion of personal experience in doing the will of God and serving the human brotherhood; Paul founded a religion in which the glorified Jesus became the object of worship and the brotherhood consisted of fellow believers in the divine Christ. In the bestowal of Jesus these two concepts were potential in his divine-human life, and it is indeed a pity that his followers failed to create a unified religion which might have given proper recognition to both the human and the divine natures of the Master as they were inseparably bound up in his earth life and so gloriously set forth in the original gospel of the kingdom.

Page 2095:3   *Paper 196:3.16*

The Hebrews had a religion of moral sublimity; the Greeks evolved a religion of beauty; Paul and his conferees founded a religion of faith, hope, and charity. **Jesus revealed and exemplified a religion of love: security in the Father's love, with joy and satisfaction consequent upon sharing this love in the service of the human brotherhood.**

Marriage
Examples given from our neighboring planet with the intent of advancing civilization and augmenting governmental evolution on Urantia.

Page 812:1–2   *Paper 72:3.8–9*

Marriage and divorce laws are uniform throughout the nation. Marriage before twenty—the age of civil enfranchisement—is not permitted. Permission to marry is only granted after one year's notice of intention, and after both bride and groom present certificates showing that they have been duly instructed in the parental schools regarding the responsibilities of married life.

Divorce regulations are somewhat lax, but decrees of separation, issued by the parental courts, may not be had until one year after application therefor has been recorded, and the year on this planet is considerably longer than on Urantia. Notwithstanding their easy divorce laws, the present rate of divorces is only one tenth that of the civilized races of Urantia.

Page 929:3   *Paper 83:7.9*

The ancients seem to have regarded marriage just about as seriously as some present-day people do. And it does not appear that many of the hasty and unsuccessful marriages of modern times are much of an improvement over the ancient practices of qualifying young men and women for mating. **The great inconsistency of modern society is to exalt love and to idealize marriage while disapproving of the fullest examination of both.**

12. The Other Nations

Spirt of Truth and Divine Fragment

Page 820:3   *Paper 72:12.5*

> See "Spirt of Truth and Divine Fragment" on page 118 for full text references. (2:0.3)

Light and Life—Early Stages

Page 820:3   *Paper 72:12.5*

> See "Light and Life—Early Stages" on page 322 for full text references. (71:8.2–14)

Paper 73: The Garden of Eden

Tabamantia—Agondonter of Finaliter Status

Page 821:3   *Paper 73:0.3*

> See "Tabamantia—Agondonter of Finaliter Status" on page 279 for full text references. (49:5.5)

1. The Nodites and the Amodonites

Mansant—a Nodite?

Page 821:6   *Paper 73:1.3*

> See "Mansant—a Nodite?" on page 262 for full text references. (45:4.2–4)

2. PLANNING FOR THE GARDEN

Lucifer Rebellion—Status of Caligastia and Daligastia

Page 822:9   *Paper 73:2.5*

See "Lucifer Rebellion—Status of Caligastia and Daligastia" on page 292 for full text references. (53:8.9)

4. ESTABLISHING THE GARDEN

Eden—Named After Edentia

Page 823:7   *Paper 73:4.1*

See "Eden—Named After Edentia" on page 256 for full text references. (43:6.2)

6. THE TREE OF LIFE

Seraphic Transport

Page 825:8   *Paper 73:6.3*

See "Seraphic Transport" on page 268 for full text references. (46:5.19)

PAPER 74: ADAM AND EVE

1. ADAM AND EVE ON JERUSEM

Material Sons and Daughters—Difficulties

The difficulties of the biologic uplifters before and after the 4th epochal revelation.

Page 829:2   *Paper 74:1.6*

And thus did Adam and Eve leave Jerusem amidst the acclaim and well-wishing of its citizens. **They went forth to their new responsibilities adequately equipped and fully instructed concerning every duty and danger to be encountered on Urantia.**

Page 839:4   *Paper 75:1.3*

Adam and Eve found themselves on a sphere wholly unprepared for the proclamation of the brotherhood of man, a world groping about in abject spiritual darkness and cursed with confusion worse confounded by the miscarriage of the mission of the preceding administration. Mind and morals were at a low level, and instead of beginning the task of effecting religious unity, they must begin all anew the work of converting the inhabitants to the most simple forms of religious belief. Instead of finding one language ready for adoption, they were confronted by the world-wide confusion of hundreds upon hundreds of local dialects. No Adam of the planetary service was ever set down on a more difficult world; the obstacles seemed insuperable and the problems beyond creature solution.

Page 1313:2   *Paper 119:3.7*

Never, since this marvelous bestowal as the Planetary Prince of a world in isolation and rebellion, have any of the Material Sons or Daughters in Nebadon been tempted to complain of their assignments or to find fault with the difficulties of their planetary missions. For all time the Material Sons know that in the Creator Son of the universe they have an understanding sovereign and a sympathetic friend, one who has in "all points been tried and tested," even as they must also be tried and tested.

---

2. ARRIVAL OF ADAM AND EVE

Early Language

Page 829:4   *Paper 74:2.2*

See "Early Language" on page 306 for full text references. (63:4.6)

---

7. LIFE IN THE GARDEN

Commandments

Page 835:14, 836:1–8   *Paper 74:7.4–12*

See "Commandments" on page 311 for full text references. (66:7.8–16)

PAPER 75: THE DEFAULT OF ADAM AND EVE

---

1. THE URANTIA PROBLEM

Material Sons and Daughters—Difficulties

Page 839:4   *Paper 75:1.3*

See "Material Sons and Daughters—Difficulties" on page 326 for full text references. (74:1.6)

PAPER 76: THE SECOND GARDEN

---

2. CAIN AND ABEL

Enoch

Page 849:3   *Paper 76:2.9*

See "Enoch" on page 264 for full text references. (45:4.13)

---

4. THE VIOLET RACE

Childbirth—Pain
Childbirth with mixed and unmixed tribes.

Page 850:8   *Paper 76:4.2*

Eve did not suffer pain in childbirth; neither did the early evolutionary races. Only the mixed races produced by the union of evolutionary man with the Nodites and later with the Adamites suffered the severe pangs of childbirth.

Page 935:7   *Paper 84:4.7*

Among the unmixed tribes, childbirth was comparatively easy, occupying only two or three hours; it is seldom so easy among the mixed races. If a woman died in childbirth, especially during the delivery of twins, she was believed to have been guilty of spirit adultery. Later on, the higher tribes looked upon death in childbirth as the will of heaven; such mothers were regarded as having perished in a noble cause.

Adam and Eve—Ascension

Page 852:4   *Paper 76:5.5*

> See "Adam and Eve—Ascension" on page 280 for full text references. (49:6.8)

Adam and Eve—Ascension

Page 853:3   *Paper 76:6.2*

> See "Adam and Eve—Ascension" on page 280 for full text references. (49:6.8)

Spiritual Economy

Page 855:4   *Paper 77:1.2*

> See "Spiritual Economy" on page 165 for full text references. (14:4.4)

Primary Midwayers—Discovery of Production

Page 855:5   *Paper 77:1.3*

> See "Primary Midwayers—Discovery of Production" on page 310 for full text references. (66:4.10)

Germ Plasm

Page 857:2   *Paper 77:2.5*

> See "Germ Plasm" on page 229 for full text references. (36:2.7)

Neighboring Universes

Page 857:3   Paper 77:2.6

See "Neighboring Universes" on page 210 for full text references. (32:2.12)

3. THE TOWER OF BABEL

Traditions of Dilmun

Page 858:2   Paper 77:3.1

After the submergence of Dalamatia the Nodites moved north and east, presently **founding the new city of Dilmun** as their racial and cultural headquarters. And about fifty thousand years after the death of Nod, when the offspring of the Prince's staff had become too numerous to find subsistence in the lands immediately surrounding their new city of Dilmun, and after they had reached out to intermarry with the Andonite and Sangik tribes adjoining their borders, it occurred to their leaders that something should be done to preserve their racial unity. Accordingly a council of the tribes was called, and after much deliberation the plan of Bablot, a descendant of Nod, was endorsed.

Page 975:7   Paper 89:2.3

Sin was ritual, not rational; an act, not a thought. **And this entire concept of sin was fostered by the lingering traditions of Dilmun** and the days of a little paradise on earth. The tradition of Adam and the Garden of Eden also lent substance to the dream of a onetime "golden age" of the dawn of the races. And all this confirmed the ideas later expressed in the belief that man had his origin in a special creation, that he started his career in perfection, and that transgression of the taboos—sin—brought him down to his later sorry plight.

4. NODITE CENTERS OF CIVILIZATION

Enoch

Page 859:8   Paper 77:4.5

See "Enoch" on page 264 for full text references. (45:4.13)

Early Language

Page 860:2   *Paper 77:4.7*

See "Early Language" on page 306 for full text references. (63:4.6)

5. *ADAMSON AND RATTA*

Secondary Midwayers—Not an Accident

Page 861:5–6   *Paper 77:5.5–6*

See "Secondary Midwayers—Not an Accident" on page 238 for full text references. (38: 9.5)

Secondary Midwayers

Page 861:5–6   *Paper 77:5.5–6*

See "Secondary Midwayers" on page 244 for full text references. (40:3.1)

6. *THE SECONDARY MIDWAYERS*

Early Language

Page 862:8   *Paper 77:6.4*

See "Early Language" on page 306 for full text references. (63:4.6)

7. *THE REBEL MIDWAYERS*

Lucifer Rebellion—Demoniacal Possession

Page 863:6, 864:1   *Paper 77:7.5,8*

See "Lucifer Rebellion—Demoniacal Possession" on page 291 for full text references. (53:8.6)

Thought Adjusters—Self Acting

Self-acting Adjusters and the function of midway creatures (rebel and loyal), and their ability to influence mortal minds.

Page 863:6   *Paper 77:7.5*

On no world can evil spirits possess any mortal mind subsequent to the life of a Paradise bestowal Son. But before the days of Christ Michael on Urantia—before the universal coming of the Thought Adjusters and the pouring out of the Master's spirit upon all flesh—these rebel midwayers were actually able to influence the minds of certain inferior mortals and somewhat to control their actions. **This was accomplished in much the same way as the loyal midway creatures function** when they serve as efficient contact guardians of the human minds of the Urantia reserve corps of destiny at those times **when the Adjuster is, in effect, detached from the personality** during a season of contact with superhuman intelligences.

Page 1196:3,7, 1197:1   *Paper 109:2.1,5,10*

You have been informed of the classification of Adjusters in relation to experience—virgin, advanced, and supreme. You should also recognize a certain functional classification—the self-acting Adjusters. **A self-acting Adjuster is one who:...**

4. **Has a subject who has been mustered into one of the reserve corps of destiny** on an evolutionary world of mortal ascension....

Supreme and self-acting Adjusters can leave the human body at will. The indwellers are not an organic or biologic part of mortal life; they are divine superimpositions thereon. In the original life plans they were provided for, but they are not indispensable to material existence. Nevertheless it should be recorded that they very rarely, even temporarily, leave their mortal tabernacles after they once take up their indwelling.

8. THE UNITED MIDWAYERS

Most Highs

Page 865:6   *Paper 77:8.13*

See "Most Highs" on page 126 for full text references. (3:1.10)

PAPER 78: THE VIOLET RACE AFTER THE DAYS OF ADAM

---

3. EARLY EXPANSIONS OF THE ADAMITES

Horse

Page 871:6   *Paper 78:3.10*

See "Horse" on page 302 for full text references. (61:3.11)

---

5. THE ANDITE MIGRATIONS

Amosad—Post-Adamic Teacher
The leadership of Amosad.

Page 872:5   *Paper 78:5.1*

For twenty thousand years the culture of the second garden persisted, but it experienced a steady decline until about 15,000 B.C., **when the regeneration of the Sethite priesthood and the leadership of Amosad inaugurated a brilliant era.** The massive waves of civilization which later spread over Eurasia immediately followed the great renaissance of the Garden consequent upon the extensive union of the Adamites with the surrounding mixed Nodites to form the Andites.

Page 1009:3   *Paper 92:5.6*

1. *The Sethite period.* **The Sethite priests, as regenerated under the leadership of Amosad, became the great post-Adamic teachers.** They functioned throughout the lands of the Andites, and their influence persisted longest among the Greeks, Sumerians, and Hindus. Among the latter they have continued to the present time as the Brahmans of the Hindu faith. The Sethites and their followers never entirely lost the Trinity concept revealed by Adam.

Early Language

Page 872:7   *Paper 78:5.3*

See "Early Language" on page 306 for full text references. (63:4.6)

8. THE SUMERIANS—LAST OF THE ANDITES

Ur and Susa
Jesus was fascinated with Ur and Susa—Parts III and IV.

Page 875:6, 876:6   Paper 78:8.2,9

It was during the floodtimes that Susa so greatly prospered. The first and lower city was inundated so that the second or higher town succeeded the lower as the headquarters for the peculiar artcrafts of that day. With the later diminution of these floods, Ur became the center of the pottery industry. About seven thousand years ago Ur was on the Persian Gulf, the river deposits having since built up the land to its present limits. These settlements suffered less from the floods because of better controlling works and the widening mouths of the rivers....

After the breakup of this Kish confederation there ensued a long period of constant warfare between these valley cities for supremacy. **And the rulership variously shifted between Sumer, Akkad, Kish, Erech, Ur, and Susa.**

Page 1427:3   Paper 130:0.3

After their stay in Rome they went overland to Tarentum, where they set sail for Athens in Greece, stopping at Nicopolis and Corinth. From Athens they went to Ephesus by way of Troas. From Ephesus they sailed for Cyprus, putting in at Rhodes on the way. They spent considerable time visiting and resting on Cyprus and then sailed for Antioch in Syria. From Antioch they journeyed south to Sidon and then went over to Damascus. From there they traveled by caravan to Mesopotamia, passing through Thapsacus and Larissa. **They spent some time in Babylon, visited Ur and other places, and then went to Susa. From Susa they journeyed to Charax, from which place Gonod and Ganid embarked for India.**

Page 1481:4   Paper 133:9.2

Jesus was much interested in the early history of Ur, the birthplace of Abraham, and he was equally fascinated with the ruins and traditions of Susa, so much so that Gonod and Ganid extended their stay in these parts three weeks in order to afford Jesus more time to conduct his investigations and also to provide the better opportunity to persuade him to go back to India with them.

2. THE ANDITE CONQUEST OF INDIA

India—Highlands

Page 879:7   *Paper 79:2.1*

See "India—Highlands" on page 308 for full text references. (64:5.1–2)

5. RED MAN AND YELLOW MAN

Sangik Races—Fifth Glacier

Page 883:3   *Paper 79:5.2*

See "Sangik Races—Fifth Glacier" on page 303 for full text references. (61:7.4)

Onamonalonton—a Red Man

Page 884:2   *Paper 79:5.8*

See "Onamonalonton—a Red Man" on page 263 for full text references. (45:4.2,5)

Polynesian Andites

Page 884:3   *Paper 79:5.9*

See "Polynesian Andites" on page 309 for full text references. (64:7.5)

PAPER 80: ANDITE EXPANSION IN THE OCCIDENT

4. THE ANDITE INVASIONS OF EUROPE

Horse

Page 892:7   *Paper 80:4.4*

See "Horse" on page 302 for full text references. (61:3.11)

8. THE DANUBIAN ANDONITES

Early Language

Page 896:7   *Paper 80:8.2*

See "Early Language" on page 306 for full text references. (63:4.6)

PAPER 81: DEVELOPMENT OF MODERN CIVILIZATION

1. THE CRADLE OF CIVILIZATION

Evolution—Slow but Effective
Evolution, terribly and unerringly effective.

Page 900:5   *Paper 81:1.3*

Climatic evolution is now about to accomplish what all other efforts had failed to do, that is, to compel Eurasian man to abandon hunting for the more advanced callings of herding and farming. **Evolution may be slow, but it is terribly effective.**

Page 957:2   *Paper 86:7.6*

Primitive religion prepared the soil of the human mind, by the powerful and awesome force of false fear, for the bestowal of a bona fide spiritual force of supernatural origin, the Thought Adjuster. And the divine Adjusters have ever since labored to transmute God-fear into God-love. **Evolution may be slow, but it is unerringly effective.**

3. CITIES, MANUFACTURE, AND COMMERCE

International Trade

Page 904:3   *Paper 81:3.7*

See "International Trade" on page 320 for full text references. (70:3.4)

336

6. THE MAINTENANCE OF CIVILIZATION

Most Highs

Page 908:8   Paper 81:6.19

See "Most Highs" on page 126 for full text references. (3:1.10)

PAPER 83: THE MARRIAGE INSTITUTION

7. THE DISSOLUTION OF WEDLOCK

Marriage

Page 929:3   Paper 83:7.9

See "Marriage" on page 324 for full text references. (72:3.8–9)

PAPER 84: MARRIAGE AND FAMILY LIFE

3. THE FAMILY UNDER FATHER DOMINANCE

Work—Roles of Men and Women

Page 934:5–6   Paper 84:3.6–7

See "Work—Roles of Men and Women" on page 318 for full text references. (69:3.3)

4. WOMAN'S STATUS IN EARLY SOCIETY

Childbirth—Pain

Page 935:7   Paper 84:4.7

See "Childbirth—Pain" on page 328 for full text references. (76:4.2)

7. The Ideals of Family Life

Patience and Forbearance

Page 941:9   *Paper 84:7.28*

See "Patience and Forbearance" on page 295 for full text references. (54:5.4)

8. Dangers of Self-Gratification

Overindulgence—Perils

Page 942:2–3, 943:1   *Paper 84:8.1–2,6*

See "Overindulgence—Perils" on page 316 for full text references. (68:2.5)

Paper 86: Early Evolution of Religion

7. The Function of Primitive Religion

Evolution—Slow but Effective

Page 957:2   *Paper 86:7.6*

See "Evolution—Slow but Effective" on page 336 for full text references. (81:1.3)

Paper 87: The Ghost Cults

2. Ghost Placation

Religious Practices—Borneans and African Tribes
Borneans and other practices of African tribes.

Page 960:2   *Paper 87:2.8*

It was customary to dispatch a large number of subjects to accompany a dead chief; slaves were killed when their master died that they might serve him in ghostland. **The Borneans still provide a courier companion;** a slave is speared to death to make the ghost journey with his deceased master. Ghosts of murdered persons were believed to be delighted to have the ghosts of their murderers as slaves; this notion motivated men to head hunting.

Page 971:1   *Paper 88:4.5*

Magic gained such a strong hold upon the savage because he could not grasp the concept of natural death. The later idea of original sin helped much to weaken the grip of magic on the race in that it accounted for natural death. It was at one time not at all uncommon for ten innocent persons to be put to death because of supposed responsibility for one natural death. **This is one reason why ancient peoples did not increase faster, and it is still true of some African tribes.** The accused individual usually confessed guilt, even when facing death.

---

4. GOOD AND BAD SPIRIT GHOSTS

Power of an Idea
The power of an idea lies...

Page 961:4   *Paper 87:4.3*

The notion of two kinds of spirit ghosts made slow but sure progress throughout the world. This new dual spiritism did not have to spread from tribe to tribe; it sprang up independently all over the world. In influencing the expanding evolutionary mind, the **power of an idea lies not in its reality or reasonableness but rather in its vividness and the universality of its ready and simple application.**

Page 1005:5   *Paper 92:3.3*

Primitive religion is nothing more nor less than the struggle for material existence extended to embrace existence beyond the grave. The observances of such a creed represented the extension of the self-maintenance struggle into the domain of an imagined ghost-spirit world. But when tempted to criticize evolutionary religion, be careful. Remember, that is *what happened*; it is a historical fact. And further recall that **the power of any idea lies, not in its certainty or truth, but rather in the vividness of its human appeal.**

Page 1090:1   *Paper 99:4.5*

After all, it is what one believes rather than what one knows that determines conduct and dominates personal performances. Purely factual knowledge exerts very little influence upon the average man unless it becomes emotionally activated. But the activation of religion is superemotional, unifying the entire human experience on transcendent levels through contact with, and release of, spiritual energies in the mortal life.

Lucifer Rebellion—Caligastic

Page 962:1   *Paper 87:4.7*

See "Lucifer Rebellion—Caligastic" on page 225 for full text references. (34:7.6)

---

7. NATURE OF CULTISM

Cults

Two papers on the cult.

Page 966:1   *Paper 87:7.6*

Regardless of the drawbacks and handicaps, every new revelation of truth has given rise to a new cult, and even the restatement of the religion of Jesus must develop a new and appropriate symbolism. Modern man must find some adequate symbolism for his new and expanding ideas, ideals, and loyalties. This enhanced symbol must arise out of religious living, spiritual experience. And this higher symbolism of a higher civilization must be predicated on the concept of the Fatherhood of God and be pregnant with the mighty ideal of the brotherhood of man.

Page 1006:1–2   *Paper 92:3.4–5*

Evolutionary religion makes no provision for change or revision; unlike science, it does not provide for its own progressive correction. Evolved religion commands respect because its followers believe it is *The Truth*; "the faith once delivered to the saints" must, in theory, be both final and infallible. **The cult resists development because real progress is certain to modify or destroy the cult itself;** therefore must revision always be forced upon it.

Only two influences can modify and uplift the dogmas of natural religion: the pressure of the slowly advancing mores and the periodic illumination of epochal revelation. And it is not strange that progress was slow; in ancient days, to be progressive or inventive meant to be killed as a sorcerer. **The cult advances slowly in generation epochs and agelong cycles.** But it does move forward. Evolutionary belief in ghosts laid the foundation for a philosophy of revealed religion which will eventually destroy the superstition of its origin.

PAPER 88: FETISHES, CHARMS, AND MAGIC

1. BELIEF IN FETISHES

Parsees

Page 967:6   *Paper 88:1.4*

See "Parsees" on page 319 for full text references. (69:6.6)

2. EVOLUTION OF THE FETISH

Moses—Kenite Traditions
Kenite traditions in regards to Moses.

Page 969:3   *Paper 88:2.5*

Moses, in the addition of the second commandment to the ancient Dalamatian moral code, made an effort to control fetish worship among the Hebrews. He carefully directed that they should make no sort of image that might become consecrated as a fetish. He made it plain, "You shall not make a graven image or any likeness of anything that is in heaven above, or on the earth beneath, or in the waters of the earth." While this commandment did much to retard art among the Jews, it did lessen fetish worship. **But Moses was too wise to attempt suddenly to displace the olden fetishes,** and he therefore consented to the putting of certain relics alongside the law in the combined war altar and religious shrine which was the ark.

Page 1058:1   *Paper 96:5.3*

Many of the advances which Moses made over and above the religion of the Egyptians and the surrounding Levantine tribes were due to the Kenite traditions of the time of Melchizedek. Without the teaching of Machiventa to Abraham and his contemporaries, the Hebrews would have come out of Egypt in hopeless darkness. Moses and his father-in-law, Jethro, gathered up the residue of the traditions of the days of Melchizedek, and these teachings, joined to the learning of the Egyptians, guided Moses in the creation of the improved religion and ritual of the Israelites. **Moses was an organizer; he selected the best in the religion and mores of Egypt and Palestine and, associating these practices with the traditions of the Melchizedek teachings, organized the Hebrew ceremonial system of worship.**

4. Magic

Necromancers

The necromancers.

Page 970:7   *Paper 88:4.2*

> The object of magic, sorcery, and **necromancy** was twofold:
> 1. To secure insight into the future.
> 2. Favorably to influence environment.

Page 987:6   *Paper 90:2.2*

> Ancient black art, both religious and secular, was called white art when practiced by either priests, seers, shamans, or medicine men. The **practitioners of the black art were called** sorcerers, magicians, wizards, witches, enchanters, **necromancers**, conjurers, and soothsayers. As time passed, all such purported contact with the supernatural was classified either as witchcraft or shamancraft.

Religious Practices—Borneans and African Tribes

Page 971:1   *Paper 88:4.5*

> See "Religious Practices—Borneans and African Tribes" on page 338 for full text references. (87:2.8)

PAPER 89: SIN, SACRIFICE, AND ATONEMENT

1. The Taboo

Commandments

Page 975:1   *Paper 89:1.4*

> See "Commandments" on page 311 for full text references. (66:7.8–16)

2. The Concept of Sin

Traditions of Dilmun

Page 975:7   *Paper 89:2.3*

See "Traditions of Dilmun" on page 330 for full text references. (77:3.1)

3. Renunciation and Humiliation

Continence Cult—Influence to Paul
The continence cult and Paul. Part III and IV.

Page 977:1   *Paper 89:3.6*

It was only natural that the cult of renunciation and humiliation should have paid attention to sexual gratification. **The continence cult originated as a ritual among soldiers prior to engaging in battle; in later days it became the practice of "saints." This cult tolerated marriage only as an evil lesser than fornication.** Many of the world's great religions have been adversely influenced by this ancient cult, but none more markedly than Christianity. **The Apostle Paul was a devotee of this cult,** and his personal views are reflected in the teachings which he fastened onto Christian theology: "It is good for a man not to touch a woman." "I would that all men were even as I myself." "I say, therefore, to the unmarried and widows, it is good for them to abide even as I." Paul well knew that such teachings were not a part of Jesus' gospel, and his acknowledgment of this is illustrated by his statement, "I speak this by permission and not by commandment." But this cult led Paul to look down upon women. And the pity of it all is that his personal opinions have long influenced the teachings of a great world religion. If the advice of the tentmaker-teacher were to be literally and universally obeyed, then would the human race come to a sudden and inglorious end. **Furthermore, the involvement of a religion with the ancient continence cult leads directly to a war against marriage and the home, society's veritable foundation and the basic institution of human progress.** And it is not to be wondered at that all such beliefs fostered the formation of celibate priesthoods in the many religions of various peoples.

Page 1679:2   *Paper 150:1.3*

It was most astounding in that day, when women were not even allowed on the main floor of the synagogue (being confined to the women's gallery), to behold them being recognized as authorized teachers of the new gospel of the kingdom. The charge which Jesus gave these ten women as he set them apart for gospel teaching and ministry was the emancipation proclamation which set free all women and for all time; no more was man to look upon woman as his spiritual inferior. This was a decided shock to even the twelve apostles. Notwithstanding they had many times heard the Master say that "in the kingdom of heaven there is neither rich nor poor, free nor bond, male nor female, all are equally the sons and daughters of God," they were literally stunned when he proposed formally to commission these ten women as religious teachers and even to permit their traveling about with them. The whole country was stirred up by this proceeding, the enemies of Jesus making great capital out of this move, but everywhere the women believers in the good news stood stanchly behind their chosen sisters and voiced no uncertain approval of this tardy acknowledgment of woman's place in religious work. And this liberation of women, giving them due recognition, was practiced by the apostles immediately after the Master's departure, **albeit they fell back to the olden customs in subsequent generations.** Throughout the early days of the Christian church women teachers and ministers were called *deaconesses* and were accorded general recognition. **But Paul, despite the fact that he conceded all this in theory, never really incorporated it into his own attitude and personally found it difficult to carry out in practice.**

PAPER 90: SHAMANISM—MEDICINE MEN AND PRIESTS

---

2. THE SHAMANISTIC PRACTICES

Necromancers

Page 987:6   *Paper 90:2.2*

See "Necromancers" on page 342 for full text references. (88:4.2)

---

3. THE SHAMANIC THEORY OF DISEASE AND DEATH

Scaffolding—Stages of Development
Three papers on different types of scaffolding presented—Parts III and IV.

Page 990:5   *Paper 90:3.10*

Evolution unerringly achieves its end: **It imbues man with that superstitious fear of the unknown and dread of the unseen which is the scaffolding for the God concept.** And having witnessed the birth of an advanced comprehension of Deity, through the co-ordinate action of revelation, this same technique of evolution then unerringly sets in motion those forces of thought which will inexorably obliterate the scaffolding, which has served its purpose.

Page 1260:3   Paper 115:1.2

Conceptual frames of the universe are only relatively true; they are serviceable scaffolding which must eventually give way before the expansions of enlarging cosmic comprehension. The understandings of truth, beauty, and goodness, morality, ethics, duty, love, divinity, origin, existence, purpose, destiny, time, space, even Deity, are only relatively true. God is much, much more than a Father, but the Father is man's highest concept of God; nonetheless, the Father-Son portrayal of Creator-creature relationship will be augmented by those supermortal conceptions of Deity which will be attained in Orvonton, in Havona, and on Paradise. Man must think in a mortal universe frame, but that does not mean that he cannot envision other and higher frames within which thought can take place.

Page 1488:6   Paper 134:5.8

The difficulty in the evolution of political sovereignty from the family to all mankind, lies in the inertia-resistance exhibited on all intervening levels. Families have, on occasion, defied their clan, while clans and tribes have often been subversive of the sovereignty of the territorial state. **Each new and forward evolution of political sovereignty is (and has always been) embarrassed and hampered by the "scaffolding stages" of the previous developments in political organization.** And this is true because human loyalties, once mobilized, are hard to change. The same loyalty which makes possible the evolution of the tribe, makes difficult the evolution of the supertribe—the territorial state. And the same loyalty (patriotism) which makes possible the evolution of the territorial state, vastly complicates the evolutionary development of the government of all mankind.

PAPER 91: THE EVOLUTION OF PRAYER

---

9. CONDITIONS OF EFFECTIVE PRAYER

Mind—Exchange for the Mind of Jesus

Page 1002:6–8   Paper 91:9.1–3

See "Mind—Exchange for the Mind of Jesus" on page 274 for full text references. (48:6.15)

PAPER 92: THE LATER EVOLUTION OF RELIGION

Thought Adjusters—Preparation for Bestowal

Page 1003:4   *Paper 92:0.4*

>   See "Thought Adjusters—Preparation for Bestowal" on page 218 for full text references. (34:5.3)

3. THE NATURE OF EVOLUTIONARY RELIGION

Power of an Idea

Page 1005:5   *Paper 92:3.3*

>   See "Power of an Idea" on page 339 for full text references. (87:4.3)

Cults

Page 1006:1–2   *Paper 92:3.4–5*

>   See "Cults" on page 340 for full text references. (87:7.6)

4. THE GIFT OF REVELATION

Revelation—Fifth Epochal Revelation
Opportunities with this revelation.

Page 1007:4   *Paper 92:4.4*

There have been many events of **religious revelation** but only five of epochal significance. These were as follows:

1. *The Dalamatian teachings....2. The Edenic teachings....3. Melchizedek of Salem....4. Jesus of Nazareth....5. The Urantia Papers....*

Page 1145:1   *Paper 104:1.13*

Not since the times of Jesus has the factual identity of the Paradise Trinity been known on Urantia (except by a few individuals to whom it was especially revealed) **until its presentation in these revelatory disclosures.** But though the Christian concept of the Trinity erred in fact, it was practically true with respect to spiritual relationships. Only in its philosophic implications and cosmological consequences did this concept suffer embarrassment: It has been difficult for many who are cosmic minded to believe that the Second Person of Deity, the second member of an infinite Trinity, once dwelt on Urantia; and while in spirit this is true, in actuality it is not a fact. The Michael Creators fully embody the divinity of the Eternal Son, but they are not the absolute personality.

Page 2090:3   *Paper 196:1.2*

The time is ripe to witness the figurative resurrection of the human Jesus from his burial tomb amidst the theological traditions and the religious dogmas of nineteen centuries. Jesus of Nazareth must not be longer sacrificed to even the splendid concept of the glorified Christ. **What a transcendent service if, through this revelation,** the Son of Man should be recovered from the tomb of traditional theology and be presented as the living Jesus to the church that bears his name, and to all other religions! Surely the Christian fellowship of believers will not hesitate to make such adjustments of faith and of practices of living as will enable it to "follow after" the Master in the demonstration of his real life of religious devotion to the doing of his Father's will and of consecration to the unselfish service of man. Do professed Christians fear the exposure of a self-sufficient and unconsecrated fellowship of social respectability and selfish economic maladjustment? Does institutional Christianity fear the possible jeopardy, or even the overthrow, of traditional ecclesiastical authority if the Jesus of Galilee is reinstated in the minds and souls of mortal men as the ideal of personal religious living? Indeed, the social readjustments, the economic transformations, the moral rejuvenations, and the religious revisions of Christian civilization would be drastic and revolutionary if the living religion of Jesus should suddenly supplant the theologic religion about Jesus.

Trinity Concept—Sethite Priests
The Trinity concept and Sethite priests.

Page 1007:6   *Paper 92:4.6*

2. *The Edenic teachings.* Adam and Eve again portrayed the concept of the Father of all to the evolutionary peoples. The disruption of the first Eden halted the course of the Adamic revelation before it had ever fully started. But the aborted teachings of Adam **were carried on by the Sethite priests,** and some of these truths have never been entirely lost to the world. The entire trend of Levantine religious evolution was modified by the teachings of the Sethites. But by 2500 B.C. mankind had largely lost sight of the revelation sponsored in the days of Eden.

Page 1009:3   *Paper 92:5.6*

1. *The Sethite period.* The Sethite priests, as regenerated under the leadership of Amosad, became the great post-Adamic teachers. They functioned throughout the lands of the Andites, and their influence persisted longest among the Greeks, Sumerians, and Hindus. Among the latter they have continued to the present time as the Brahmans of the Hindu faith. **The Sethites and their followers never entirely lost the Trinity concept revealed by Adam.**

Revelation—Written Word
Revelation and the written word.

Page 1008:2   *Paper 92:4.9*

5. *The Urantia Papers.* The papers, of which this is one, constitute the most recent presentation of truth to the mortals of Urantia. **These papers** differ from all previous revelations, for they are not the work of a single universe personality but a composite presentation by many beings. But no revelation short of the attainment of the Universal Father can ever be complete. All other celestial ministrations are no more than partial, transient, and practically adapted to local conditions in time and space. While such admissions as this may possibly detract from the immediate force and authority of all revelations, the time has arrived on Urantia when it is advisable to make such frank statements, even at the risk of weakening the future influence and authority of this, the most recent of the revelations of truth to the mortal races of Urantia.

Page 1106:3   *Paper 101:2.4*

There are two basic reasons for believing in a God who fosters human survival:
1. Human experience, personal assurance, the somehow registered hope and trust initiated by the indwelling Thought Adjuster.
2. The revelation of truth, whether by direct personal ministry of the Spirit of Truth, by the world bestowal of divine Sons, **or through the revelations of the written word.**

Page 1109:3   *Paper 101:4.2*

Mankind should understand that we who participate in the revelation of truth are very rigorously limited by the instructions of our superiors. We are not at liberty to anticipate the scientific discoveries of a thousand years. Revelators must act in accordance with the instructions which form a part of the revelation mandate. We see no way of overcoming this difficulty, either now or at any future time. We full well know that, **while the historic facts and religious truths of this series of revelatory presentations** will stand on the records of the ages to come, within a few short years many of our statements regarding the physical sciences will stand in need of revision in consequence of additional scientific developments and new discoveries. These new developments we even now foresee, but we are forbidden to include such humanly undiscovered facts in the revelatory records. Let it be made clear that revelations are not necessarily inspired. The cosmology of these revelations is *not inspired*. It is limited by our permission for the co-ordination and sorting of present-day knowledge. While divine or spiritual insight is a gift, *human wisdom must evolve.*

Page 1115:2   Paper 101:9.1

No professed revelation of religion could be regarded as authentic if it failed to recognize the duty demands of ethical obligation which had been created and fostered by preceding evolutionary religion. **Revelation unfailingly enlarges the ethical horizon of evolved religion while it simultaneously and unfailingly expands the moral obligations of all prior revelations.**

---

5. THE GREAT RELIGIOUS LEADERS

Onamonalonton—a Red Man

Page 1008:8   Paper 92:5.3

See "Onamonalonton—a Red Man" on page 263 for full text references. (45:4.2,5)

Amosad—Post-Adamic Teacher

Page 1009:3   Paper 92:5.6

See "Amosad—Post-Adamic Teacher" on page 333 for full text references. (78:5.1)

Trinity Concept—Sethite Priests

Page 1009:3   Paper 92:5.6

See "Trinity Concept—Sethite Priests" on page 347 for full text references. (92:4.6)

Religious Awakening—6th Century B.C.
6th Century B.C.

Page 1009:6   Paper 92:5.9

4. *The sixth century before Christ.* Many men arose to proclaim truth in this, one of the greatest centuries of religious awakening ever witnessed on Urantia. Among these should be recorded Gautama, Confucius, Lao-tse, Zoroaster, and the Jainist teachers. The teachings of Gautama have become widespread in Asia, and he is revered as the Buddha by millions. Confucius was to Chinese morality what Plato was to Greek philosophy, and while there were religious repercussions to the teachings of both, strictly speaking, neither was a religious teacher; Lao-tse envisioned more of God in Tao than did Confucius in humanity or Plato in idealism. Zoroaster, while much affected by the prevalent concept of dual spiritism, the good and the bad, at the same time definitely exalted the idea of one eternal Deity and of the ultimate victory of light over darkness.

Page 1033:4   Paper 94:6.1

About six hundred years before the arrival of Michael, it seemed to Melchizedek, long since departed from the flesh, that the purity of his teaching on earth was being unduly jeopardized by general absorption into the older Urantia beliefs. It appeared for a time that his mission as a forerunner of Michael might be in danger of failing. **And in the sixth century before Christ, through an unusual co-ordination of spiritual agencies, not all of which are understood even by the planetary supervisors, Urantia witnessed a most unusual presentation of manifold religious truth.** Through the agency of several human teachers the Salem gospel was restated and revitalized, and as it was then presented, much has persisted to the times of this writing.

7. *THE FURTHER EVOLUTION OF RELIGION*

Parent-Child Relationship
Parent-child relationships—Parts III and IV

Page 1013:6   Paper 92:7.7

Religious meanings progress in self-consciousness when the child transfers his ideas of omnipotence from his parents to God. **And the entire religious experience of such a child is largely dependent on whether fear or love has dominated the parent-child relationship.** Slaves have always experienced great difficulty in transferring their master-fear into concepts of God-love. Civilization, science, and advanced religions must deliver mankind from those fears born of the dread of natural phenomena. And so should greater enlightenment deliver educated mortals from all dependence on intermediaries in communion with Deity.

Page 1921:6   Paper 177:2.2

"I know you will prove loyal to the gospel of the kingdom because I can depend upon your present faith and love when these qualities are grounded upon such an early training as has been your portion at home. You are the product of a home where the parents bear each other a sincere affection, and therefore you have not been overloved so as injuriously to exalt your concept of self-importance. Neither has your personality suffered distortion in consequence of your parents' loveless maneuvering for your confidence and loyalty, the one against the other. You have enjoyed that parental love which insures laudable self-confidence and which fosters normal feelings of security. **But you have also been fortunate in that your parents possessed wisdom as well as love;** and it was wisdom which led them to withhold most forms of indulgence and many luxuries which wealth can buy while they sent you to the synagogue school along with your neighborhood playfellows, and they also encouraged you to learn how to live in this world by permitting you to have original experience. You came over to the Jordan, where we preached and John's disciples baptized, with your young friend Amos. Both of you desired to go with us. When you returned to Jerusalem, your parents consented; Amos's parents refused; they loved their son so much that they denied him the blessed experience which you have had, even such as you this day enjoy. By running away from home, Amos could have joined us, but in so doing he would have wounded love and sacrificed loyalty. Even if such a course had been wise, it would have been a terrible price to pay for experience, independence, and liberty. Wise parents, such as yours, see to it that their children do not have to wound love or stifle loyalty in order to develop independence and enjoy invigorating liberty when they have grown up to your age.

PAPER 93: MACHIVENTA MELCHIZEDEK

---

2. THE SAGE OF SALEM

Three Concentric Circles

Page 1015:5   Paper 93:2.5

See "Three Concentric Circles" on page 288 for full text references. (53:5.4)

Bestowals and Offspring

Page 1015:6   *Paper 93:2.6*

> See "Bestowals and Offspring" on page 184 for full text references. (20:6.2)

---

3. MELCHIZEDEK'S TEACHINGS

Lanaforge

Page 1016:4   *Paper 93:3.2*

> See "Lanaforge" on page 261 for full text references. (45:2.4)

Three Concentric Circles

Page 1016:5   *Paper 93:3.3*

> See "Three Concentric Circles" on page 288 for full text references. (53:5.4)

Bestowals and Offspring

Page 1017:1   *Paper 93:3.7*

> See "Bestowals and Offspring" on page 184 for full text references. (20:6.2)

---

4. THE SALEM RELIGION

Commandments

Page 1017:8–15   *Paper 93:4.6–13*

> See "Commandments" on page 311 for full text references. (66:7.8–16)

Melchizedek Bestowal—Truth of One God

One of the purposes of Melchizedek's incarnation.

Page 1018:2   *Paper 93:4.15*

Like Jesus, Melchizedek attended strictly to the fulfillment of the mission of his bestowal. He did not attempt to reform the mores, to change the habits of the world, nor to promulgate even advanced sanitary practices or scientific truths. **He came to achieve two tasks:** to keep alive on earth the truth of the one God and to prepare the way for the subsequent mortal bestowal of a Paradise Son of that Universal Father.

Page 1052:2   *Paper 96:0.2*

The Salem religion was revered as a tradition by the Kenites and several other Canaanite tribes. **And this was one of the purposes of Melchizedek's incarnation:** That a religion of one God should be so fostered as to prepare the way for the earth bestowal of a Son of that one God. Michael could hardly come to Urantia until there existed a people believing in the Universal Father among whom he could appear.

5. THE SELECTION OF ABRAHAM

Melchizedek Bestowal—Abraham and Brother

Leaving Ur—Parts III and IV.

Page 1019:1   *Paper 93:5.4*

Terah and his whole family were halfhearted converts to the Salem religion, which had been preached in Chaldea; they learned of Melchizedek through the preaching of Ovid, a Phoenician teacher who proclaimed the Salem doctrines in Ur. **They left Ur intending to go directly through to Salem, but Nahor,** Abraham's brother, not having seen Melchizedek, was lukewarm and persuaded them to tarry at Haran. And it was a long time after they arrived in Palestine before they were willing to destroy *all* of the household gods they had brought with them; they were slow to give up the many gods of Mesopotamia for the one God of Salem.

Page 1598:5   *Paper 142:3.4*

2. *The Most High.* This concept of the Father in heaven was proclaimed by Melchizedek to Abraham and was carried far from Salem by those who subsequently believed in this enlarged and expanded idea of Deity. **Abraham and his brother left Ur because of the establishment of sun worship,** and they became believers in Melchizedek's teaching of El Elyon—the Most High God. Theirs was a composite concept of God, consisting in a blending of their older Mesopotamian ideas and the Most High doctrine.

6. MELCHIZEDEK'S COVENANT WITH ABRAHAM

Melchizedek Bestowal—Divine Covenant
The abrogated divine covenant—three papers in Parts III and IV.

Page 1020:4,6   Paper 93:6.1,3

Abraham envisaged the conquest of all Canaan. His determination was only weakened by the fact that Melchizedek would not sanction the undertaking. But Abraham had about decided to embark upon the enterprise when the thought that he had no son to succeed him as ruler of this proposed kingdom began to worry him. He arranged another conference with Melchizedek; and it was in the course of this interview that the priest of Salem, **the visible Son of God,** persuaded Abraham to abandon his scheme of material conquest and temporal rule in favor of the spiritual concept of the kingdom of heaven....

And Melchizedek **made a formal covenant** with Abraham at Salem. Said he to Abraham: "Look now up to the heavens and number the stars if you are able; so numerous shall your seed be." And Abraham believed Melchizedek, "and it was counted to him for righteousness." And then Melchizedek told Abraham the story of the future occupation of Canaan by his offspring after their sojourn in Egypt.

Page 1910:1   Paper 175:3.2

From this time on the Jews were left to finish their brief and short lease of national life wholly in accordance with their purely human status among the nations of Urantia. **Israel had repudiated the Son of the God who made a covenant with Abraham,** and the plan to make the children of Abraham the light-bearers of truth to the world had been shattered. **The divine covenant** had been abrogated, and the end of the Hebrew nation drew on apace.

Page 1969:4   Paper 182:3.9

The experience of parting with the apostles was a great strain on the human heart of Jesus; this sorrow of love bore down on him and made it more difficult to face such a death as he well knew awaited him. He realized how weak and how ignorant his apostles were, and he dreaded to leave them. He well knew that the time of his departure had come, but his human heart longed to find out whether there might not possibly be some legitimate avenue of escape from this terrible plight of suffering and sorrow. And when it had thus sought escape, and failed, it was willing to drink the cup. The divine mind of Michael knew he had done his best for the twelve apostles; but the human heart of Jesus wished that more might have been done for them before they should be left alone in the world. Jesus' heart was being crushed; he truly loved his brethren. He was isolated from his family in the flesh; one of his chosen associates was betraying him. **His father Joseph's people had rejected him and thereby sealed their doom as a people with a special mission on earth.** His soul was tortured by baffled love and rejected mercy. It was just one of those awful human moments when everything seems to bear down with crushing cruelty and terrible agony.

---

7. THE MELCHIZEDEK MISSIONARIES

Melchizedek Bestowal—Missionaries
Missionaries of Machiventa Melchizedek—Parts III and IV

Page 1021:5   *Paper 93:7.1*

Melchizedek continued for some years to instruct his students and to train the Salem missionaries, who penetrated to all the surrounding tribes, especially to Egypt, Mesopotamia, and Asia Minor. And as the decades passed, these teachers journeyed farther and farther from Salem, carrying with them Machiventa's gospel of belief and faith in God.

Page 1027:1   *Paper 94:0.1*

The early teachers of the Salem religion penetrated to the remotest tribes of Africa and Eurasia, ever preaching Machiventa's gospel of man's faith and trust in the one universal God as the only price of obtaining divine favor. Melchizedek's covenant with Abraham was the pattern for all the early propaganda that went out from Salem and other centers. **Urantia has never had more enthusiastic and aggressive missionaries of any religion than these noble men and women who carried the teachings of Melchizedek over the entire Eastern Hemisphere.** These missionaries were recruited from many peoples and races, and they largely spread their teachings through the medium of native converts. They established training centers in different parts of the world where they taught the natives the Salem religion and then commissioned these pupils to function as teachers among their own people.

Page 1442:1   *Paper 131:0.1*

During the Alexandrian sojourn of Jesus, Gonod, and Ganid, the young man spent much of his time and no small sum of his father's money making a collection of the teachings of the world's religions about God and his relations with mortal man. Ganid employed more than threescore learned translators in the making of this abstract of the religious doctrines of the world concerning the Deities. **And it should be made plain in this record that all these teachings portraying monotheism were largely derived, directly or indirectly, from the** preachments of the missionaries of Machiventa Melchizedek, who went forth from their Salem headquarters to spread the doctrine of one God—the Most High—to the ends of the earth.

Page 1444:1   *Paper 131:2.1*

The Kenites of Palestine salvaged much of the teaching of Melchizedek, **and from these records,** as preserved and modified by the Jews, Jesus and Ganid made the following selection:

10. Present Status of Machiventa Melchizedek

Four and Twenty Counselors

Page 1025:1   *Paper 93:10.5*

See "Four and Twenty Counselors" on page 197 for full text references. (24:5.3)

Paper 94: The Melchizedek Teachings in the Orient

Melchizedek Bestowal—Missionaries

Page 1027:1   *Paper 94:0.1*

See "Melchizedek Bestowal—Missionaries" on page 355 for full text references. (93:7.1)

3. Brahmanic Philosophy

Deity Function—Levels

Page 1030:3   *Paper 94:3.3*

See "Deity Function—Levels" on page 101 (0:1.3–10)

Religious Awakening—6th Century B.C.

Page 1033:4   Paper 94:6.1

See "Religious Awakening—6th Century B.C." on page 350 for full text references. (92:5.9)

Jesus' Gospel—Hour Is Striking
Hour is striking—two times in Parts III and IV.

Page 1041:5   Paper 94:12.7

All Urantia is waiting for the proclamation of the ennobling message of Michael, unencumbered by the accumulated doctrines and dogmas of nineteen centuries of contact with the religions of evolutionary origin. **The hour is striking for presenting to Buddhism, to Christianity, to Hinduism, even to the peoples of all faiths,** not the gospel about Jesus, **but the living, spiritual reality of the gospel of Jesus.**

Page 2083:1   Paper 195:9.5

The modern age will refuse to accept a religion which is inconsistent with facts and out of harmony with its highest conceptions of truth, beauty, and goodness. **The hour is striking for a rediscovery of the true and original foundations of present-day distorted and compromised Christianity—the real life and teachings of Jesus.**

Efficacy of Spittle—Belief In
Efficacy of spittle—two times in Parts III and IV.

Page 1044:5   Paper 95:2.6

The **superstitions of these times** are well illustrated by the general belief in the **efficacy of spittle as a healing agent,** an idea which had its origin in Egypt and spread therefrom to Arabia and Mesopotamia. In the legendary battle of Horus with Set the young god lost his eye, but after Set was vanquished, this eye was restored by the wise god Thoth, who spat upon the wound and healed it.

Page 1812:5   *Paper 164:3.11*

This is one of the strangest of all the Master's miracles. This man did not ask for healing. He did not know that the Jesus who had directed him to wash at Siloam, and who had promised him vision, was the prophet of Galilee who had preached in Jerusalem during the feast of tabernacles. **This man had little faith that he would receive his sight, but the people of that day had great faith in the efficacy of the spittle of a great or holy man;** and from Jesus' conversation with Nathaniel and Thomas, Josiah had concluded that his would-be benefactor was a great man, a learned teacher or a holy prophet; accordingly he did as Jesus directed him.

---

4. THE TEACHINGS OF AMENEMOPE

God-consciousness—Prelude to Worship

The secret of Jesus' unparalleled religious life was this "prelude to true worship"—Parts III and IV.

Page 1046:3   *Paper 95:4.2*

Amenemope taught that riches and fortune were the gift of God, and this concept thoroughly colored the later appearing Hebrew philosophy. **This noble teacher believed that God-consciousness was the determining factor in all conduct; that every moment should be lived in the realization of the presence of, and responsibility to, God.** The teachings of this sage were subsequently translated into Hebrew and became the sacred book of that people long before the Old Testament was reduced to writing. The chief preachment of this good man had to do with instructing his son in uprightness and honesty in governmental positions of trust, and these noble sentiments of long ago would do honor to any modern statesman.

Page 1133:1   *Paper 103:4.1*

The characteristic difference between a social occasion and a religious gathering is that in contrast with the secular the religious is pervaded by the atmosphere of *communion.* In this way human association generates a feeling of fellowship with the divine, and this is the beginning of group worship. Partaking of a common meal was the earliest type of social communion, and so did early religions provide that some portion of the ceremonial sacrifice should be eaten by the worshipers. Even in Christianity the Lord's Supper retains this mode of communion. The atmosphere of the communion provides a refreshing and comforting period of truce in the conflict of the self-seeking ego with the altruistic urge of the indwelling spirit Monitor. **And this is the prelude to true worship—the practice of the presence of God which eventuates in the emergence of the brotherhood of man.**

Page 2087:3, 2088:5   *Paper 196:0.3,10*

Jesus did not cling to faith in God as would a struggling soul at war with the universe and at death grips with a hostile and sinful world; he did not resort to faith merely as a consolation in the midst of difficulties or as a comfort in threatened despair; faith was not just an illusory compensation for the unpleasant realities and the sorrows of living. In the very face of all the natural difficulties and the temporal contradictions of mortal existence, he experienced the tranquillity of supreme and unquestioned trust in God and felt the tremendous thrill of living, **by faith, in the very presence of the heavenly Father.** And this triumphant faith was a living experience of actual spirit attainment. Jesus' great contribution to the values of human experience was not that he revealed so many new ideas about the Father in heaven, but rather that he so magnificently and humanly demonstrated a new and higher type of *living faith in God.* Never on all the worlds of this universe, in the life of any one mortal, did God ever become such a *living reality* as in the human experience of Jesus of Nazareth....

Jesus brought to God, as a man of the realm, the greatest of all offerings: the consecration and dedication of his own will to the majestic service of doing the divine will. Jesus always and consistently interpreted religion wholly in terms of the Father's will. When you study the career of the Master, as concerns prayer or any other feature of the religious life, look not so much for what he taught as for what he did. Jesus never prayed as a religious duty. To him prayer was a sincere expression of spiritual attitude, a declaration of soul loyalty, a recital of personal devotion, an expression of thanksgiving, an avoidance of emotional tension, a prevention of conflict, an exaltation of intellection, an ennoblement of desire, a vindication of moral decision, an enrichment of thought, an invigoration of higher inclinations, a consecration of impulse, a clarification of viewpoint, a declaration of faith, a transcendental surrender of will, a sublime assertion of confidence, a revelation of courage, the proclamation of discovery, a confession of supreme devotion, the validation of consecration, a technique for the adjustment of difficulties, and the mighty mobilization of the combined soul powers to withstand all human tendencies toward selfishness, evil, and sin. He lived just such a life of prayerful consecration to the doing of his Father's will and ended his life triumphantly with just such a prayer. **The secret of his unparalleled religious life was this consciousness of the presence of God;** and he attained it by intelligent prayer and sincere worship—unbroken communion with God—and not by leadings, voices, visions, or extraordinary religious practices.

5. The Remarkable Ikhnaton

Iknaton and the Angels of the Churches
The Iknaton connection with the "angels of the churches".

Page 1047:2   *Paper 95:5.2*

Since the disappearance of Melchizedek in the flesh, no human being up to that time had possessed such an amazingly clear concept of the revealed religion of Salem as Ikhnaton. In some respects this young Egyptian king is one of the most remarkable persons in human history. During this time of increasing spiritual depression in Mesopotamia, he kept alive the doctrine of El Elyon, the One God, in Egypt, thus maintaining the philosophic monotheistic channel which was vital to the religious background of the then future bestowal of Michael. **And it was in recognition of this exploit, among other reasons, that the child Jesus was taken to Egypt, where some of the spiritual successors of Ikhnaton saw him and to some extent understood certain phases of his divine mission to Urantia.**

Page 1255:6   *Paper 114:6.7*

3. *The religious guardians.* **These are the "angels of the churches," the earnest contenders for that which is and has been.** They endeavor to maintain the ideals of that which has survived for the sake of the safe transit of moral values from one epoch to another. They are the checkmates of the angels of progress, all the while seeking to translate from one generation to another the imperishable values of the old and passing forms into the new and therefore less stabilized patterns of thought and conduct. These angels do contend for spiritual forms, but they are not the source of ultrasectarianism and meaningless controversial divisions of professed religionists. The corps now functioning on Urantia is the fifth thus to serve.

Page 1258:2   *Paper 114:7.7*

The reservists unconsciously act as conservators of essential planetary information. Many times, upon the death of a reservist, a transfer of certain vital data from the mind of the dying reservist to a younger successor is made by a liaison of the two Thought Adjusters. The Adjusters undoubtedly function in many other ways unknown to us, in connection with these reserve corps.

Page 1355:3   *Paper 123:0.3*

Throughout the two years of their **sojourn at Alexandria,** Jesus enjoyed good health and continued to grow normally. **Aside from a few friends and relatives no one was told about Jesus' being a "child of promise."** One of Joseph's relatives revealed this to a few friends in Memphis, **descendants of the distant Ikhnaton,** and they, with a small group of Alexandrian believers, assembled at the palatial home of Joseph's relative-benefactor a short time before the return to Palestine to wish the Nazareth family well and to pay their respects to the child. On this occasion the assembled friends presented Jesus with a complete copy of the Greek translation of the Hebrew scriptures. But this copy of the Jewish sacred writings was not placed in Joseph's hands until both he and Mary had finally declined the invitation of their Memphis and Alexandrian friends to remain in Egypt. These believers insisted that the child of destiny would be able to exert a far greater world influence as a resident of Alexandria than of any designated place in Palestine. These persuasions delayed their departure for Palestine for some time after they received the news of Herod's death.

6. THE SALEM DOCTRINES IN IRAN

Mithraic Cult
The Mithraic cult and the Ahura-Mazda.

Page 1049:5   *Paper 95:6.2*

This founder of a new religion was a virile and adventurous youth, who, on his first pilgrimage to Ur in Mesopotamia, had learned of the traditions of the Caligastia and the Lucifer rebellion—along with many other traditions—all of which had made a strong appeal to his religious nature. Accordingly, as the result of a dream while in Ur, he settled upon a program of returning to his northern home to undertake the remodeling of the religion of his people. He had imbibed the Hebraic idea of a God of justice, the Mosaic concept of divinity. The idea of a supreme God was clear in his mind, and he set down all other gods as devils, consigned them to the ranks of the demons of which he had heard in Mesopotamia. **He had learned of the story of the Seven Master Spirits as the tradition lingered in Ur, and, accordingly, he created a galaxy of seven supreme gods with Ahura-Mazda at its head.** These subordinate gods he associated with the idealization of Right Law, Good Thought, Noble Government, Holy Character, Health, and Immortality.

Page 1082:4   *Paper 98:5.3*

The Mithraic cult portrayed a militant god taking origin in a great rock, engaging in valiant exploits, and causing water to gush forth from a rock struck with his arrows. There was a flood from which one man escaped in a specially built boat and a last supper which Mithras celebrated with the sun-god before he ascended into the heavens. This sun-god, or Sol Invictus, **was a degeneration of the Ahura-Mazda** deity concept of Zoroastrianism. Mithras was conceived as the surviving champion of the sun-god in his struggle with the god of darkness. And in recognition of his slaying the mythical sacred bull, Mithras was made immortal, being exalted to the station of intercessor for the human race among the gods on high.

Parsees

Page 1050:4   *Paper 95:6.8*

See "Parsees" on page 319 for full text references. (69:6.6)

PAPER 96: YAHWEH—GOD OF THE HEBREWS

Melchizedek Bestowal—Truth of One God

Page 1052:2   *Paper 96:0.2*

See "Melchizedek Bestowal—Truth of One God" on page 353 for full text references. (93:4.15)

1. DEITY CONCEPTS AMONG THE SEMITES

Growth of the God Concept In Jewish Theology
Growth of the God concept among the Jews—Parts III and IV.

Page 1052:5-Page 1053:1–7   *Paper 96:1.2–9*

The progress of the Hebrews from polytheism through henotheism to monotheism was not an unbroken and continuous conceptual development. They experienced many retrogressions in the evolution of their Deity concepts, while during any one epoch there existed varying ideas of God among different groups of Semite believers. **From time to time numerous terms were applied to their concepts of God, and in order to prevent confusion these various Deity titles will be defined as they pertain to the evolution of Jewish theology:**

1. *Yahweh* **was the god** of the southern Palestinian tribes, who associated this concept of deity with Mount Horeb, the Sinai volcano. Yahweh was merely one of the hundreds and thousands of nature gods which held the attention and claimed the worship of the Semitic tribes and peoples.

2. *El Elyon.* For centuries after Melchizedek's sojourn at Salem his doctrine of Deity persisted in various versions but was generally connoted by the term El Elyon, **the Most High God of heaven.** Many Semites, including the immediate descendants of Abraham, at various times worshiped both Yahweh and El Elyon.

3. *El Shaddai.* It is difficult to explain what El Shaddai stood for. This idea of God was a composite derived from the teachings of Amenemope's Book of Wisdom modified by Ikhnaton's doctrine of Aton and further influenced by Melchizedek's teachings embodied in the concept of El Elyon. But as the concept of El Shaddai permeated the Hebrew mind, it became thoroughly colored with the Yahweh beliefs of the desert.

   One of the dominant ideas of the religion of this era was the Egyptian concept of divine Providence, the teaching that material prosperity was a reward for serving El Shaddai.

4. *El.* Amid all this confusion of terminology and haziness of concept, many devout believers sincerely endeavored to worship all of these evolving ideas of divinity, and there grew up the practice of referring to this composite Deity as El. And this term included still other of the Bedouin nature gods.

5. *Elohim.* In Kish and Ur there long persisted Sumerian-Chaldean groups who **taught a three-in-one God concept founded on the traditions of the days of Adam and Melchizedek.** This doctrine was carried to Egypt, where this Trinity was worshiped under the name of Elohim, or in the singular as Eloah. The philosophic circles of Egypt and later Alexandrian teachers of Hebraic extraction taught this unity of pluralistic Gods, and many of Moses' advisers at the time of the exodus believed in this Trinity. But the concept of the trinitarian Elohim never became a real part of Hebrew theology until after they had come under the political influence of the Babylonians.

6. *Sundry names.* The Semites disliked to speak the name of their Deity, and they therefore resorted to numerous appellations from time to time, such as: The Spirit of God, The Lord, The Angel of the Lord, The Almighty, The Holy One, The Most High, Adonai, The Ancient of Days, The Lord God of Israel, **The Creator of Heaven and Earth,** Kyrios, Jah, The Lord of Hosts, and **The Father in Heaven.**

Page 1598:3–9 Paper 142:3.2–8

Jesus mildly upbraided the twelve, in substance saying: Do you not know the traditions of Israel relating to the growth of the idea of Yahweh, and are you ignorant of the teaching of the Scriptures concerning the doctrine of God? And then did the Master proceed to instruct the apostles about the evolution of the concept of Deity throughout the course of the development of the Jewish people. **He called attention to the following phases of the growth of the God idea:**

1. *Yahweh*—the god of the Sinai clans. This was the primitive concept of Deity which Moses exalted to the higher level of the Lord God of Israel. The Father in heaven never fails to accept the sincere worship of his children on earth, no matter how crude their concept of Deity or by what name they symbolize his divine nature.

2. *The Most High.* This concept of the Father in heaven was proclaimed by Melchizedek to Abraham and was carried far from Salem by those who subsequently believed in this enlarged and expanded idea of Deity. **Abraham and his brother left Ur because of the establishment of sun worship,** and they became believers in Melchizedek's teaching of **El Elyon**—the Most High God. Theirs was a composite concept of God, consisting in a blending of their older Mesopotamian ideas and the Most High doctrine.

3. *El Shaddai.* During these early days many of the Hebrews worshiped El Shaddai, the Egyptian concept of the God of heaven, which they learned about during their captivity in the land of the Nile. Long after the times of Melchizedek all three of these concepts of God became joined together to form the doctrine of the creator Deity, the Lord God of Israel.

4. *Elohim.* From the times of Adam the teaching of the Paradise Trinity has persisted. Do you not recall how the Scriptures begin by asserting that "In the beginning the Gods created the heavens and the earth"? This indicates that when that record was made the Trinity concept of three Gods in one had found lodgment in the religion of our forebears.

5. *The Supreme Yahweh.* By the times of **Isaiah** these beliefs about God had expanded into the concept of a Universal Creator who was simultaneously all-powerful and all-merciful. And this evolving and enlarging concept of God virtually supplanted all previous ideas of Deity in our fathers' religion.

6. *The Father in heaven.* And now do we know God as our Father in heaven. Our teaching provides a religion wherein the believer *is* a son of God. That is the good news of the gospel of the kingdom of heaven. Coexistent with the Father are the Son and the Spirit, **and the revelation of the nature and ministry of these Paradise Deities will continue to enlarge and brighten throughout the endless ages of**

the eternal spiritual progression of the ascending sons of God. At all times and during all ages the true worship of any human being—as concerns individual spiritual progress—is recognized by the indwelling spirit as homage rendered to the Father in heaven.

---

4. THE PROCLAMATION OF YAHWEH

Commandments

Page 1056:6   *Paper 96:4.4*

See "Commandments" on page 311 for full text references. (66:7.8–16)

---

5. THE TEACHINGS OF MOSES

Moses—Kenite Traditions

Page 1058:1   *Paper 96:5.3*

See "Moses—Kenite Traditions" on page 341 for full text references. (88:2.5)

---

6. THE GOD CONCEPT AFTER MOSES' DEATH

Joshua and Joshua ben Joseph
Joshua ben Joseph's namesake?—Parts III and IV.

Page 1059:4   *Paper 96:6.3*

Desperately Joshua sought to hold the concept of a supreme Yahweh in the minds of the tribesmen, causing it to be proclaimed: "As I was with Moses, so will I be with you; I will not fail you nor forsake you." Joshua found it necessary to preach a stern gospel to his disbelieving people, people all too willing to believe their old and native religion but unwilling to go forward in the religion of faith and righteousness. **The burden of Joshua's teaching became:** "Yahweh is a holy God; he is a jealous God; he will not forgive your transgressions nor your sins." The highest concept of this age pictured Yahweh as a "God of power, judgment, and justice."

Page 1346:4   *Paper 122:3.1*

One evening about sundown, before Joseph had returned home, Gabriel appeared to Mary by the side of a low stone table and, after she had recovered her composure, said: "I come at the bidding of one who is my Master and whom you shall love and nurture. To you, Mary, I bring glad tidings when I announce that the conception within you is ordained by heaven, **and that in due time you will become the mother of a son; you shall call him Joshua,** and he shall inaugurate the kingdom of heaven on earth and among men. Speak not of this matter save to Joseph and to Elizabeth, your kinswoman, to whom I have also appeared, and who shall presently also bear a son, whose name shall be John, and who will prepare the way for the message of deliverance which your son shall proclaim to men with great power and deep conviction. And doubt not my word, Mary, for this home has been chosen as the mortal habitat of the child of destiny. My benediction rests upon you, the power of the Most Highs will strengthen you, and the Lord of all the earth shall overshadow you."

Page 1375:1   *Paper 124:6.7*

The third day they passed by two villages which had been recently built by Herod and noted their superior architecture and their beautiful palm gardens. By nightfall they reached Jericho, where they remained until the morrow. **That evening Joseph, Mary, and Jesus walked a mile and a half to the site of the ancient Jericho, where Joshua, for whom Jesus was named, had performed his renowned exploits, according to Jewish tradition.**

Page 1501:4, 1502:5   *Paper 135:6.1,7*

Early in the month of March, A.D. 25, John journeyed around the western coast of the Dead Sea and up the river Jordan to opposite Jericho, **the ancient ford over which Joshua and the children of Israel passed when they first entered the promised land;** and crossing over to the other side of the river, he established himself near the entrance to the ford and began to preach to the people who passed by on their way back and forth across the river. This was the most frequented of all the Jordan crossings....

John was a heroic but tactless preacher. One day when he was preaching and baptizing on the west bank of the Jordan, a group of Pharisees and a number of Sadducees came forward and presented themselves for baptism. Before leading them down into the water, John, addressing them as a group said: "Who warned you to flee, as vipers before the fire, from the wrath to come? I will baptize you, but I warn you to bring forth fruit worthy of sincere repentance if you would receive the remission of your sins. Tell me not that Abraham is your father. I declare that God is able of these twelve stones here before you to raise up worthy children for Abraham. And even now is the ax laid to the very roots of the trees. Every tree that brings not forth good fruit is destined to be cut down and cast into the fire." **(The twelve stones to which he referred were the reputed memorial stones set up by Joshua to commemorate the crossing of the "twelve tribes" at this very point when they first entered the promised land.)**

PAPER 97: EVOLUTION OF THE GOD CONCEPT AMONG THE HEBREWS

---

2. ELIJAH AND ELISHA

Elijah

Page 1064:3   *Paper 97:2.2*

See "Elijah" on page 265 for full text references. (45:4.15)

---

3. YAHWEH AND BAAL

Elijah

Page 1065:2–3   *Paper 97:3.5–6*

See "Elijah" on page 265 for full text references. (45:4.15)

---

5. THE FIRST ISAIAH

"The Spirit of the Lord Is Upon Me"
The spirit of the Lord is upon me—Parts III and IV.

Page 1066:7   *Paper 97:5.3*

Speaking to the fear-ridden and soul-hungry Hebrews, this prophet said: "Arise and shine, for your light has come, and the glory of the Lord has risen upon you." **"The spirit of the Lord is upon me because he has anointed me** to preach good tidings to the meek; he has sent me to bind up the brokenhearted, to proclaim liberty to the captives and the opening of the prison to those who are bound." "I will greatly rejoice in the Lord, my soul shall be joyful in my God, for he has clothed me with the garments of salvation and has covered me with his robe of righteousness." "In all their afflictions he was afflicted, and the angel of his presence saved them. In his love and in his pity he redeemed them."

Page 1391:5–6   *Paper 126:4.1–2*

With the coming of his fifteenth birthday, Jesus could officially occupy the synagogue pulpit on the Sabbath day. Many times before, in the absence of speakers, Jesus had been asked to read the Scriptures, but now the day had come when, according to law, he could conduct the service. Therefore on the first Sabbath after his fifteenth birthday the chazan arranged for Jesus to conduct the morning service of the synagogue. **And when all the faithful in Nazareth had assembled, the young man, having made his selection of Scriptures, stood up and began to read:**

"**The spirit of the Lord God is upon me, for the Lord has anointed me;** he has sent me to bring good news to the meek, to bind up the brokenhearted, to proclaim liberty to the captives, and to set the spiritual prisoners free; to proclaim the year of God's favor and the day of our God's reckoning; to comfort all mourners, to give them beauty for ashes, the oil of joy in the place of mourning, a song of praise instead of the spirit of sorrow, that they may be called trees of righteousness, the planting of the Lord, wherewith he may be glorified.

PAPER 98: THE MELCHIZEDEK TEACHINGS IN THE OCCIDENT

---

5. THE CULT OF MITHRAS

Mithraic Cult

Page 1082:4   *Paper 98:5.3*

See "Mithraic Cult" on page 361 for full text references. (95:6.2)

---

7. THE CHRISTIAN RELIGION

Death on the Cross

Death on the cross was to stimulate man's *realization* of Father's eternal love—Parts III and IV.

Page 1083:6   *Paper 98:7.1*

A Creator Son **did not incarnate** in the likeness of mortal flesh and bestow himself upon the humanity of Urantia **to reconcile** an angry God but rather to win all mankind to the recognition of the Father's love and to the realization of their sonship with God. After all, even the great advocate of the atonement doctrine realized something of this truth, for he declared that "**God was in Christ reconciling the world to himself.**"

Page 2019:6   *Paper 188:5.13*
We know that the death on the cross was not to effect man's reconciliation to God but to stimulate man's *realization* of the Father's eternal love and his Son's unending mercy, and to broadcast these universal truths to a whole universe.

PAPER 99: THE SOCIAL PROBLEMS OF RELIGION

---

1. RELIGION AND SOCIAL RECONSTRUCTION

Revelation—Unrevealed Concepts

Page 1086:4   *Paper 99:1.1*
> See "Revelation—Unrevealed Concepts" on page 203 for full text references. (30:0.2)

Jesus—Teachings During Current Epoch of Social Readjustment
The teachings of Jesus during current epoch of social readjustment.

Page 1086:6   *Paper 99:1.3*
Urantia society can never hope to settle down as in past ages. The social ship has steamed out of the sheltered bays of established tradition and has begun its cruise upon the high seas of evolutionary destiny; and the soul of man, as never before in the world's history, needs carefully to scrutinize its charts of morality and painstakingly to observe the compass of religious guidance. **The paramount mission of religion as a social influence is to stabilize the ideals of mankind during these dangerous times of transition from one phase of civilization to another, from one level of culture to another.**

Page 2082:7–8   *Paper 195:9.2–3*
But paganized and socialized Christianity stands in need of new contact with the uncompromised teachings of Jesus; it languishes for lack of a new vision of the Master's life on earth. A new and fuller revelation of the religion of Jesus is destined to conquer an empire of materialistic secularism and to overthrow a world sway of mechanistic naturalism. Urantia is now quivering on the very brink of one of its most amazing **and enthralling epochs of social readjustment,** moral quickening, and spiritual enlightenment.

The teachings of Jesus, even though greatly modified, survived the mystery cults of their birthtime, the ignorance and superstition of the dark ages, **and are even now slowly triumphing over the materialism, mechanism, and secularism of the twentieth century.** And such times of great testing and threatened defeat are always times of great revelation.

---

4. TRANSITION DIFFICULTIES

Power of an Idea

Page 1090:1   *Paper 99:4.5*

See "Power of an Idea" on page 339 for full text references. (87:4.3)

---

5. SOCIAL ASPECTS OF RELIGION

Feelings That Lie Too Deep for Words

Page 1091:8   *Paper 99:5.9*

Primitive man made little effort to put his religious convictions into words. His religion was danced out rather than thought out. Modern men have thought out many creeds and created many tests of religious faith. Future religionists must live out their religion, dedicate themselves to the wholehearted service of the brotherhood of man. **It is high time that man had a religious experience so personal and so sublime that it could be realized and expressed only by "feelings that lie too deep for words."**

Page 1122:8   *Paper 102:3.12*

The pursuit of knowledge constitutes science; the search for wisdom is philosophy; the love for God is religion; the hunger for truth *is* a revelation. **But it is the indwelling Thought Adjuster that attaches the feeling of reality to man's spiritual insight into the cosmos.**

PAPER 100: RELIGION IN HUMAN EXPERIENCE

1. RELIGIOUS GROWTH

Parents—Influence Upon Children
Adult loyalties and parental influence.

Page 1094:6   *Paper 100:1.4*

Children are permanently impressed **only by the loyalties of their adult associates;** precept or even example is not lastingly influential. Loyal persons are growing persons, and growth is an impressive and inspiring reality. Live loyally today—grow—and tomorrow will attend to itself. The quickest way for a tadpole to become a frog is to live loyally each moment as a tadpole.

Page 1922:3   *Paper 177:2.5*

For more than an hour Jesus and John continued this discussion of home life. The Master went on to explain to John how a child is wholly dependent on his parents and the associated home life for all his early concepts of everything intellectual, social, moral, and even spiritual since the family represents to the young child all that he can first know of either human or divine relationships. **The child must derive his first impressions of the universe from the mother's care; he is wholly dependent on the earthly father for his first ideas of the heavenly Father.** The child's subsequent life is made happy or unhappy, easy or difficult, in accordance with his early mental and emotional life, conditioned by these social and spiritual relationships of the home. A human being's entire afterlife is enormously influenced by what happens during the first few years of existence.

Changing Responses to Urges

Page 1095:1   *Paper 100:1.6*

Religious experience is markedly influenced by physical health, inherited temperament, and social environment. But these temporal conditions do not inhibit inner spiritual progress by a soul dedicated to the doing of the will of the Father in heaven. There are present in all normal mortals certain innate drives toward growth and self-realization which function if they are not specifically inhibited. The certain technique of fostering this constitutive endowment of the potential of spiritual growth is to maintain an attitude of wholehearted devotion to supreme values.

Page 1572:8   *Paper 140:4.8*

An effective philosophy of living is formed by a combination of cosmic insight and the total of one's emotional reactions to the social and economic environment. Remember: **While inherited urges cannot be fundamentally modified, emotional responses to such urges can be changed; therefore the moral nature can be modified, character can be improved.** In the strong character emotional responses are integrated and co-ordinated, and thus is produced a unified personality. Deficient unification weakens the moral nature and engenders unhappiness.

Spiritual Economy

Page 1095:3   *Paper 100:1.8*

See "Spiritual Economy" on page 165 for full text references. (14:4.4)

3. CONCEPTS OF SUPREME VALUE

Personality—Civil War
Civil war in the personality.

Page 1097:2   *Paper 100:3.5*

Values can never be static; reality signifies change, growth. Change without growth, expansion of meaning and exaltation of value, is valueless—**is potential evil.** The greater the quality of cosmic adaptation, the more of meaning any experience possesses. Values are not conceptual illusions; they are real, but always they depend on the fact of relationships. Values are always both actual and potential—not what was, but what is and is to be.

Page 1097:5   *Paper 100:4.1*

Religious living is devoted living, and devoted living is creative living, original and spontaneous. New religious insights arise out of conflicts which initiate the choosing of new and better reaction habits in the place of older and inferior reaction patterns. **New meanings only emerge amid conflict; and conflict persists only in the face of refusal to espouse the higher values connoted in superior meanings.**

Page 1220:10   *Paper 111:4.11*

This is the problem: If freewill man is endowed with the powers of creativity in the inner man, then must we recognize that freewill creativity embraces the potential of freewill destructivity. And when creativity is turned to destructivity, you are face to face with the devastation of evil and sin—oppression, war, and destruction. Evil is a partiality of creativity which tends toward disintegration and eventual destruction. **All conflict is evil in that it inhibits the creative function of the inner life—it is a species of civil war in the personality.**

4. PROBLEMS OF GROWTH

Spiritual Wisdom—Divine Destinies to Human Origins Perspective

Page 1097:5   *Paper 100:4.1*

See "Spiritual Wisdom—Divine Destinies to Human Origins Perspective" on page 138 for full text references. (3:6.3)

Personality—Civil War

Page 1097:5   *Paper 100:4.1*

See "Personality—Civil War" on page 372 for full text references. (100:3.5)

6. MARKS OF RELIGIOUS LIVING

Cosmic Citizenship

Page 1100:5   *Paper 100:6.3*

See "Cosmic Citizenship" on page 241 for full text references. (39:4.9)

PAPER 101: THE REAL NATURE OF RELIGION

1. TRUE RELIGION

True Religion

Page 1104:4   *Paper 101:1.1*

True religion is not a system of philosophic belief which can be reasoned out and substantiated by natural proofs, neither is it a fantastic and mystic experience of indescribable feelings of ecstasy which can be enjoyed only by the romantic devotees of mysticism. Religion is not the product of reason, but viewed from within, it is altogether reasonable. Religion is not derived from the logic of human philosophy, but as a mortal experience it is altogether logical. **Religion is the experiencing of divinity in the consciousness of a moral being of evolutionary origin; it represents true experience with eternal realities in time, the realization of spiritual satisfactions while yet in the flesh.**

Page 1728:2   *Paper 155:4.2*

While pausing for lunch under the shadow of an overhanging ledge of rock, near Luz, **Jesus delivered one of the most remarkable addresses which his apostles ever listened to throughout all their years of association with him.** No sooner had they seated themselves to break bread than Simon Peter asked Jesus: "Master, since the Father in heaven knows all things, and since his spirit is our support in the establishment of the kingdom of heaven on earth, why is it that we flee from the threats of our enemies? Why do we refuse to confront the foes of truth?" But before Jesus had begun to answer Peter's question, Thomas broke in, asking: "Master, I should really like to know just what is wrong with the religion of our enemies at Jerusalem. **What is the real difference between their religion and ours?** Why is it we are at such diversity of belief when we all profess to serve the same God?" And when Thomas had finished, Jesus said: "While I would not ignore Peter's question, knowing full well how easy it would be to misunderstand my reasons for avoiding an open clash with the rulers of the Jews at just this time, still it will prove more helpful to all of you if I choose rather to answer Thomas's question. And that I will proceed to do when you have finished your lunch."

Page 1728:3–7   *Paper 155:5.1–5*

This memorable discourse on religion, summarized and restated in modern phraseology, gave expression to the following truths:
While the religions of the world have a double origin—natural and revelatory—at any one time and among any one people there are to be found three distinct forms of religious devotion. And these three manifestations of the religious urge are:
1. *Primitive religion.* The seminatural and instinctive urge to fear mysterious energies and worship superior forces, chiefly a religion of the physical nature, the religion of fear.
2. *The religion of civilization.* The advancing religious concepts and practices of the civilizing races—the religion of the mind—the intellectual theology of the authority of established religious tradition.

3. *True religion*—**the religion of revelation.** The revelation of supernatural values, a partial insight into eternal realities, a glimpse of the goodness and beauty of the infinite character of the Father in heaven—**the religion of the spirit as demonstrated in human experience.**

2. THE FACT OF RELIGION

Spirt of Truth and Divine Fragment

Page 1106:3   *Paper 101:2.4*

See "Spirt of Truth and Divine Fragment" on page 118 for full text references. (2:0.3)

Revelation—Written Word

Page 1106:3   *Paper 101:2.4*

See "Revelation—Written Word" on page 348 for full text references. (92:4.9)

Truth Sensitivity

Page 1106:9   *Paper 101:2.8*

See "Truth Sensitivity" on page 275 for full text references. (48: 6.17)

3. THE CHARACTERISTICS OF RELIGION

Cosmic Mind

Page 1108:1   *Paper 101:3.2*

See "Cosmic Mind" on page 174 for full text references. (16:9.1–2)

4. THE LIMITATIONS OF REVELATION

Revelation—Unrevealed Concepts

Page 1109:3   *Paper 101:4.2*

See "Revelation—Unrevealed Concepts" on page 203 for full text references. (30:0.2)

Revelation—Written Word

Page 1109:3   *Paper 101:4.2*

See "Revelation—Written Word" on page 348 for full text references. (92:4.9)

6. PROGRESSIVE RELIGIOUS EXPERIENCE

Cosmic Citizenship

Page 1112:4   *Paper 101:6.8*

See "Cosmic Citizenship" on page 241 for full text references. (39:4.9)

9. RELIGION AND MORALITY

Revelation—Written Word

Page 1115:2   *Paper 101:9.1*

See "Revelation—Written Word" on page 348 for full text references. (92:4.9)

3. KNOWLEDGE, WISDOM, AND INSIGHT

Revelation—Substitute for Morontia Mota

Page 1122:1   *Paper 102:3.5*

See "Revelation—Substitute for Morontia Mota" on page 139 for full text references. (4:2.7)

Feelings That Lie Too Deep for Words

Page 1122:8   *Paper 102:3.12*

See "Feelings That Lie Too Deep for Words" on page 370 for full text references. (99:5.9)

4. THE FACT OF EXPERIENCE

Mind—Exchange for the Mind of Jesus

Page 1123:1   *Paper 102:4.1*

See "Mind—Exchange for the Mind of Jesus" on page 274 for full text references. (48:6.15)

6. THE CERTAINTY OF RELIGIOUS FAITH

Kingdom Within

Page 1124:3   *Paper 102:6.1*

See "Kingdom Within" on page 219 for full text references. (34:6.7)

Supreme Mind

Page 1125:5   *Paper 102:6.10*

See "Supreme Mind" on page 298 for full text references. (56:2.3)

PAPER 103: THE REALITY OF RELIGIOUS EXPERIENCE

2. RELIGION AND THE INDIVIDUAL

Personality—When Is it Bestowed?

Page 1130:6   *Paper 103:2.1*

> See "Personality—When Is it Bestowed?" on page 195 for full text references. (24:2.8)

Cooperation or Resistance to Divine Leading— Survival of Immortal Soul

Page 1130:6   *Paper 103:2.1*

> See "Cooperation or Resistance to Divine Leading—Survival of Immortal Soul" on page 205 for full text references. (30:4.8)

Born of the Spirit
Spiritual birthday and Rodan's views on gaining entrance into the brotherhood.

Page 1130:6   *Paper 103:2.1*

Religion is functional in the human mind and has been realized in experience prior to its appearance in human consciousness. A child has been in existence about nine months before it experiences *birth*. **But the "birth" of religion is not sudden; it is rather a gradual emergence. Nevertheless, sooner or later there is a "birth day." You do not enter the kingdom of heaven unless you have been "born again"—born of the Spirit.** Many spiritual births are accompanied by much anguish of spirit and marked psychological perturbations, as many physical births are characterized by a "stormy labor" and other abnormalities of "delivery." Other spiritual births are a natural and normal growth of the recognition of supreme values with an enhancement of spiritual experience, albeit no religious development occurs without conscious effort and positive and individual determinations. Religion is never a passive experience, a negative attitude. What is termed the "birth of religion" is not directly associated with so-called conversion experiences which usually characterize religious episodes occurring later in life as a result of mental conflict, emotional repression, and temperamental upheavals.

Page 1774:6   *Paper 160:1.14*

In a continually changing world, in the midst of an evolving social order, it is impossible to maintain settled and established goals of destiny. Stability of personality can be experienced only by those who have discovered and embraced the living God as the eternal goal of infinite attainment. **And thus to transfer one's goal from time to eternity, from earth to Paradise, from the human to the divine, requires that man shall become regenerated, converted, be born again; that he shall become the re-created child of the divine spirit; that he shall gain entrance into the brotherhood of the kingdom of heaven.** All philosophies and religions which fall short of these ideals are immature. The philosophy which I teach, linked with the gospel which you preach, represents the new religion of maturity, the ideal of all future generations. And this is true because our ideal is final, infallible, eternal, universal, absolute, and infinite.

---

4. Spiritual Communion

God-consciousness—Prelude to Worship

Page 1133:1   *Paper 103:4.1*

See "God-consciousness—Prelude to Worship" on page 358 for full text references. (95:4.2)

Sense of Guilt
Sense of guilt; we are human.

Page 1133:2–3   *Paper 103:4.2–3*

When primitive man felt that his communion with God had been interrupted, he resorted to sacrifice of some kind in an effort to make atonement, to restore friendly relationship. The hunger and thirst for righteousness leads to the discovery of truth, and truth augments ideals, and this creates new problems for the individual religionists, **for our ideals tend to grow by geometrical progression, while our ability to live up to them is enhanced only by arithmetical progression.**

The sense of guilt (not the consciousness of sin) comes either from interrupted spiritual communion or from the lowering of one's moral ideals. Deliverance from such a predicament can only come through the realization that one's highest moral ideals are not necessarily synonymous with the will of God. **Man cannot hope to live up to his highest ideals, but he can be true to his purpose of finding God and becoming more and more like him.**

Page 1739:3   *Paper 156:5.8*

Do not become discouraged by the discovery that you are human. Human nature may tend toward evil, but it is not inherently sinful. Be not downcast by your failure wholly to forget some of your regrettable experiences. The mistakes which you fail to forget in time will be forgotten in eternity. Lighten your burdens of soul by speedily acquiring a long-distance view of your destiny, a universe expansion of your career.

---

5. THE ORIGIN OF IDEALS

"Whosoever Saves His Life Selfishly..."

Page 1134:4   *Paper 103:5.6*

The attempt to secure equal good for the self and for the greatest number of other selves presents a problem which cannot always be satisfactorily resolved in a time-space frame. Given an eternal life, such antagonisms can be worked out, but in one short human life they are incapable of solution. Jesus referred to such a paradox when he said: **"Whosoever shall save his life shall lose it, but whosoever shall lose his life for the sake of the kingdom, shall find it."**

Page 1760:2   *Paper 158:7.5*

After they had recovered from the first shock of Jesus' stinging rebuke, and before they resumed their journey, the Master spoke further: "If any man would come after me, let him disregard himself, take up his responsibilities daily, and follow me. **For whosoever would save his life selfishly, shall lose it, but whosoever loses his life for my sake and the gospel's, shall save it.** What does it profit a man to gain the whole world and lose his own soul? What would a man give in exchange for eternal life? Be not ashamed of me and my words in this sinful and hypocritical generation, even as I will not be ashamed to acknowledge you when in glory I appear before my Father in the presence of all the celestial hosts. Nevertheless, many of you now standing before me shall not taste death till you see this kingdom of God come with power."

---

6. PHILOSOPHIC CO-ORDINATION

Truth Sensitivity

Page 1136:3   *Paper 103:6.8*

See "Truth Sensitivity" on page 275 for full text references. (48: 6.17)

PAPER 104: GROWTH OF THE TRINITY CONCEPT

1. URANTIAN TRINITY CONCEPTS

Three Concentric Circles

Page 1143:6 *Paper 104:1.3*

See "Three Concentric Circles" on page 288 for full text references. (53:5.4)

Revelation—Fifth Epochal Revelation

Page 1145:1 *Paper 104:1.13*

See "Revelation—Fifth Epochal Revelation" on page 346 for full text references. (92:4.4)

4. THE SEVEN TRIUNITIES

Seven Triunities

Page 1147:11, 1148:1–2 *Paper 104:4.1–3*

See "Seven Triunities" on page 163 for full text references. (10:5.7)]

5. TRIODITIES

The Triodities
Non-Father triune relationships, the triodities.

Page 1151:1–11 *Paper 104:5.1–5*

There are certain other triune relationships which are non-Father in constitution, but they are not real triunities, and they are always distinguished from the Father triunities. They are called variously, associate triunities, co-ordinate triunities, and *triodities.* They are consequential to the existence of the triunities. **Two of these associations are constituted as follows:**

The Triodity of Actuality. **This triodity consists in the interrelationship of the three absolute actuals:**

1. The Eternal Son.
2. The Paradise Isle.
3. The Conjoint Actor.

The Eternal Son is the absolute of spirit reality, the absolute personality. The Paradise Isle is the absolute of cosmic reality, the absolute pattern. The Conjoint Actor is the absolute of mind reality, the co-ordinate of absolute spirit reality, and the existential Deity synthesis of personality and power. This triune association eventuates the co-ordination of the sum total of actualized reality—spirit, cosmic, or mindal. It is unqualified in actuality.

The Triodity of Potentiality. **This triodity consists in the association of the three Absolutes of potentiality:**

1. The Deity Absolute.
2. The Universal Absolute.
3. The Unqualified Absolute.

Thus are interassociated the infinity reservoirs of all latent energy reality—spirit, mindal, or cosmic. This association yields the integration of all latent energy reality. It is infinite in potential.

Page 1165:5   *Paper 106:2.8*

Within the completed power-personality synthesis of the Supreme Being there will be associated all of the absoluteness of the several triodities which could be so associated, and this majestic personality of evolution will be experientially attainable and understandable by all finite personalities. When ascenders attain the postulated seventh stage of spirit existence, they will therein experience the realization of a new meaning-**value of the absoluteness and infinity of the triodities** as such is revealed on subabsolute levels in the Supreme Being, who is experiencible. But the attainment of these stages of maximum development will probably await the co-ordinate settling of the entire grand universe in light and life.

Page 1262:2,4–5   *Paper 115:3.5,7–8*

One basic conception of the absolute level involves a postulate of three phases:

2. *The Actual.* The union of the three Absolutes of actuality, the Second, Third, and Paradise Sources and Centers. **This triodity** of the Eternal Son, the Infinite Spirit, and the Paradise Isle constitutes the actual revelation of the originality of the First Source and Center.

3. *The Potential.* The union of the three Absolutes of potentiality, the Deity, Unqualified, and Universal Absolutes. **This triodity** of existential potentiality constitutes the potential revelation of the originality of the First Source and Center.

PAPER 105: DEITY AND REALITY

---

1. THE PHILOSOPHIC CONCEPT OF THE I AM

I AM

Page 1152:4–5, Page 1153:2   *Paper 105:1.1–2,5*

See "I AM" on page 102 for full text references. (0:3.13, 15)

Prereality

Page 1153:2   *Paper 105:1.5*

See "Prereality" on page 252 for full text references. (42:2.5)

---

2. THE I AM AS TRIUNE AND AS SEVENFOLD

Prereality

Page 1154:2   *Paper 105:2.3*

See "Prereality" on page 252 for full text references. (42:2.5)

Supergravity Presence

Page 1155:3   *Paper 105:2.10*

See "Supergravity Presence" on page 108 for full text references. (0:11.7)

---

3. THE SEVEN ABSOLUTES OF INFINITY

Unqualified Absolute—Space Presence

Page 1156:4   *Paper 105:3.7*

See "Unqualified Absolute—Space Presence" on page 190 for full text references. (21:2.12)

5. Promulgation of Finite Reality

Tertiary Maximums and Finites

Page 1158:10   *Paper 105:5.8*

We speak of the perfect and the perfected as primary and secondary maximums, but there is still another type: Trinitizing and other relationships between the primaries **and the secondaries result in the appearance of** *tertiary maximums*—things, meanings, and values that are neither perfect nor perfected yet are co-ordinate with both ancestral factors.

Page 1163:14   *Paper 106:1.1*

The primary or spirit-origin phases of finite reality find immediate expression on creature levels as perfect personalities and on universe levels as the perfect Havona creation. Even experiential Deity is thus expressed in the spirit person of God the Supreme in Havona. But the secondary, evolutionary, time-and-matter-conditioned phases of the finite become cosmically integrated only as a result of growth and attainment. Eventually all secondary or perfecting finites are to attain a level equal to that of primary perfection, but such destiny is subject to a time delay, a constitutive superuniverse qualification which is not genetically found in the central creation. (**We know of the existence of tertiary finites, but the technique of their integration is as yet unrevealed.**)

7. Eventuation of Transcendentals

Transcendence of Finite Limitations

Page 1160:2   *Paper 105:7.4*

Among those realities which are associated with the **transcendental level** are the following:
1. The Deity presence of the Ultimate.
2. The concept of the master universe.
3. The Architects of the Master Universe.
4. The two orders of Paradise force organizers.
5. **Certain modifications in space potency.**
6. **Certain values of spirit.**
7. **Certain meanings of mind.**
8. **Absonite qualities and realities.**
9. **Omnipotence, omniscience, and omnipresence.**
10. **Space.**

Page 1281:5–6   *Paper 117:3.3–4*

Said Jesus: "I am the living way," and so he is the living way from the material level of self-consciousness to the spiritual level of God-consciousness. And even as he is this living way of ascension from the self to God, so is the Supreme the living way from finite consciousness to **transcendence of consciousness, even to the insight of absonity.**

Your Creator Son can actually be such a living channel from humanity to divinity since he has personally experienced the fullness of the traversal of this universe path of progression, from the true humanity of Joshua ben Joseph, the Son of Man, to the Paradise divinity of Michael of Nebadon, the Son of the infinite God. **Similarly can the Supreme Being function as the universe approach to the transcendence of finite limitations, for he is the actual embodiment and personal epitome of all creature evolution, progression, and spiritualization.** Even the grand universe experiences of the descending personalities from Paradise are that part of his experience which is complemental to his summation of the ascending experiences of the pilgrims of time.

Page 1291:10   *Paper 117:7.6*

It may be that on the upper limits of the finite, where time conjoins transcended time, there is some sort of blurring and blending of sequence. **It may be that the Supreme is able to forecast his universe presence onto these supertime levels and then to a limited degree anticipate future evolution by reflecting this future forecast back to the created levels as the Immanence of the Projected Incomplete.** Such phenomena may be observed wherever finite makes contact with superfinite, as in the experiences of human beings who are indwelt by Thought Adjusters that are veritable predictions of man's future universe attainments throughout all eternity.

PAPER 106: UNIVERSE LEVELS OF REALITY

Spiritual Wisdom—Divine Destinies to Human Origins Perspective

Page 1162:1   *Paper 106:0.1*

See "Spiritual Wisdom—Divine Destinies to Human Origins Perspective" on page 138 for full text references. (3:6.3)

1. PRIMARY ASSOCIATION OF FINITE FUNCTIONALS

Tertiary Maximums and Finites

Page 1163:14   *Paper 106:1.1*

See "Tertiary Maximums and Finites" on page 384 for full text references. (105:5.8)

2. SECONDARY SUPREME FINITE INTEGRATION

The Triodities

Page 1165:5   *Paper 106:2.8*

See "The Triodities" on page 381 for full text references. (104:5.1–5)

4. ULTIMATE QUARTAN INTEGRATION

Qualified Vicegerents of the Ultimate

Page 1166:6   *Paper 106:4.3*

See "Qualified Vicegerents of the Ultimate" on page 171 for full text references. (15:10.7)

5. COABSOLUTE OR FIFTH-PHASE ASSOCIATION

Consummator of Universe Destiny

Page 1167:2   *Paper 106:5.1*

See "Consummator of Universe Destiny" on page 109 for full text references. (0:12.7)

Consummator of Universe Destiny

Page 1169:2   *Paper 106:7.3*

See "Consummator of Universe Destiny" on page 109 for full text references. (0:12.7)

Consummator of Universe Destiny

Page 1171:4   *Paper 106:8.8*

See "Consummator of Universe Destiny" on page 109 for full text references. (0:12.7)

Majeston

Page 1172:5   *Paper 106:8.17*

See "Majeston" on page 176 for full text references. (17:2.2)

Kingdom Within

Page 1175:1   *Paper 106:9.12*

See "Kingdom Within" on page 219 for full text references. (34:6.7)

Irrevocable Choice

Page 1182:4   *Paper 107:6.2*

See "Irrevocable Choice" on page 143 for full text references. (5:1.11)

7. ADJUSTERS AND PERSONALITY

Father Contact

Page 1183:3   *Paper 107:7.1*

Page 1183:6   *Paper 107:7.4*

See "Father Contact" on page 114 for full text references. (1:7.1)

Impersonal Entities, Mind, and Mortal Will

Page 1183:6   *Paper 107:7.4*

See "Impersonal Entities, Mind, and Mortal Will" on page 159 for full text references. (9:5.4)

PAPER 108: MISSION AND MINISTRY OF THOUGHT ADJUSTERS

God—Foreknowledge

Page 1185:2   *Paper 108:0.2*

See "God—Foreknowledge" on page 133 for full text references. (3:2.6)

2. PREREQUISITES OF ADJUSTER INDWELLING

Thought Adjusters—Preparation for Bestowal

Page 1187:1   *Paper 108:2.2*

See "Thought Adjusters—Preparation for Bestowal" on page 218 for full text references. (34:5.3)

Thought Adjusters—Preparation for Bestowal

Page 1187:4   *Paper 108:2.5*

See "Thought Adjusters—Preparation for Bestowal" on page 218 for full text references. (34:5.3)

3. ORGANIZATION AND ADMINISTRATION

Andon and Fonta—Receiving Universe Names

Page 1188:5   *Paper 108:3.3*

See "Andon and Fonta—Receiving Universe Names" on page 304 for full text references. (63:0.2–3)

Tabamantia—Agondonter of Finaliter Status

Page 1189:1   *Paper 108:3.5*

See "Tabamantia—Agondonter of Finaliter Status" on page 279 for full text references. (49:5.5)

PAPER 109: RELATION OF ADJUSTERS TO UNIVERSE CREATURES

2. SELF-ACTING ADJUSTERS

Spiritual Economy

Page 1196:9   *Paper 109:2.7*

See "Spiritual Economy" on page 165 for full text references. (14:4.4)

Thought Adjusters—Self Acting

Page 1196:3,7, 1197:1   *Paper 109:2.1,5,10*

See "Thought Adjusters—Self Acting" on page 332 for full text references. (77:7.5)

3. RELATION OF ADJUSTERS TO MORTAL TYPES

Thought Adjusters—Preparation for Bestowal

Page 1198:2   *Paper 109:3.7*

See "Thought Adjusters—Preparation for Bestowal" on page 218 for full text references. (34:5.3)

7. DESTINY OF PERSONALIZED ADJUSTERS

Most Highs on Urantia—Times of Crisis

Page 1201:8   Paper 109:7.7

See "Most Highs on Urantia—Times of Crisis" on page 226 for full text references. (35:5.5–6)

PAPER 110: RELATION OF ADJUSTERS TO INDIVIDUAL MORTALS

Impersonal Entities, Mind, and Mortal Will

Page 1203:2   Paper 110:0.2

See "Impersonal Entities, Mind, and Mortal Will" on page 159 for full text references. (9:5.4)

---

2. ADJUSTERS AND HUMAN WILL

Predestination

Page 1204:5   Paper 110:2.1

See "Predestination" on page 136 for full text references. (3:5.3)

---

3. CO-OPERATION WITH THE ADJUSTER

Cooperation or Resistance to Divine Leading— Survival of Immortal Soul

Page 1206:3   Paper 110:3.5

See "Cooperation or Resistance to Divine Leading—Survival of Immortal Soul" on page 205 for full text references. (30:4.8)

Survival—Co-operation with Indwelling Adjuster

Page 1206:3   Paper 110:3.5

See "Survival—Co-operation with Indwelling Adjuster" on page 245 for full text references. (40:5.16)

Cosmic Citizenship

Page 1206:4–8   *Paper 110:3.6–10*
    See "Cosmic Citizenship" on page 241 for full text references. (39:4.9)

6. THE SEVEN PSYCHIC CIRCLES

Cosmic Citizenship

Page 1211:1   *Paper 110:6.16*
    See "Cosmic Citizenship" on page 241 for full text references. (39:4.9)

PAPER 111: THE ADJUSTER AND THE SOUL

4. THE INNER LIFE

Personality

Page 1220:5   *Paper 111:4.6*
    See "Personality" on page 148 for full text references. (5:6.10)

Personality—Civil War

Page 1220:10   *Paper 111:4.11*
    See "Personality—Civil War" on page 372 for full text references. (100:3.5)

PAPER 112: PERSONALITY SURVIVAL

Gift to God

Page 1225:2,10   *Paper 112:0.2,10*
    See "Gift to God" on page 111 for full text references. (1:1.2)

Personality

Page 1225:2–16   *Paper 112:0.2–16*

    See "Personality" on page 148 for full text references. (5:6.10)

2. THE SELF

Personality Relationships

Page 1228:3   *Paper 112:2.4*

    See "Personality Relationships" on page 248 for full text references. (40:9.7)

3. THE PHENOMENON OF DEATH

Irrevocable Choice

Page 1229:9   *Paper 112:3.2*

    See "Irrevocable Choice" on page 143 for full text references. (5:1.11)

4. ADJUSTERS AFTER DEATH

Third Day—Third Period

Page 1232:1   *Paper 112:4.6*

    See "Third Day—Third Period" on page 204 for full text references. (30:4.4–5)

5. SURVIVAL OF THE HUMAN SELF

Personality—When Is it Bestowed?

Page 1232:5   *Paper 112:5.4*

    See "Personality—When Is it Bestowed?" on page 195 for full text references. (24:2.8)

Transition Mortals

Page 1233:3   *Paper 112:5.7*

See "Transition Mortals" on page 269 for full text references. (47:1.1)

Cooperation or Resistance to Divine Leading—Survival of Immortal Soul

Page 1233:5   *Paper 112:5.9*

See "Cooperation or Resistance to Divine Leading—Survival of Immortal Soul" on page 205 for full text references. (30:4.8)

Survival—Co-operation with Indwelling Adjuster

Page 1233:5   *Paper 112:5.9*

See "Survival—Co-operation with Indwelling Adjuster" on page 245 for full text references. (40:5.16)

Personality Relationships

Page 1235:4   *Paper 112:5.22*

See "Personality Relationships" on page 248 for full text references. (40:9.7)

7. ADJUSTER FUSION

Irrevocable Choice

Page 1237:7   *Paper 112:7.5*

See "Irrevocable Choice" on page 143 for full text references. (5:1.11)

Corps of the Finality—Future Destiny

Page 1239:5   *Paper 112:7.16*

See "Corps of the Finality—Future Destiny" on page 209 for full text references. (31:10.12)

PAPER 113: SERAPHIC GUARDIANS OF DESTINY

3. RELATION TO OTHER SPIRIT INFLUENCES

Ministering Spirits from the Conjoint Actor— Relation with Humans

Page 1244:2   *Paper 113:3.1*

See "Ministering Spirits from the Conjoint Actor—Relation with Humans" on page 145 for full text references. (5:3.5)

5. SERAPHIC MINISTRY TO MORTALS

Human Will—Sovereignty

Page 1245:7   *Paper 113:5.1*

See "Human Will—Sovereignty" on page 147 for full text references. (5:6.8)

6. GUARDIAN ANGELS AFTER DEATH

Third Day—Third Period

Page 1247:2   *Paper 113:6.4*

See "Third Day—Third Period" on page 204 for full text references. (30:4.4–5)

PAPER 114: SERAPHIC PLANETARY GOVERNMENT

2. THE BOARD OF PLANETARY SUPERVISORS

Lucifer Rebellion—Panoptians

Page 1252:2   *Paper 114:2.4*

See "Lucifer Rebellion—Panoptians" on page 289 for full text references. (53:7.1)

4. THE MOST HIGH OBSERVER

Race Commissioners

Page 1253:5   *Paper 114:4.2*

See "Race Commissioners" on page 273 for full text references. (48:6.11)

Most Highs on Urantia—Times of Crisis

Page 1253:7   *Paper 114:4.4*

See "Most Highs on Urantia—Times of Crisis" on page 226 for full text references. (35:5.5–6)

6. THE MASTER SERAPHIM OF PLANETARY SUPERVISION

Most Highs

Page 1255:3,7   *Paper 114:6.4,8*

See "Most Highs" on page 126 for full text references. (3:1.10)

Iknaton and the Angels of the Churches

Page 1255:6   *Paper 114:6.7*

See "Iknaton and the Angels of the Churches" on page 360 for full text references. (95:5.2)

Education—Purpose

Page 1255:7   *Paper 114:6.8*

See "Education—Purpose" on page 126 for full text references.(2:7.12)

Race Commissioners

Page 1255:8   *Paper 114:6.9*

See "Race Commissioners" on page 273 for full text references. (48:6.11)

3. Original, Actual, and Potential

The Triodities

Page 1262:2,4–5   *Paper 115:3.5,7–8*
See "The Triodities" on page 381 for full text references. (104:5.1–5)

God—Foreknowledge

Page 1262:3   *Paper 115:3.6*

Page 1262:9   *Paper 115:3.12*
See "God—Foreknowledge" on page 133 for full text references. (3:2.6)

Paper 116: The Almighty Supreme

1. The Supreme Mind

Supreme Mind

Page 1269:1–2,4   *Paper 116:1.2–3,5*
See "Supreme Mind" on page 298 for full text references. (56:2.3)

4. The Almight and the Supreme Creators

Supreme Mind

Page 1272:2   *Paper 116:4.3*
See "Supreme Mind" on page 298 for full text references. (56:2.3)

3. SIGNIFICANCE OF THE SUPREME TO UNIVERSE CREATURES

Transcendence of Finite Limitations

Page 1281:5–6   Paper 117:3.3–4

See "Transcendence of Finite Limitations" on page 384 for full text references. (105:7.4)

4. THE FINITE GOD

Cosmic Citizenship

Page 1284:5–6   Paper 117:4.9–10

See "Cosmic Citizenship" on page 241 for full text references. (39:4.9)

6. THE QUEST FOR THE SUPREME

Personality Circuit as Divine Love

Page 1289:3   Paper 117:6.10

See "Personality Circuit as Divine Love" on page 149 for full text references. (5:6.12)

7. THE FUTURE OF THE SUPREME

Qualified Vicegerents of the Ultimate

Page 1291:7–8   Paper 117:7.3–4

See "Qualified Vicegerents of the Ultimate" on page 171 for full text references. (15:10.7)

Transcendence of Finite Limitations

Page 1291:10   *Paper 117:7.6*

See "Transcendence of Finite Limitations" on page 384 for full text references. (105:7.4)

Corps of the Finality—Allegiance to the Paradise Trinity

Page 1292:1   *Paper 117:7.7*

See "Corps of the Finality—Allegiance to the Paradise Trinity" on page 208 for full text references. (31:0.6)

Relationship Between the Supreme and Superuniverse Citizens

Page 1292:10   *Paper 117:7.13*

See "Relationship Between the Supreme and Superuniverse Citizens" on page 166 for full text references. (14:4.6)

PAPER 118: SUPREME AND ULTIMATE—TIME AND SPACE

---

1. TIME AND ETERNITY

Patience and Forbearance

Page 1295:6   *Paper 118:1.6*

See "Patience and Forbearance" on page 295 for full text references. (54:5.4)

---

3. TIME-SPACE RELATIONSHIPS

Pattern and Space

Page 1297:8   *Paper 118:3.7*

See "Pattern and Space" on page 158 for full text references. (9:4.4)

4. Primary and Secondary Causation

Unqualified Absolute—Space Presence

Page 1298:6   *Paper 118:4.6*

See "Unqualified Absolute—Space Presence" on page 190 for full text references. (21:2.12)

7. Omniscience and Predestination

God—Foreknowledge

Page 1300:5   *Paper 118:7.1*

See "God—Foreknowledge" on page 133 for full text references. (3:2.6)

Sin

Page 1301:1   *Paper 118:7.4*

See "Sin" on page 293 for full text references. (54:0.1)

9. Universe Mechanisms

Supreme Mind

Page 1303:5   *Paper 118:9.4*

See "Supreme Mind" on page 298 for full text references. (56:2.3)

Paper 119:The Bestowals of Christ Michael

3. The Third Bestowal

Material Sons and Daughters—Difficulties

Page 1313:2   *Paper 119:3.7*

See "Material Sons and Daughters—Difficulties" on page 326 for full text references. (74:1.6)

4. THE FOURTH BESTOWAL

Creator Son—Fourth Creature Bestowal

Page 1313:5–5   *Paper 119:4.2–3*

    See "Creator Son—Fourth Creature Bestowal" on page 240 for full text references. (39:1.15)

7. THE SEVENTH AND FINAL BESTOWAL

Bestowals and Offspring

Page 1316:6   *Paper 119:7.3*

    See "Bestowals and Offspring" on page 184 for full text references. (20:6.2)

PAPER 120: THE BESTOWAL OF MICHAEL ON URANTIA

Order of "Days"

Page 1324:3   *Paper 120:0.6*

    See "Order of "Days"" on page 178 for full text references. (18:6.1)

1. THE SEVENTH BESTOWAL COMMISSION

Gabriel and Father Melchizedek

Page 1326:1   *Paper 120:1.4*

    See "Gabriel and Father Melchizedek" on page 216 for full text references. (33:6.2)

Creator Sons—Sovereignty

Page 1326:3   *Paper 120:1.6*

    See "Creator Sons—Sovereignty" on page 191 for full text references. (21:3.2)

401

3. FURTHER COUNSEL AND ADVICE

Bestowals and Offspring

Page 1330:3   *Paper 120:3.8*

See "Bestowals and Offspring" on page 184 for full text references. (20:6.2)

4. THE INCARNATION—MAKING TWO ONE

God and Man—Combined Being

Page 1331:2   *Paper 120:4.2*

See "God and Man—Combined Being" on page 179 for full text references. (19:3.7)

Incarnational Bestowals of Paradise Sons— Mysterious and Miraculous

Page 1331:5   *Paper 120:4.5*

See "Incarnational Bestowals of Paradise Sons—Mysterious and Miraculous" on page 183 for full text references. (20:6.1)

PAPER 122: BIRTH AND INFANCY OF JESUS

Creator Sons—Selection of Bestowal Race

Page 1344:2   *Paper 122:0.2*

See "Creator Sons—Selection of Bestowal Race" on page 192 for full text references. (21:4.3)

2. GABRIEL APPEARS TO ELIZABETH

John the Baptist—Distant Cousins

Page 1346:1   *Paper 122:2.6*

See "John the Baptist—Distant Cousins" on page 265 for full text references. (45:4.17)

3. GABRIEL'S ANNOUNCEMENT TO MARY

Joshua and Joshua ben Joseph

Page 1346:4   *Paper 122:3.1*

See "Joshua and Joshua ben Joseph" on page 365 for full text references. (96:6.3)

PAPER 123: THE EARLY CHILDHOOD OF JESUS

Iknaton and the Angels of the Churches

Page 1355:3   *Paper 123:0.3*

See "Iknaton and the Angels of the Churches" on page 360 for full text references. (95:5.2)

PAPER 124: THE LATER CHILDHOOD OF JESUS

6. THE JOURNEY TO JERUSALEM

Joshua and Joshua ben Joseph

Page 1375:1   *Paper 124:6.7*

See "Joshua and Joshua ben Joseph" on page 365 for full text references. (96:6.3)

"Be About Your Father's Business"

Page 1376:1   Paper 124:6.15

See ""Be About Your Father's Business"" on page 232 for full text references. (37:2.8)

PAPER 126: THE TWO CRUCIAL YEARS

3. THE FIFTEENTH YEAR (A.D. 9)

"Be About Your Father's Business"

Page 1389:8   Paper 126:3.5

See ""Be About Your Father's Business"" on page 232 for full text references. (37:2.8)

4. FIRST SERMON IN THE SYNAGOGUE

"The Spirit of the Lord Is Upon Me"

Page 1391:5–6   Paper 126:4.1–2

See ""The Spirit of the Lord Is Upon Me"" on page 367 for full text references. (97:5.3)

Commandments

Page 1392:7   Paper 126:4.9

See "Commandments" on page 311 for full text references. (66:7.8–16)

PAPER 127: THE ADOLESCENT YEARS

6. HIS TWENTIETH YEAR (A.D. 14)

Bestowals and Offspring

Page 1404:7   *Paper 127:6.8*

> See "Bestowals and Offspring" on page 184 for full text references. (20:6.2)

Jesus—Perfecting Mortal Existence

Page 1405:4   *Paper 127:6.12*

> See "Jesus—Perfecting Mortal Existence" on page 276 for full text references. (48:7.2,7,18,24,30)

PAPER 130: ON THE WAY TO ROME

Ur and Susa

Page 1427:3   *Paper 130:0.3*

> See "Ur and Susa" on page 334 for full text references. (78:8.2)

2. AT CAESAREA

Subjective Self-consciousness

Page 1431:5   *Paper 130:2.10*

> See "Subjective Self-consciousness" on page 161 for full text references. (9:8.6)

4. DISCOURSE ON REALITY

Kingdom Within

Page 1434:2   *Paper 130:4.3*

> See "Kingdom Within" on page 219 for full text references. (34:6.7)

Relativity

Page 1436:1   *Paper 130:4.15*

See "Relativity" on page 152 for full text references. (7:1.2)

PAPER 131: THE WORLD'S RELIGIONS

Melchizedek Bestowal—Missionaries

Page 1442:1   *Paper 131:0.1*

See "Melchizedek Bestowal—Missionaries" on page 355 for full text references. (93:7.1)

---

2. JUDAISM

Melchizedek Bestowal—Missionaries

Page 1444:1   *Paper 131:2.1*

See "Melchizedek Bestowal—Missionaries" on page 355 for full text references. (93:7.1)

Commandments

Page 1446:1   *Paper 131:2.12*

See "Commandments" on page 311 for full text references. (66:7.8–16)

PAPER 132: THE SOJOURN AT ROME

2. GOOD AND EVIL

Goodness and Evil

Page 1458:2–5   *Paper 132:2.5–8*

See "Goodness and Evil" on page 140 for full text references. (4:3.6)

4. Personal Ministry

Courts and Judges—Integrity

Page 1462:1   *Paper 132:4.8*

See "Courts and Judges—Integrity" on page 320 for full text references. (70:11.10)

Paper 133: The Return from Rome

6. At Ephesus—Discourse on the Soul

Soul

Page 1478:4   *Paper 133:6.5*

See "Soul" on page 104 for full text references. (0:5.10)

7. The Sojourn at Cyprus—Discourse on Mind

Happiness

Page 1480:4   *Paper 133:7.12*

See "Happiness" on page 124 for full text references. (2:7.6)

9. In Mesopotamia

Ur and Susa

Page 1481:4   *Paper 133:9.2*

See "Ur and Susa" on page 334 for full text references. (78:8.2)

PAPER 134: THE TRANSITION YEARS

---

5. POLITICAL SOVEREIGNTY

Scaffolding—Stages of Development

Page 1488:6   *Paper 134:5.8*

See "Scaffolding—Stages of Development" on page 344 for full text references. (90:3.10)

---

6. LAW, LIBERTY, AND SOVEREIGNTY

Global Viewpoints

Page 1491:3,5   *Paper 134:6.9,11*

See "Global Viewpoints" on page 296 for full text references. (55:3.1)

PAPER 135: JOHN THE BAPTIST

---

6. JOHN BEGINS TO PREACH

Joshua and Joshua ben Joseph

Page 1501:4, 1502:5   *Paper 135:6.1,7*

See "Joshua and Joshua ben Joseph" on page 365 for full text references. (96:6.3)

PAPER 136: BAPTISM AND THE FORTY DAYS

---

2. THE BAPTISM OF JESUS

Forty Days

Page 1512:2   *Paper 136:2.6*

See "Forty Days" on page 271 for full text references. (47:8.5)

3. THE FORTY DAYS

Forty Days

Page 1512:6   *Paper 136:3.2*

See "Forty Days" on page 271 for full text references. (47:8.5)

5. THE FIRST GREAT DECISION

Angels—Twelve Legions

Page 1516:1   *Paper 136:5.1*

See "Angels—Twelve Legions" on page 237 for full text references. (38:6.2)

8. THE FOURTH DECISION

Kingdom Within

Page 1521:1   *Paper 136:8.6*

See "Kingdom Within" on page 219 for full text references. (34:6.7)

9. THE FIFTH DECISION

Kingdom Within

Page 1522:1   *Paper 136:9.2*

Page 1522:5   *Paper 136:9.6*

See "Kingdom Within" on page 219 for full text references. (34:6.7)

2. CHOOSING PHILIP AND NATHANIEL

Jesus—Follow Me
Full significance of one supreme requirement.

Page 1526:5   *Paper 137:2.5*

It suddenly dawned on Philip that Jesus was a really great man, possibly the Messiah, and he decided to abide by Jesus' decision in this matter; and he went straight to him, asking, "Teacher, shall I go down to John or shall I join my friends who follow you?" **And Jesus answered, "Follow me."** Philip was thrilled with the assurance that he had found the Deliverer.

Page 2089:3   *Paper 196:0.13*

Jesus does not require his disciples to believe in him but rather to believe *with* him, believe in the reality of the love of God and in full confidence accept the security of the assurance of sonship with the heavenly Father. The Master desires that all his followers should fully share his transcendent faith. Jesus most touchingly challenged his followers, not only to believe *what* he believed, but also to believe *as* he believed. **This is the full significance of his one supreme requirement, "Follow me."**

Page 2090:3–4   *Paper 196:1.2–3*

The time is ripe to witness the figurative resurrection of the human Jesus from his burial tomb amidst the theological traditions and the religious dogmas of nineteen centuries. Jesus of Nazareth must not be longer sacrificed to even the splendid concept of the glorified Christ. What a transcendent service if, through this revelation, the Son of Man should be recovered from the tomb of traditional theology and be presented as the living Jesus to the church that bears his name, and to all other religions! **Surely the Christian fellowship of believers will not hesitate to make such adjustments of faith and of practices of living as will enable it to "follow after"** the Master in the demonstration of his real life of religious devotion to the doing of his Father's will and of consecration to the unselfish service of man. Do professed Christians fear the exposure of a self-sufficient and unconsecrated fellowship of social respectability and selfish economic maladjustment? Does institutional Christianity fear the possible jeopardy, or even the overthrow, of traditional ecclesiastical authority if the Jesus of Galilee is reinstated in the minds and souls of mortal men as the ideal of personal religious living? Indeed, the social readjustments, the economic transformations, the moral rejuvenations, and the religious revisions of Christian civilization would be drastic and revolutionary if the living religion of Jesus should suddenly supplant the theologic religion about Jesus.

To "follow Jesus" means to personally share his religious faith and to enter into the spirit of the Master's life of unselfish service for man. One of the most important things in human living is to find out what Jesus believed, to discover his ideals, and to strive for the achievement of his exalted life purpose. Of all human knowledge, that which is of greatest value is to know the religious life of Jesus and how he lived it.

7. FOUR MONTHS OF TRAINING

Kingdom Within

Page 1535:5   *Paper 137:7.13*

See "Kingdom Within" on page 219 for full text references. (34:6.7)

8. SERMON ON THE KINGDOM

Kingdom Within

Page 1537:2   *Paper 137:8.15*

See "Kingdom Within" on page 219 for full text references. (34:6.7)

PAPER 139: THE TWELVE APOSTLES

2. SIMON PETER

Seventy Times and Seven

Page 1551:1   *Paper 139:2.5*

See "Seventy Times and Seven" on page 247 for full text references. (40:8.2)

Paper 140: The Ordination of the Twelve

---

1. Preliminary Instruction

Kingdom Within

Page 1568:5   *Paper 140:1.2*

See "Kingdom Within" on page 219 for full text references. (34:6.7)

---

4. You Are the Salt of the Earth

Changing Responses to Urges

Page 1572:8   *Paper 140:4.8*

See "Changing Responses to Urges" on page 371 for full text references. (100:1.6)

---

5. Fatherly and Brotherly Love

Patience and Forbearance

Page 1574:4   *Paper 140:5.11*

See "Patience and Forbearance" on page 295 for full text references. (54:5.4)

Paper 141: Beginning the Public Work

---

2. God's Law and the Father's Will

Kingdom Within

Page 1588:4   *Paper 141:2.1*

See "Kingdom Within" on page 219 for full text references. (34:6.7)

PAPER 142: THE PASSOVER AT JERUSALEM

---

3. THE CONCEPT OF GOD

Growth of the God Concept In Jewish Theology

Page 1598:3–9   *Paper 142:3.2–8*

See "Growth of the God Concept In Jewish Theology" on page 362 for full text references. (96:1.2–9)

Melchizedek Bestowal—Abraham and Brother

Page 1598:5   *Paper 142:3.4*

See "Melchizedek Bestowal—Abraham and Brother" on page 353 for full text references. (93:5.4)

Commandments

Page 1599:2–13   *Paper 142:3.10–21*

See "Commandments" on page 311 for full text references. (66:7.8–16)

PAPER 143: GOING THROUGH SAMARIA

---

2. LESSON ON SELF-MASTERY

Kingdom Within

Page 1609:5   *Paper 143:2.4*

See "Kingdom Within" on page 219 for full text references. (34:6.7)

2. AT JOTAPATA

Mind—Exchange for the Mind of Jesus

Page 1639:5   *Paper 146:2.10*

See "Mind—Exchange for the Mind of Jesus" on page 274 for full text references. (48:6.15)

PAPER 148: TRAINING EVANGELISTS AT BETHSAIDA

4. EVIL, SIN, AND INIQUITY

Sin

Page 1660:3   *Paper 148:4.4*

See "Sin" on page 293 for full text references. (54:0.1)

PAPER 149: THE SECOND PREACHING TOUR

5. LESSON REGARDING CONENTMENT

Happiness

Page 1674:3–4,6   *Paper 149:5.1–2,4*

See "Happiness" on page 124 for full text references. (2:7.6)

PAPER 150: THE THIRD PREACHING TOUR

1. THE WOMEN'S EVANGELISTIC CORPS

Continence Cult—Influence to Paul

Page 1679:2   *Paper 150:1.3*

See "Continence Cult—Influence to Paul" on page 343 for full text references. (89:3.6)

PAPER 154: LAST DAYS AT CAPERNAUM

2. A WEEK OF REST

Inevitabilities

Page 1719:1   *Paper 154:2.5*

   See "Inevitabilities" on page 136 for full text references. (3:5.5)

PAPER 155: FLEEING THROUGH NORTHERN GALILEE

4. ON THE WAY TO PHOENICIA

True Religion

Page 1728:2   *Paper 155:4.2*

   See "True Religion" on page 373 for full text references. (101:1.1)

5. THE DISCOURSE ON TRUE RELIGION

True Religion

Page 1728:3–7   *Paper 155:5.1–5*

   See "True Religion" on page 373 for full text references. (101:1.1)

6. THE SECOND DISCOURSE ON RELIGION

Survival—Co-operation with Indwelling Adjuster

Page 1733:5   *Paper 155:6.17*

   See "Survival—Co-operation with Indwelling Adjuster" on page 245
   for full text references. (40:5.16)

PAPER 156: THE SOJOURN AT TYRE AND SIDON

5. JESUS' TEACHING AT TYRE

Sense of Guilt

Page 1739:3   *Paper 156:5.8*

See "Sense of Guilt" on page 379 for full text references. (103:4.2)

PAPER 158: THE MOUNT OF TRANSFIGURATION

1. THE TRANSFIGURATION

Gabriel and Father Melchizedek

Page 1753:4   *Paper 158:1.8*

See "Gabriel and Father Melchizedek" on page 216 for full text references. (33:6.2)

7. PETER'S PROTEST

"Whosoever Saves His Life Selfishly..."

Page 1760:2   *Paper 158:7.5*

See ""Whosoever Saves His Life Selfishly..."" on page 380 for full text references. (103:5.6)

PAPER 159: THE DECAPOLS TOUR

1. THE SERMON ON FORGIVENESS

Seventy Times and Seven

Page 1763:1   *Paper 159:1.4*

See "Seventy Times and Seven" on page 247 for full text references. (40:8.2

PAPER 160: RODAN OF ALEXANDRIA

---

1. RODAN'S GREEK PHILOSOPHY

Cooperation or Resistance to Divine Leading— Survival of Immortal Soul

Page 1774:6   *Paper 160:1.14*

See "Cooperation or Resistance to Divine Leading—Survival of Immortal Soul" on page 205 for full text references. (30:4.8)

Born of the Spirit

Page 1774:6   *Paper 160:1.14*

See "Born of the Spirit" on page 378 for full text references. (103:2.1)

---

4. THE BALANCE OF MATURITY

Ability

Page 1779:3   *Paper 160:4.5*

See "Ability" on page 259 for full text references. (44:8.3)

PAPER 162: AT THE FEAST OF TABERNACLES

---

5. SERMON ON THE LIGHT OF THE WORLD

Cooperation or Resistance to Divine Leading— Survival of Immortal Soul

Page 1795:2   *Paper 162:5.3*

See "Cooperation or Resistance to Divine Leading—Survival of Immortal Soul" on page 205 for full text references. (30:4.8)

PAPER 164: AT THE FEAST OF DEDICATION

3. HEALING THE BLIND BEGGAR

Efficacy of Spittle—Belief In

Page 1812:5   *Paper 164:3.11*

See "Efficacy of Spittle—Belief In" on page 357 for full text references. (95:2.6)

PAPER 165: THE PEREAN MISSION BEGINS

1. AT THE PELLA CAMP

Jesus—Thief at the Pella Camp

Page 1817:5   *Paper 165:1.1*

By the middle of January more than twelve hundred persons were gathered together at **Pella, and Jesus taught this multitude at least once each day when he was in residence at the camp,** usually speaking at nine o'clock in the morning if not prevented by rain. Peter and the other apostles taught each afternoon. The evenings Jesus reserved for the usual sessions of questions and answers with the twelve and other advanced disciples. The evening groups averaged about fifty.

Page 2002:1   *Paper 186:4.5*

As soon as the thieves could be made ready, they were led into the courtyard, where they gazed upon Jesus, one of them for the first time, **but the other had often heard him speak, both in the temple and many months before at the Pella camp.**

5. TALKS TO THE APOSTLES ON WEALTH

Anxiety

Page 1823:2   *Paper 165:5.2*

See "Anxiety" on page 277 for full text references. (48:7.21)

PAPER 170: THE KINGDOM OF HEAVEN

---

5. LATER IDEAS OF THE KINGDOM

Kingdom Within

Page 1865:1   *Paper 170:5.11*

Page 1866:2   *Paper 170:5.19*
  See "Kingdom Within" on page 219 for full text references. (34:6.7)

PAPER 174: TUESDAY MORNING IN THE TEMPLE

---

4. THE GREAT COMMANDMENT

Commandments

Page 1901:2   *Paper 174:4.2*
  See "Commandments" on page 311 for full text references. (66:7.8)

PAPER 175: THE LAST TEMPLE DISCOURSE

---

3. THE FATEFUL SANHEDRIM MEETING

Melchizedek Bestowal—Divine Covenant

Page 1910:1   *Paper 175:3.2*
  See "Melchizedek Bestowal—Divine Covenant" on page 354 for full
  text references. (93:6.1,3)

PAPER 177: WEDNESDAY, THE REST DAY

2. EARLY HOME LIFE

Parent-Child Relationship

Page 1921:6   *Paper 177:2.2*

    See "Parent-Child Relationship" on page 350 for full text references. (92:7.7)

Parents—Influence Upon Children

Page 1922:3   *Paper 177:2.5*

    See "Parents—Influence Upon Children" on page 371 for full text references. (100:1.4)

PAPER 179: THE LAST SUPPER

3. WASHING THE APOSTLES' FEET

Commandments

Page 1939:4-5   *Paper 179:3.6-7*

    See "Commandments" on page 311 for full text references. (66:7.8)

PAPER 180: THE FAREWELL DISCOURSE

1. THE NEW COMMANDMENT

Commandments

Page 1944:4   *Paper 180:1.1*

    See "Commandments" on page 311 for full text references. (66:7.8)

Cosmic Mind

Page 1949:5   *Paper 180:5.3*

"Cosmic Mind" on page 174 (16:9.1–2)

Kingdom Within

Page 1951:2   *Paper 180:6.1*

See "Kingdom Within" on page 219 for full text references. (34:6.7)

Melchizedek Bestowal—Divine Covenant

Page 1969:4   *Paper 182:3.9*

See "Melchizedek Bestowal—Divine Covenant" on page 354 for full text references. (93:6.1,3)

Lucifer Rebellion—Satan

Page 1972:1   *Paper 183:1.2*

See "Lucifer Rebellion—Satan" on page 254 for full text references. (43:4.9)

---

3. The Master's Arrest

Angels—Twelve Legions

Page 1974:5   *Paper 183:3.7*

See "Angels—Twelve Legions" on page 237 for full text references. (38:6.2)

Paper 184: Before the Sanhedrin Court

---

4. The Hour of Humiliation

"Prophesy to us..."
"Prophesy to us"—two papers.

Page 1984:2   *Paper 184:4.1*

The Jewish law required that, in the matter of passing the death sentence, there should be two sessions of the court. This second session was to be held on the day following the first, and the intervening time was to be spent in fasting and mourning by the members of the court. But these men could not await the next day for the confirmation of their decision that Jesus must die. They waited only one hour. In the meantime Jesus was left in the audience chamber in the custody of the temple guards, who, with the servants of the high priest, amused themselves by heaping every sort of indignity upon the Son of Man. They mocked him, spit upon him, and cruelly buffeted him. **They would strike him in the face with a rod and then say, "Prophesy to us, you the Deliverer, who it was that struck you."** And thus they went on for one full hour, reviling and mistreating this unresisting man of Galilee.

Page 2000:2   *Paper 186:2.10*

His love for ignorant mortals is fully disclosed by his patience and great self-possession in the face of the jeers, blows, and buffetings of the coarse soldiers and the unthinking servants. **He was not even angry when they blindfolded him and, derisively striking him in the face, exclaimed: "Prophesy to us who it was that struck you."**

PAPER 186: JUST BEFORE THE CRUCIFIXION

---

2. THE MASTER'S ATTITUDE

"Prophesy to us..."

Page 2000:2   *Paper 186:2.10*

See ""Prophesy to us..."" on page 422 for full text references. (184:4.1)

---

4. PREPARATION FOR THE CRUCIFIXION

Jesus—Thief at the Pella Camp

Page 2002:1   *Paper 186:4.5*

See "Jesus—Thief at the Pella Camp" on page 418 for full text references. (165:1.1)

PAPER 187: THE CRUCIFIXION

---

5. LAST HOUR ON THE CROSS

Creator Son Final Bestowal—"It is Finished"

Page 2011:2   *Paper 187:5.6*

See "Creator Son Final Bestowal—"It is Finished"" on page 193 for full text references. (21:4.5)

PAPER 188: THE TIME OF THE TOMB

---

4. MEANING OF THE DEATH ON THE CROSS

Atonement Doctrine

Page 2016:6   *Paper 188:4.1*

Page 2016:8–10   *Paper 188:4.3–5*

Page 2017:3   *Paper 188:4.8*
  See "Atonement Doctrine" on page 123 for full text references. (2:6.5)

---

5. LESSONS FROM THE CROSS

Death on the Cross

Page 2019:6   *Paper 188:5.13*
  See "Death on the Cross" on page 368 for full text references. (98:7.1)

PAPER 189: THE RESURRECTION

---

1. THE MORONTIA TRANSIT

Bestowals—Resurrection on Third Day

Page 2021:3   *Paper 189:1.4*
  See "Bestowals—Resurrection on Third Day" on page 185 for full text
  references. (20:6.6)

---

3. THE DISPENSATIONAL RESURRECTION

Archangels' Circuit on Urantia

Page 2024:4   *Paper 189:3.2*
  See "Archangels' Circuit on Urantia" on page 233 for full text
  references. (37:3.3)

PAPER 190: MORONTIA APPEARANCES OF JESUS

Jesus—Morontia

Page 2029:3   *Paper 190:0.3*

See "Jesus—Morontia" on page 253 for full text references. (43:1.3)

5. *THE WALK-WITH TWO BROTHERS*

Jesus—Morontia

Page 2035:2   *Paper 190:5.5*

See "Jesus—Morontia" on page 253 for full text references. (43:1.3)

PAPER 192: APPEARANCES IN GALILEE

1. *APPEARANCE BY THE LAKE*

Jesus—Morontia

Page 2047:1   *Paper 192:1.8*

See "Jesus—Morontia" on page 253 for full text references. (43:1.3)

PAPER 193: FINAL APPEARANCES AND ASCENSION

God Conscious—Positive Proof

Page 2052:3   *Paper 193:0.3*

See "God Conscious—Positive Proof" on page 112 for full text references. (1:2.5)

PAPER 194: BESTOWAL OF THE SPIRIT OF TRUTH

Spirit of Truth—Bestowal

Page 2059:1   *Paper 194:0.1*

See "Spirit of Truth—Bestowal" on page 186 for full text references. (20:6.9)

2. THE SIGNIFICANCE OF PENTECOST

Kingdom Within

Page 2061:6   *Paper 194:2.8*

See "Kingdom Within" on page 219 for full text references. (34:6.7)

3. WHAT HAPPENED AT PENTECOST

Spirit of Truth—Bestowal

Page 2065:2   *Paper 194:3.14*

Page 2065:5   *Paper 194:3.17*

See "Spirit of Truth—Bestowal" on page 186 for full text references. (20:6.9)

Self-assertion

Page 2065:6   *Paper 194:3.18*

See "Self-assertion" on page 294 for full text references. (54:1.6)

PAPER 195: AFTER PENTECOST

7. THE VULNERABILITY OF MATERIALISM

Relativity

Page 2078:8 *Paper 195:7.5*

See "Relativity" on page 152 for full text references. (7:1.2)

9. CHRISTIANITY'S PROBLEM

Jesus—Teachings During Current Epoch of Social Readjustment

Page 2082:7–8 *Paper 195:9.2–3*

See "Jesus—Teachings During Current Epoch of Social Readjustment" on page 369 for full text references. (99:1.3)

Cosmic Citizenship

Page 2082:9 *Paper 195:9.4*

See "Cosmic Citizenship" on page 241 for full text references. (39:4.9)

Jesus' Gospel—Hour Is Striking

Page 2083:1 *Paper 195:9.5*

See "Jesus' Gospel—Hour Is Striking" on page 357 for full text references. (94:12.7)

10. THE FUTURE

Education—Purpose

Page 2086:3 *Paper 195:10.17*

See "Education—Purpose" on page 126 for full text references. (2:7.12)

427

PAPER 196: THE FAITH OF JESUS

God-consciousness—Prelude to Worship

Page 2087:3, 2088:5   *Paper 196:0.3,10*

See "God-consciousness—Prelude to Worship" on page 358 for full text references. (95:4.2)

Kingdom Within

Page 2088:3–4   *Paper 196:0.8–9*

See "Kingdom Within" on page 219 for full text references. (34:6.7)

Jesus—Follow Me

Page 2089:3   *Paper 196:0.13*

See "Jesus—Follow Me" on page 410 for full text references. (137:2.5)

1. JESUS—THE MAN

Revelation—Fifth Epochal Revelation

Page 2090:3   *Paper 196:1.2*

See "Revelation—Fifth Epochal Revelation" on page 346 for full text references. (92:4.4)

Jesus—Follow Me

Page 2090:3–4   *Paper 196:1.2–3*

See "Jesus—Follow Me" on page 410 for full text references. (137:2.5)

"Be About Your Father's Business"

Page 2091:2   *Paper 196:1.6*

See ""Be About Your Father's Business"" on page 232 for full text references. (37:2.8)

2. THE RELIGION OF JESUS

Spirt of Truth and Divine Fragment

Page 2092:2   Paper 196:2.4

See "Spirt of Truth and Divine Fragment" on page 118 for full text references. (2:0.3)

Religion—Will of God and Serving Human Brotherhood

Page 2092:4   Paper 196:2.6

See "Religion—Will of God and Serving Human Brotherhood" on page 323 for full text references. (72:3.5)

3. THE SUPREMACY OF RELIGION

Religion—Will of God and Serving Human Brotherhood

Page 2095:3   Paper 196:3.16

See "Religion—Will of God and Serving Human Brotherhood" on page 323 for full text references. (72:3.5)

Subjective Self-consciousness

Page 2095:5   Paper 196:3.18

See "Subjective Self-consciousness" on page 161 for full text references. (9:8.6)

God Conscious—Positive Proof

Page 2097:2   Paper 196:3.31

See "God Conscious—Positive Proof" on page 112 for full text references. (1:2.5)

Highest Human Concept

Page 2097:3   Paper 196:3.32

See "Highest Human Concept" on page 117 for full text references. (2:0.1)

Index to Chapter 7—Study Companion

Appendix

*JJ's intrastate cartage company in Phoenix, AZ during the early 1990s.
JJ drove these mobile billboards around Arizona with the words:*

THE

URANTIA

BOOK

YOU'VE GOT TO READ IT.

*Ask JJ about the responses he got
when you see him at a study group or conference.*

Testimonials

For three decades my path has crossed many truth seekers. Most of them I do not hear about or run into again. I cherish and love each one of those personal encounters and have fond memories of sharing loving experiences. On rare occasions I get an e-mail or letter from these beautiful spiritual brothers and sisters. I share the following with you:

"I just want to tell you how important our meeting was for me. All I learned from you has inspired me to go on reading, and of course, spreading the word with more energy."
—Bea, Montevideo, Uruguay

"Especially you introduced me, my family, my cousins, and my friends the Treasure of Life—the beautiful gift of life that changed our lives. Now we have some purpose of life. Before we were like dead bodes; we were empty inside. You are really an angel for us. JJ, when sometimes I am writing e-mails to you, or reading The Urantia Book, I am having tears in my eyes that how fortunate I am that I have something special with me, and how much I am thankful to you that you shared it with me."
—Nadia, Pakistan

"We want to thank you again for your outreach message. Your sharing efforts made the most difference in our individual lives, more than any other. So, you see, there is even more positive impact from your outreach than you ever knew."
—Bob and Linda, Missouri, USA

"Many have done so much in such a short time to make The Urantia Book available in Korea, perhaps Sy most of all. Bottom line—your visit was SO important and so motivating during this early time of the Korean translation. Everyone was energized by your visit and so impressed by you."
—Barry and Kay, East Coast, USA

"Thanks to all of you and JJ Johnson's great page in the Honolulu Star Bulletin—Captain Cook edition—entitled "Man's Gift to God—Wholehearted Desire to Do His Will." ... So full of feelings." (1978, Honolulu, Hawaii)
—Bjorn Lovell Bjornson, British Columbia, Canada

Testimonials

Read what others have said about JJ:

"Up Close and Personal *is just like having JJ, with his years of cross-references and insights, right with you at your study group."*
—Dick Johnson, Phoenix, Arizona, retired attorney and 35-year student of *The Urantia Book*

A true renaissance man ......
—Will B./State Department - Foreign Service

"JJ, you are truly a cosmic ambassador, preparing the ground for a higher order of spiritual and social living."
—David Kantor, IT webmaster and active in International Fellowship of Readers of *The Urantia Book*

Comments to JJ about The Urantia Book

"I've read it several times, the first time about 25 years ago. As you suggested, it is a tremendously insightful book and still find it very interesting."
—Dr. Stephen Covey, author of best-seller *7 Habits*

"My bold statement in Garden of Ediacara on continental drift as it relates to The Urantia Book has infuriated some of my closest and most eminent colleagues (even though they agree with all the science in Garden). Nevertheless, I am prepared to stand my ground. Please consider me a friend of the Urantia Fellowship."
—Dr. Mark A.S. McMenamin, author of *The Garden of Ediacara.* **Columbia University Press, New York.**

"JJ's remarkable observation and unsurpassed knowledge of The Urantia Papers has helped make this effort much more than it would have been without him."
—Excerpt from *A History of the Urantia Papers* **by Larry Mullins with Dr. Meredith Justin Sprunger**

Author Biography

JJ Johnson is a dedicated and devoted Urantia student whose zeal in supporting the Fifth Epochal Revelation's messages has been apostolic in nature. As a Founding Member and Past President of the Grand Canyon Society for Readers of *The Urantia Book*, he served for many years in a number of capacities including Chairman of the Education committee.

Johnson's ombudsman service to the Urantia movement is truly international in scope. He visited and placed *Urantia Books* in both national and university libraries in more than 20 countries, as well as catalyzing new readers and attending study groups all over the world and throughout the 50 states. From 2001 to 2005, Johnson facilitated study groups from Islamabad, Pakistan, to Rangoon, Burma, introducing spiritually hungry fellow agondonters to concepts that provided a way to the truth they were searching for. After meeting a blind reader at an Urantia conference, Johnson's dogged insistence and resolute persistence achieved the publication of the very first *Urantia Book* translated into Braille.

Secretary of State Condoleezza Rice thanking JJ for his service and wishing him well on his upcoming assignment. (April 2007)

JJ's recent assignment afforded him the opportunity to place Urantia Books in National Libraries and Universities around the globe during his off hours. That, plus meeting new and long-time readers from around the world, was one of the highlights of his tour.

www.ingramcontent.com/pod-product-compliance
Lightning Source LLC
Chambersburg PA
CBHW062031090426
42740CB00016B/2883